NOISE AND MILITARY SERVICE

Implications for Hearing Loss and Tinnitus

Committee on Noise-Induced Hearing Loss and Tinnitus Associated with Military Service from World War II to the Present

Medical Follow-up Agency

Larry E. Humes, Lois M. Joellenbeck, and Jane S. Durch, *Editors*

INSTITUTE OF MEDICINE
OF THE NATIONAL ACADEMIES

THE NATIONAL ACADEMIES PRESS
Washington, DC
www.nap.edu

THE NATIONAL ACADEMIES PRESS • 500 Fifth Street, N.W. • Washington, DC 20001

NOTICE: The project that is the subject of this report was approved by the Governing Board of the National Research Council, whose members are drawn from the councils of the National Academy of Sciences, the National Academy of Engineering, and the Institute of Medicine. The members of the committee responsible for the report were chosen for their special competences and with regard for appropriate balance.

This study was supported by Contract No. V101(93)P-1637 #29 between the National Academy of Sciences and the Department of Veterans Affairs. Any opinions, findings, conclusions, or recommendations expressed in this publication are those of the author(s) and do not necessarily reflect the view of the organizations or agencies that provided support for this project.

Library of Congress Cataloging-in-Publication Data

Noise and military service : implications for hearing loss and tinnitus / Committee on Noise-Induced Hearing Loss and Tinnitus Associated with Military Service from World War II to the Present, Medical Follow-up Agency ; Larry E. Humes, Lois M. Joellenbeck, and Jane S. Durch, editors.
p. ; cm.
Includes bibliographical references.
ISBN 0-309-09949-8 — ISBN 0-309-65307-X
1. Deafness—Etiology. 2. Tinnitus—Etiology. 3. Deafness, Noise induced. 4. Acoustic trauma. 5. Soldiers—Health and hygiene. 6. Veterans—Health and hygiene. 7. Noise—Health aspects. I. Humes, Larry. II. Joellenbeck, Lois M. (Lois Mary), 1963- . III. Durch, Jane. IV. Institute of Medicine (U.S.). Committee on Noise-Induced Hearing Loss and Tinnitus Associated with Military Service from World War II to the Present.
[DNLM: 1. Noise—adverse effects. 2. Hearing Loss, Noise-Induced. 3. Military Personnel. 4. Tinnitus. WV 270 N7836 2005]
RF291.5.S65N65 2005
617.8'07—dc22

2005030236

Additional copies of this report are available from the National Academies Press, 500 Fifth Street, N.W., Lockbox 285, Washington, DC 20055; (800) 624-6242 or (202) 334-3313 (in the Washington metropolitan area); Internet, http://www.nap.edu.

For more information about the Institute of Medicine, visit the IOM home page at: **www.iom.edu.**

Printed in the United States of America.

The serpent has been a symbol of long life, healing, and knowledge among almost all cultures and religions since the beginning of recorded history. The serpent adopted as a logotype by the Institute of Medicine is a relief carving from ancient Greece, now held by the Staatliche Museen in Berlin.

"Knowing is not enough; we must apply.
Willing is not enough; we must do."
—Goethe

INSTITUTE OF MEDICINE
OF THE NATIONAL ACADEMIES

Advising the Nation. Improving Health.

THE NATIONAL ACADEMIES

Advisers to the Nation on Science, Engineering, and Medicine

The **National Academy of Sciences** is a private, nonprofit, self-perpetuating society of distinguished scholars engaged in scientific and engineering research, dedicated to the furtherance of science and technology and to their use for the general welfare. Upon the authority of the charter granted to it by the Congress in 1863, the Academy has a mandate that requires it to advise the federal government on scientific and technical matters. Dr. Ralph J. Cicerone is president of the National Academy of Sciences.

The **National Academy of Engineering** was established in 1964, under the charter of the National Academy of Sciences, as a parallel organization of outstanding engineers. It is autonomous in its administration and in the selection of its members, sharing with the National Academy of Sciences the responsibility for advising the federal government. The National Academy of Engineering also sponsors engineering programs aimed at meeting national needs, encourages education and research, and recognizes the superior achievements of engineers. Dr. Wm. A. Wulf is president of the National Academy of Engineering.

The **Institute of Medicine** was established in 1970 by the National Academy of Sciences to secure the services of eminent members of appropriate professions in the examination of policy matters pertaining to the health of the public. The Institute acts under the responsibility given to the National Academy of Sciences by its congressional charter to be an adviser to the federal government and, upon its own initiative, to identify issues of medical care, research, and education. Dr. Harvey V. Fineberg is president of the Institute of Medicine.

The **National Research Council** was organized by the National Academy of Sciences in 1916 to associate the broad community of science and technology with the Academy's purposes of furthering knowledge and advising the federal government. Functioning in accordance with general policies determined by the Academy, the Council has become the principal operating agency of both the National Academy of Sciences and the National Academy of Engineering in providing services to the government, the public, and the scientific and engineering communities. The Council is administered jointly by both Academies and the Institute of Medicine. Dr. Ralph J. Cicerone and Dr. Wm. A. Wulf are chair and vice chair, respectively, of the National Research Council.

www.national-academies.org

COMMITTEE ON NOISE-INDUCED HEARING LOSS AND TINNITUS ASSOCIATED WITH MILITARY SERVICE FROM WORLD WAR II TO THE PRESENT

LARRY E. HUMES (*Chair*), Professor, Department of Speech and Hearing Sciences, Indiana University

ELLIOTT H. BERGER, Senior Scientist, Auditory Research, E•A•R/ Aearo Company, Indianapolis, Indiana

BARBARA A. BOHNE, Professor, Department of Otolaryngology, Washington University School of Medicine

ANTHONY T. CACACE, Associate Professor, The Neuroscience Institute and Advanced Imaging Research Center, Albany Medical College

KAREN J. CRUICKSHANKS, Professor, Department of Ophthalmology and Visual Sciences and Department of Population Health Sciences, University of Wisconsin Medical School

RICHARD W. DANIELSON, Manager for Audiology and Hearing Conservation, National Space Biomedical Research Institute, Baylor College of Medicine

JUDY R. DUBNO, Professor, Department of Otolaryngology–Head and Neck Surgery, Medical University of South Carolina

GEORGE A. GATES, Emeritus Director/Professor, Virginia Merrill Bloedel Hearing Research Center, Department of Otolaryngology–Head and Neck Surgery, University of Washington School of Medicine

DONALD HENDERSON, Professor, Center for Hearing and Deafness, State University of New York at Buffalo

JOHN H. MILLS, Professor, Department of Otolaryngology–Head and Neck Surgery, Medical University of South Carolina

MARK R. STEPHENSON, Senior Research Audiologist, Division of Applied Research and Technology, National Institute for Occupational Safety and Health, Centers for Disease Control and Prevention, Cincinnati, Ohio

RICHARD S. TYLER, Professor and Director of Audiology, Department of Otolaryngology–Head and Neck Surgery, and Department of Speech Pathology and Audiology, The University of Iowa

ROBERT B. WALLACE, Irene Ensminger Stecher Professor of Epidemiology and Internal Medicine, Department of Epidemiology, The University of Iowa

Project Staff

LOIS JOELLENBECK, Senior Program Officer (Study Director)

JANE DURCH, Senior Program Officer

WILLIAM PAGE, Senior Program Officer

SUSAN VAN HEMEL, Senior Program Officer, Board on Behavioral, Cognitive, and Sensory Science, Division of Behavioral and Social Sciences and Education
KRISTEN GILBERTSON, Research Assistant
ALLISON BERGER, Program Assistant
KAREN KAZMERZAK, Research Associate
RICK ERDTMANN, Director, Medical Follow-up Agency
PAMELA RAMEY-MCCRAY, Administrative Assistant
ANDREA COHEN, Financial Associate

Data Operations
HARRIET CRAWFORD, IT Project Manager
JOHN LARSON, Records Manager, St. Louis Office
NOREEN STEVENSON, Research Assistant
AL MATTEI, Staff Assistant

Reviewers

This report has been reviewed in draft form by individuals chosen for their diverse perspectives and technical expertise, in accordance with procedures approved by the National Research Council's Report Review Committee. The purpose of this independent review is to provide candid and critical comments that will assist the institution in making its published report as sound as possible and to ensure that the report meets institutional standards for objectivity, evidence, and responsiveness to the study charge. The review comments and draft manuscript remain confidential to protect the integrity of the deliberative process. We wish to thank the following individuals for their review of this report:

Bobby R. Alford, Department of Otolaryngology–Head and Neck Surgery, Baylor College of Medicine
Robert A. Dobie, Department of Otolaryngology, School of Medicine, University of California, Davis
Roger P. Hamernik, Auditory Research Laboratories, State University of New York at Plattsburgh
Richard N. Miller, Institute of Medicine (retired)
Theresa Y. Schulz, James H. Quillen VA Medical Center, Mountain Home, Tennessee
James B. Snow, Jr., National Institute on Deafness and Other Communication Disorders, National Institutes of Health (retired)
Evelyn E. Talbott, Department of Epidemiology, University of Pittsburgh
William A. Yost, Parmly Hearing Institute and the Graduate School, Loyola University of Chicago

Although the reviewers listed above have provided many constructive comments and suggestions, they were not asked to endorse the conclusions or recommendations nor did they see the final draft of the report before its release. The review of this report was overseen by Hyla Napadensky, Napadensky Energetics Inc. (retired), and Linda D. Cowan, Health Sciences Center, University of Oklahoma. Appointed by the National Research Council and Institute of Medicine, they were responsible for making certain that an independent examination of this report was carried out in accordance with institutional procedures and that all review comments were carefully considered. Responsibility for the final content of this report rests entirely with the authoring committee and the institution.

Preface

This report represents the collective efforts of many people over a period of approximately 15 months. First and foremost are the committee members, who devoted countless hours of volunteer service to address the committee's charge. The numerous discussions held, both face to face and via conference calls, were thorough and, when opinions differed, were always conducted with respect for divergent views. The report reflects the broad input and consensus of all committee members and is much stronger as a result. Clearly, the charge could not be addressed without a panel having sufficiently broad areas of expertise, but also a willingness to listen to sometimes differing perspectives with an open mind before moving to consensus. I truly appreciate not only the amount of expertise available through the committee, but the manner in which each member shared that expertise and worked together to meet our collective responsibilities.

Another group central to the committee's work included the staff members from the Institute of Medicine (IOM) and National Research Council (NRC) who supported our efforts. Lois Joellenbeck, who served as study director, and Jane Durch were the committee's principal guides. We were ably aided as well by research assistant Kristen Gilbertson, whose contributions included not only research and administrative tasks, but also participating in the collection and entry of data from veterans' medical records. Program assistant Allison Berger managed the logistics of many of the committee's earlier meetings. Karen Kazmerzak helped initiate our research efforts during the first 3 months of the committee's work. Susan Van Hemel, a senior program officer with the NRC, contributed advice throughout the study.

This study included the collection and analysis of data from the service medical records of some 3,500 veterans by members of the staff of the Medical Follow-up Agency of the IOM. The large data collection and data management task was overseen by Harriet Crawford, with the work of data collection and entry performed by Noreen Stevenson, John Larson, and Al Mattei. The data analysis was performed by William Page, a biostatistician with the Medical Follow-up Agency. Identifying and obtaining these records required the assistance of personnel from the Department of Veterans Affairs, the Department of Defense, and the National Archives and Records Administration. On behalf of the committee and staff, I want to thank the many individuals from those agencies for their assistance, and to offer special thanks to Lynda Russell, Yvonne Hamilton, and James White from the Department of Veterans Affairs.

The work of the committee was also supported by other staff members from the Medical Follow-up Agency, particularly the director, Rick Erdtmann, and the administrative assistant, Pamela Ramey-McCray. In addition, on behalf of the committee and staff, I wish to thank Liesl Peters, Clyde Behney, Janice Mehler, Jennifer Bitticks, Jennifer Otten, and Andrea Cohen from the IOM and NRC staffs, who participated in the report review, preproduction, dissemination, and financial management.

I would like to acknowledge the assistance of Lauren Strauser in the preparation and analyses of the data from the many cross-sectional studies of noise-induced hearing loss in military personnel that are the basis for Figures 3-3 through 3-6. In addition, the time needed for these analyses, as well as for my writing and editorial duties, would not have been available were it not for a generous release from teaching responsibilities granted to me by Indiana University for the spring 2005 semester.

Finally, on behalf of the committee, I wish to express our gratitude to the individuals who made presentations to the committee at its three information-gathering meetings: Mary Schoelen from the Senate Committee on Veterans' Affairs; COL David Chandler and Douglas Ohlin from the U.S. Army; John Page and Kurt Yankaskas from the U.S. Navy; Maj. Joseph Narrigan and Maj. Robert Pluta from the U.S. Air Force; Al Lillibridge from the U.S. Marine Corps; CDR Wade McConnell from the U.S. Coast Guard; Michael Hodgson, Kyle Dennis, Judy Schafer, and Bradley Flohr from the Department of Veterans Affairs; and veterans Arnold Mathias, Ansell (Mac) MacMillan, Carroll Williams, John Warwick, Rick Weidman, and David Kayal. Our project officer from the Department of Veterans Affairs, Michael Hodgson, provided valuable guidance on the scope of the study as the committee began its work. Also at the Department of Veterans Affairs, Michael Wells responded to several requests for data on veterans receiving disability compensation. Throughout the study, representatives of the military services provided ongoing assistance by responding to numerous requests for information. In particular, we want to acknowledge the assistance of the following individuals in this regard: COL David Chandler, Douglas Ohlin, Kevin Wisniewski, Felix Sachs, and COL Mark Rubertone from the U.S. Army; John Page, Leslie Crowder, Kurt Yankaskas, and Valerie Bjorn from the U.S. Navy; Maj. Robert Pluta and Lt. Col. Wade Weisman from the U.S. Air Force; and CDR Wade McConnell from the U.S. Coast Guard.

Larry E. Humes
Committee Chair

Contents

Figures, Tables, and Boxes

FIGURES

TABLES

BOXES

Executive Summary

People serving in the military will, at some point, be exposed to high-intensity noise of various types. Some may develop hearing loss, especially for high-frequency sounds, or tinnitus ("ringing in the ears"), or both, as a result of their noise exposure. Hearing loss or tinnitus incurred or aggravated during military service may qualify veterans for services and financial compensation from the Department of Veterans Affairs (VA). Since World War II, the human and financial costs associated with hearing loss among military veterans have repeatedly drawn attention to noise, hearing loss, and the need for hearing conservation in military settings. In recent years, tinnitus has emerged as a significant concern as well.

VA reported that the 2.5 million veterans receiving disability compensation at the end of fiscal year 2003 had approximately 6.8 million separate disabilities related to their military service.[1] Disabilities of the auditory system, including tinnitus and hearing loss, were the third most common type, accounting for nearly 10 percent of the total number of disabilities among these veterans. For the roughly 158,000 veterans who began receiving compensation in 2003, auditory disabilities were the second most common type of disability. These veterans had approximately 75,300 disabilities of the auditory system out of a total of some 485,000 disabilities. At the

[1]Veterans may have hearing loss and other disabilities that have been determined to have been incurred during or aggravated by military service but that do not qualify for disability compensation payments (a "zero percent" service-connected disability). Veterans with service-connected hearing loss who do not qualify for any disability compensation payments are not included in the VA data on numbers of disabilities or numbers of veterans with disabilities. All veterans with service-connected tinnitus qualify for compensation payments.

end of 2004, the monthly compensation payments to veterans with hearing loss as their major form of disability represented an annualized cost of some $660 million. The corresponding compensation payments to veterans with tinnitus as their major disability were close to $190 million on an annualized basis.

Determining whether hearing loss or tinnitus, evident at the time a claim is filed by a veteran, is attributable to prior military service can pose challenges for VA. After the fact, hearing loss or tinnitus incurred as a result of military service cannot be distinguished with certainty from subsequent noise-induced hearing loss or tinnitus resulting from work in a noisy industry or from participation in a variety of noisy recreational activities, such as hunting. Furthermore, high-frequency hearing losses are seen not only with noise exposure, but also at older ages (presbycusis), although the specific patterns of loss are generally distinguishable until 60–70 years of age (see Chapter 2). Tinnitus may also develop in response to factors other than noise exposure (e.g., head injury, brain tumors, middle ear diseases, certain medications) and can occur with or without hearing loss. If documentation of hearing thresholds or tinnitus during military service is not available, even a detailed case history from the veteran may leave considerable uncertainty about the association between a current hearing loss or tinnitus and prior military service.

Concerns about the noise hazards associated with military service and questions about the relationship between noise exposure and hearing loss or tinnitus led Congress to direct VA to contract with the National Academies for a study of these issues.[2] The committee convened by the Institute of Medicine of the National Academies to conduct this study was charged with reviewing the following for the period from World War II to the present: (1) the available data on hearing loss that could be expected among members of the armed forces; (2) sources of hazardous noise exposure during military service; (3) the levels of noise exposure necessary to cause hearing loss or tinnitus; (4) the course of hearing loss following noise exposure, including whether onset can be delayed; (5) risk factors for noise-induced hearing loss and tinnitus; and (6) compliance by the military services with requirements for audiometric testing and the adequacy of the services' hearing conservation programs to protect the hearing of service members.

APPROACH TO THE STUDY

The committee's considerations included noise-induced hearing loss, which most commonly results from repeated exposures to hazardous noise

[2]The study was called for in Section 104 of the Veterans Benefits Act of 2002 (P.L. 107-330).

for a period of several years, and the phenomenon of acoustic trauma, which is the sudden loss of hearing following a single exposure to very hazardous noise. Hearing loss was assessed primarily on the basis of data on hearing thresholds measured in humans by pure-tone audiometry. Noise-induced hearing loss is often characterized by a "notch" in the audiogram, reflecting worse hearing at frequencies between 3000 and 6000 Hz than at lower and higher frequencies. The specific pattern of changes in pure-tone thresholds can vary depending on the type of noise exposure. The committee focused most of its attention on permanent changes in those thresholds. In adults, hearing loss is typically considered to be present when pure-tone thresholds are worse than 25 dB HL at any frequency usually tested.[3]

The study also included consideration of noise-induced tinnitus. Tinnitus is the perception of sound (e.g., ringing, buzzing, whistling) that cannot be attributed to an external sound source and is perceivable only by the person who is experiencing it. This subjective phenomenon is distinct from perceived sound that can be generated by events in the head or neck and that may be perceptible by an observer. The presence of tinnitus is determined primarily by self-report, but perceptual attributes, such as its pitch and loudness, can be established reliably under controlled conditions (psychoacoustic testing). The mechanisms underlying tinnitus are not fully understood. Some people with tinnitus experience serious problems associated with emotional well-being, sleep, hearing, and concentration. No current treatment will eliminate tinnitus, but some treatments may reduce its adverse impact. Promising treatments may include counseling, counseling combined with sound therapies, medications, and electrical and magnetic stimulation.

The committee reviewed material from peer-reviewed journals, books, reports prepared by or for the military services, and documents and data provided by the military services at the committee's request. The committee's information gathering also included testimony and presentations from veterans and representatives of the military services. Published peer-reviewed reports generally carried the most weight. Ideally, the committee would like to have drawn on data from reports of longitudinal, population-based studies of noise-induced hearing loss and tinnitus in humans in military settings. There are few such studies, and therefore, the committee was compelled to turn to other sources of evidence to address its charge.

The committee's findings and conclusions concerning each element of its charge are summarized here. (See Box ES-1 for a complete listing of findings.) Also summarized are needs the committee identified for opera-

---

[3]A standardized value representing the average thresholds measured in a large group of young normal-hearing adults at a given frequency is said to be 0 dB "hearing level" or 0 dB HL. Hearing thresholds are commonly measured at the following frequencies: 250, 500, 1000, 2000, 3000, 4000, 6000, and 8000 Hz.

tional changes and further research. These proposals for operational changes and research are aimed at improving hearing protection, preventing hearing loss and noise-induced tinnitus during military service, and gaining a better understanding of noise hazards, noise-induced hearing loss, and tinnitus, especially among military personnel.

SOURCES OF NOISE DURING MILITARY SERVICE

Many sources of potentially damaging noise have long existed in military settings. For the period addressed by this report—World War II to the present—some of these sources include weapons systems (e.g., handguns, rifles, artillery pieces, rockets), wheeled and tracked vehicles, fixed- and rotary-wing aircraft, ships, and communications devices (Chapter 3). Service members may encounter these noise sources through training, standard military operations, and combat. Exposure to combat-related noise may be unpredictable in onset and duration. Service members may also be exposed to hazardous noise through activities that are not unique to the military environment, including various engineering, industrial, construction, or maintenance tasks.

Throughout the period since World War II, the military services have collected data on noise levels associated with various kinds of equipment and activities, but a complete catalog of noise sources and the noise levels they produce is not feasible. The committee compiled an illustrative listing of documents reporting on sound levels in military settings (see Chapter 3 and Appendix F).

HAZARDOUS NOISE LEVELS

The specific noise levels that cause noise-induced hearing loss vary with the duration of the exposure, the type of noise, and the frequency content of the noise, as well as the susceptibility of the exposed individual (Chapters 1 and 2). Time-weighted average noise exposures of approximately 85 dBA for 8 hours per day for a 40-hour work week, or the equivalent, are considered to be hazardous, but a person must be so exposed for a number of years before developing noise-induced hearing loss. On the other hand, impulse noise with peak levels exceeding approximately 140 dB SPL may be hazardous even for a single exposure. With regard to noise-induced tinnitus, specific parameters of hazardous noise exposure have not been defined, but noise levels associated with hearing loss are also likely to be associated with tinnitus (Chapter 4).

HEARING LOSS AND TINNITUS AMONG MILITARY PERSONNEL

In the more than 60 years since the U.S. entrance into World War II, over 25 million people have served in the U.S. armed forces. Their experiences, in five different services and at least five major conflicts, as well as peacetime eras, have exposed many to loud noise. The total number who experienced noise-induced hearing loss or tinnitus by the time their military service ended may be substantial, but the available data provide no basis for a valid estimate of the number (Chapters 3 and 4). The abundant clinical evidence that noise exposure can result in hearing loss or tinnitus is not sufficient to determine in a particular group of people the extent to which such conditions have actually occurred or to establish that exposure to noise during military service was the cause of observed hearing losses or tinnitus.

Over the decades since World War II, noise exposures are likely to have varied widely, even within similar occupational specialties and eras. Data and analyses to document and quantify noise exposures of military personnel during this period, as well as to document and quantify their hearing thresholds and permanent changes in those thresholds over the course of military service, are rarely available. The committee found only a limited number of studies of hearing loss on which to base its findings, and those studies were primarily for the period since 1970. Among these were cross-sectional studies showing patterns of hearing loss consistent with noise exposure, but no longitudinal studies that could provide reliable data on changes in individuals' hearing thresholds over the course of military service (Chapters 1 and 3). The available studies were not designed to be representative of a service as a whole and only rarely of a particular military occupational group. Furthermore, the variability of individual responses to noise exposure precludes using the average hearing thresholds reported in many studies to estimate the hearing loss of individuals.

No epidemiological studies of tinnitus among U.S. military personnel were identified, and the services' hearing conservation programs do not include surveillance for tinnitus. Limited tinnitus surveillance was introduced in 2003 with post-deployment health assessment questionnaires.

Together, these factors made it impossible to generalize findings from the available studies to broader populations of military service members or veterans. In particular, it was not possible to estimate the proportion of a given military population that developed noise-induced hearing loss or tinnitus during military service, the amount of hearing loss incurred, or the relative risk of noise-induced hearing loss or tinnitus for a given individual, based on his or her branch of military service, occupational speciality, or service era.

ONSET AND COURSE OF HEARING LOSS

There is little evidence available to address whether noise-induced hearing loss or tinnitus progresses after noise exposure ends or whether noise-induced hearing loss can develop several months or years after the noise exposure has ended (Chapters 2 and 4). No longitudinal studies have examined patterns of hearing loss in noise-exposed humans or laboratory animals who did not develop hearing loss at the time of noise exposure. The committee's understanding of the mechanisms and processes involved in the recovery from noise exposure suggests, however, that a prolonged delay in the onset of noise-induced hearing loss is unlikely.

When hearing loss is known to have occurred as a result of a noise exposure, it has generally been thought that hearing loss for pure tones does not worsen following the cessation of a given noise exposure. However, there are no longitudinal data from humans who developed noise-induced hearing loss in early adulthood and were followed into their 60s, 70s, or 80s. Data from a few longitudinal studies of older adults, which differed in the way prior noise exposure was documented, have not produced conclusive results.

RISK FACTORS FOR NOISE-INDUCED HEARING LOSS AND TINNITUS

It is well established that individuals vary in their responses to noise exposure, but the factors that account for this variability are still poorly understood. Evidence from studies in humans was not sufficient to determine whether noise exposure combined with specific endogenous or exogenous factors was associated with additional risk for noise-induced hearing loss or tinnitus (Chapters 2 and 4). Studies of several endogenous factors—older age, gender, race, eye color, and prior hearing loss—have shown little association with noise-induced hearing loss. Conclusive results have not emerged from investigations of the effects of noise exposure in combination with the following exogenous risk factors: aminoglycoside antibiotics, cisplatin, diuretics, salicylates, solvents, carbon disulfide, carbon monoxide, cigarette smoking, whole-body vibration, body temperature, exercise, or electromagnetic fields. Some of these medications and chemicals are recognized ototoxins that may induce hearing loss unrelated to noise exposure.

The committee identified only one study in humans that had investigated the association between tinnitus and combined exposures to noise and other factors. Tinnitus risk factors, independent of noise exposure, include hearing loss, head injury, middle ear disease, and certain medications (e.g., salicylates, aminoglycoside antibiotics).

MILITARY HEARING CONSERVATION PROGRAMS

Data analyzed by the committee led to the conclusion that military hearing conservation programs, dating from the late 1970s, were not adequate to protect the hearing of service members. The committee concluded that hearing conservation activities from World War II through the 1970s would have been even less adequate to protect the hearing of service members than the programs in place since the late 1970s, because only early hearing protection devices of limited effectiveness were available and mandatory hearing conservation measures were in place only in the Air Force (Chapter 5).

Given that engineering measures to reduce noise levels and administrative measures to reduce noise exposures may not be compatible with requirements for military operations, use of hearing protection devices is often the primary defense against noise-induced hearing loss for military personnel. The effectiveness of these devices depends, in large measure, on how well and how often they are used. Data on the use of hearing protection by military personnel are limited, but a handful of reports over the past 30 years suggests that in some settings, only about half of those who should have been using hearing protection devices were doing so.

The services' hearing conservation programs require annual audiometric testing for personnel enrolled in the program. The percentage of service members tested each year who have a significant shift in hearing thresholds currently ranges from about 10 percent to 18 percent, which is two to five times higher than rates considered appropriate in industrial hearing conservation programs. Testing will not prevent noise-induced hearing loss, but it may serve to limit the loss if the detection of temporary hearing losses or small permanent losses results in increased use of hearing protection or the reassignment of individuals to lower noise environments. Available data showed, however, that some personnel may not be receiving the required tests, and discussions with personnel from the hearing conservation programs suggest that some test results may not be reaching the central hearing conservation registry system.

DOCUMENTATION OF AUDIOMETRIC TESTING DURING MILITARY SERVICE

A review of service medical records for veterans who left military service during the period from World War II to 2002 suggests that documented audiometric testing at entrance into and separation from service has not been adequate, throughout the period, to evaluate changes in hearing associated with military service for the majority of service members (Chapter 6). As argued repeatedly in this report, it is critical to obtain an audiogram at entry into and exit from military service to clearly establish whether noise-induced hearing loss developed during military service. The service medical records audited revealed that about 30 percent of personnel who left the Navy and Marine Corps during the period from the early 1980s to 2002 had both an entry and separation audiogram within ±60 days of entry or separation, whereas the percentages were even lower, typically less than 12 percent, for personnel who had served in either the Army or the Air Force. As expected, the percentage of service medical records containing audiograms of any type was lowest for the period before 1950, except for the Air Force, an early leader in requiring the collection of audiograms.

OPERATIONAL NEEDS SUGGESTED BY THE REPORT

The current irreversibility of noise-induced hearing loss and tinnitus means that preventing these problems, or limiting their progression, is especially important. From the review of information on noise exposure in military settings, hearing loss and tinnitus experienced by some service members, and the hearing conservation activities of the military services, the committee identified several steps that may enhance hearing protection for service members and improve the effectiveness of the services' hearing conservation programs. Although this report was prepared for the Department of Veterans Affairs, it is the Department of Defense and the individual military services that can take these important steps to minimize the adverse effects of noise exposure on military personnel and better document hearing loss or tinnitus when either occurs during military service. The committee strongly recommends that the following practices be implemented:

1. Work to achieve more extensive and consistent use of hearing protection by military personnel.
2. Include questions about the presence and severity of tinnitus in each ear on all audiometric records obtained from enlistment through the end of military service. (In the remaining suggestions, audiograms and audiometric records are assumed to include responses to questions about the presence and severity of tinnitus.)
3. Enforce requirements for audiograms prior to noise exposure for all new military service members at *all* basic training sites.
4. Enforce, and establish where they do not presently exist, requirements for audiograms at the completion of military service to ensure that any hearing loss or tinnitus arising during military service is adequately documented. The Department of Defense and the Department of Veterans Affairs should explore whether resources are available within the VA system to aid the military services in conducting audiometric tests and tinnitus assessments for personnel completing their military service.
5. Given the likely occurrence of maximum noise-induced hearing loss at 6000 Hz, include the measurement of hearing thresholds at 8000 Hz in all audiograms to allow for detection of the noise-notch pattern of hearing loss associated with noise exposure.

6. Enforce hearing conservation requirements for annual monitoring audiograms, as well as for follow-up audiograms if a significant threshold shift is detected in annual monitoring audiograms.
7. Continue to develop the Defense Occupational and Environmental Health Readiness System (DOEHRS) to improve its reporting capabilities to match and exceed those available with the services' previous systems. Further development of this system should include modification of the hearing conservation component (DOEHRS-HC) to track reports of tinnitus. It should also include implementation of the industrial hygiene component (DOEHRS-IH) to provide information on exposures to hazardous noise and other chemical, physical, biological, and ergonomic hazards.
8. Develop mechanisms to provide VA personnel access to records from DOEHRS-HC for review of disability claims for hearing loss or tinnitus that are not otherwise supported by audiometric records in the service medical record.

RESEARCH NEEDS SUGGESTED BY THE REPORT

The committee also saw areas where further research would be valuable for improving understanding of broad scientific questions concerning the relationship between noise exposure and hearing loss and tinnitus. Research could also address more targeted questions concerning noise exposure, hearing loss, tinnitus, and hearing conservation measures related to military service.

Two broad scientific areas were of interest to the committee:

1. Further investigate, both in laboratory animals and humans, exposures to fluctuating noise, impulse/impact noise, and combinations of noise, as well as intermittent exposures to steady-state noise, to determine the acoustic parameters associated with noise-induced hearing loss and tinnitus.
2. Further investigate the mechanisms, natural history, epidemiology, measurement, and treatment of noise-induced hearing loss and tinnitus.

Several avenues of research specifically related to military settings and military personnel could be considered. Many are offered as a means to fill the void for prospective, longitudinal, epidemiological data on noise-induced hearing loss and tinnitus in military personnel.

1. Obtain valid estimates of the incidence, prevalence, and severity of noise-induced hearing loss and tinnitus among military personnel, including gender-specific estimates. If the reporting ability and completeness of existing databases, such as DOEHRS-HC, improve, greater use might be

made of their data for analyses for personnel enrolled in hearing conservation programs.

2. Establish cohorts of military veterans with various documented noise exposures, immediately on discharge, and survey them periodically for ototoxic exposures, subsequent nonmilitary noise exposures, and hearing function, as well as presence and severity of tinnitus, in order to determine whether there is a delay in the effects of military noise exposure. These cohorts will need to be followed through the remainder of members' lifetimes, but this longitudinal study will reveal elements of the natural history of noise-induced hearing loss and tinnitus that otherwise will not be determined. The Millennium Cohort Study, which is designed to evaluate the long-term health of people who have served in the military, might provide a mechanism for conducting a longitudinal investigation of hearing health.

3. Conduct randomized trials of interventions within each military branch to determine with greater certainty which approaches to hearing conservation—including efforts to increase the use and effectiveness of hearing protection devices, compliance with requirements for audiometric testing, and the use of otoprotective medications—lead to lower incidence of noise-induced hearing loss and tinnitus.

4. On a sample basis, determine noise levels for modern military activities and also determine, with standard industrial hygiene methods, the noise dose experienced by individual military personnel where dosimetry has not been done.

5. Conduct real-world studies in military settings, including field and garrison conditions, to assess the noise attenuation and utilization rates of hearing protection devices, including the recently introduced earplugs that provide level-dependent sound attenuation.

BOX ES-1
Compilation of Report Findings

Chapter 2: Noise-Induced Hearing Loss

- The evidence from laboratory studies in humans and animals is sufficient to conclude that the most pronounced effects of a given noise exposure on pure-tone thresholds are measurable immediately following the exposure, with the length of recovery, whether partial or complete, related to the level, duration, and type of noise exposure. Most recovery to stable hearing thresholds occurs within 30 days.
- There is not sufficient evidence from longitudinal studies in laboratory animals or humans to determine whether permanent noise-induced hearing loss can develop much later in one's lifetime, long after the cessation of that noise exposure. Although the definitive studies to address this issue have not been performed, based on the anatomical and physiological data available on the recovery process following noise exposure, it is unlikely that such delayed effects occur.
- Nonacoustic factors may interact with the effects of noise to increase the measured noise-induced hearing loss. For many exogenous factors, evidence in animal models reveals that the effects of drugs or chemical agents may combine in an additive or synergistic manner with the effects of noise to increase noise-induced hearing loss. In particular, aminoglycosides, cisplatin, and solvents (toluene and styrene) interact in laboratory animals with noise presented simultaneously or sequentially to increase the amount of noise-induced hearing loss. However, there is not sufficient evidence to confirm this finding in humans. In particular, the evidence is not conclusive in humans with regard to additive or synergistic effects of noise and the following exogenous factors on hearing: aminoglycosides, cisplatin, diuretics, salicylates, solvents, carbon disulfide, carbon monoxide, cigarette smoking, whole-body vibration, body temperature, exercise, and electromagnetic fields.
- Several endogenous factors have been examined, including (old) age, gender, race, eye color, and prior hearing loss, but there is not sufficient evidence in humans to conclude that any of these factors predicts susceptibility to noise-induced hearing loss.
- The evidence from cross-sectional studies of noise-induced hearing loss in humans is sufficient to conclude that daily time-weighted average noise exposures greater than approximately 85 dBA for 8 hours for periods of many years pose a hazard to human hearing and that the hazard increases as the time-weighted average exposure exceeds this value.
- The evidence is not sufficient to determine the probability of acquiring a noise-induced hearing loss, or to estimate the magnitude of the noise-induced hearing loss, that a specific *individual* is likely to experience from a given noise exposure.

Chapter 3: Noise and Noise-Induced Hearing Loss in the Military

- The evidence is sufficient to conclude that hazardous noise levels are and have been present in many military settings.
- Extensive collections of data on sound pressure levels produced by equipment and activities in military settings are available from World War II to the present. Many estimates of noise exposures (doses) from specific activities also

are available from more restricted time periods. However, because of the changing nature of assignments in the military, the unpredictable aspects of military training and combat, the intermittent nature of many military noise exposures, and the sporadic use of hearing protection while in the military, these data do not provide a sufficient basis for estimating cumulative noise exposures over the course of military service for individuals or for subgroups (e.g., occupational specialties, branches, or eras).

- The evidence is sufficient to conclude that certain military personnel from World War II to the present have exhibited hearing thresholds while in the military that are typical of noise-induced hearing loss.
- The evidence is not sufficient to reach conclusions regarding the number or proportion of service members, overall or in specific occupational groups or eras since World War II, who have experienced noise-induced hearing loss while in the military.
- The evidence is not sufficient to determine the probability of acquiring noise-induced hearing loss associated with service in the military, or in specific branches of the military, for a given individual. The probability of acquiring noise-induced hearing loss can only be determined precisely with well-controlled, longitudinal epidemiological studies.
- The evidence is sufficient to conclude that, in the absence of audiograms obtained at the beginning and end of military service, it is difficult or impossible to determine with certainty how much of a specific individual's hearing loss was acquired during military service.

Chapter 4: Tinnitus

- The evidence is sufficient to conclude that noise doses associated with hearing loss are likely to be associated with tinnitus.
- The evidence was not sufficient to reach conclusions regarding the specific number or proportion of service members, overall or in specific branches or occupational groups, who report that tinnitus began or was exacerbated by noise exposure during military service.
- There is limited or suggestive evidence that exposure to impulse noise is associated with a greater likelihood of having tinnitus compared with exposure to steady-state noise.
- The evidence is sufficient to conclude that hearing loss (hearing thresholds greater than 25 dB HL at one or more audiometric frequencies between 250 and 8000 Hz) is associated with a higher prevalence of tinnitus.
- The evidence is not sufficient to determine precisely the magnitude of the risk of tinnitus associated with hearing loss.

Chapter 5: Responding to Noise Risks: Hearing Conservation Programs in the Military

- Compliance with requirements for use of hearing protection devices is crucial for an effective hearing conservation program. There is limited or suggestive evidence to conclude that use of hearing protection devices and the level of real-world hearing protection these devices provide have been and remain not adequate in military hearing conservation programs. However, the studies conducted in U.S. military personnel are generally consistent with studies from other settings that provide additional evidence that the use and real-world protection of hearing protection devices are not adequate.

• Results of annual audiograms are available for approximately half of military service members in hearing conservation programs reporting compliance with testing requirements during the period 1988–2003. Incomplete reporting, lack of compliance with requirements for annual audiograms, or both, severely limit the usefulness of the centralized database and the conclusions that can be drawn from it regarding hearing conservation program effectiveness.

• The evidence reviewed by the committee—including information on the effectiveness of available hearing protection devices and indicators regarding use of hearing protection, the completeness of audiometric monitoring, and compliance with requirements for entrance and separation audiograms—was sufficient to conclude that hearing conservation programs in the military are currently not adequate to protect the hearing of military service members, and have not been adequate for the period since World War II. This has important human health, personnel readiness, and financial implications.

Chapter 6: Reports of Audiometric Testing in Service Medical Records of Military Veterans

• Review of a sample of service medical records of military veterans indicates that compliance with requirements for audiometric testing at entrance into service has been limited, even in the most recent eras, and did not exceed 70 percent in any branch or era when using a ±60-day window for analysis.

• Review of a sample of service medical records of military veterans indicates that audiometric testing at separation from service has been limited, even in the most recent eras, and did not exceed 54 percent in any branch or era when using a ±60-day window for analysis.

• Review of a sample of service medical records of military veterans indicates that audiometric testing at both entrance into and separation from service has been extremely limited, even in the most recent eras, and did not exceed 34 percent in any branch or era when using a ±60-day window for analysis.

1

Overview of the Problem and Introduction

People serving in the military, especially those in areas of combat, will at some point be exposed to high-intensity noise of various types. Two possible consequences of such exposures are the development of a hearing loss, most prominent for high-frequency sounds, and tinnitus, typically referred to as "a ringing in the ears." Depending on a variety of factors, these effects may be either temporary or permanent consequences of such an exposure.

If documentation of the existence of hearing loss or tinnitus at discharge from the military is missing, it is nearly impossible to determine whether hearing loss or tinnitus detected by audiometric testing later in life is the result of noise exposure during prior military service. Both noise and aging, for example, result in similar high-frequency hearing loss, although the specific patterns of hearing loss resulting from each are generally distinguishable until 60–70 years of age (see Chapter 2). This adds to the challenge of determining the cause of the hearing loss when the only existing documentation consists of hearing thresholds measured late in life and many years after military service. In addition, it is quite likely that an individual might have experienced other hazardous noise exposures subsequent to discharge from military service that could result in significant noise-induced hearing loss or tinnitus. After the fact, for example, there are no means currently available to distinguish the hearing loss resulting from several years of military service from the noise-induced hearing loss resulting from subsequent work in a noisy industry or from participation in a wide variety of recreational activities, such as hunting (e.g., Clark, 1991). This serves to underscore the importance of measuring hearing thresholds at enlistment and at discharge, with annual measurements in between for those most at risk for noise-induced hearing loss and tinnitus.

These uncertainties regarding noise-induced hearing loss and tinnitus have placed the Department of Veterans Affairs (VA) in a quandary. Frequently, VA personnel are called on to determine whether the hearing loss measured in a 70- or 80-year-old veteran is due to this individual's prior military service. Furthermore, this assessment frequently must be done in the absence of documentation of the measurement of hearing thresholds at or around the time of military service (see Chapter 6). Even with a detailed case history from the veteran, it is next to impossible to draw a conclusion, with any degree of certainty, regarding the association of hearing loss in an older person with prior military service unless audiometric data acquired at entrance into and separation from military service are available.

VA reported that the 2.5 million veterans receiving disability compensation at the end of fiscal year 2003 had approximately 6.8 million separate disabilities related to their military service (Veterans Benefits Administration, 2004).[1] Disabilities of the auditory system, including tinnitus and hearing loss, were the third most common type, accounting for nearly 10 percent of the total number of disabilities among these veterans. For the 157,935 veterans who began receiving compensation in 2003, auditory disabilities were the second most common type of disability. These veterans had 75,316 disabilities of the auditory system out of a total of some 485,000 disabilities of all types. At the end of 2004, the monthly compensation payments to veterans with hearing loss as their major form of disability represented an annualized cost of some $660 million (Department of Veterans Affairs, 2005a). The corresponding compensation payments to veterans with tinnitus as their major disability were close to $190 million on an annualized basis (Department of Veterans Affairs, 2005b). Such staggering human and financial costs have served as the rationale for many reports examining hearing loss among military service members over the past several decades (e.g., Johnson, 1957; Yarington, 1968; Walden et al., 1971; Edwards and Price, 1989; Donahue and Ohlin, 1993; Rench et al., 2001).

CHARGE TO THE COMMITTEE

The charge to this committee arose from Public Law 107-330, which required VA to contract with the National Academies to review and evalu-

[1]Veterans may have hearing loss and other disabilities that have been determined to have been incurred during or aggravated by military service but that do not qualify for disability compensation payments (a "zero percent" service-connected disability). Veterans with service-connected hearing loss who do not qualify for any disability compensation payments are not included in the VA data on numbers of disabilities or numbers of veterans with disabilities. All veterans determined to have service-connected tinnitus qualify for compensation payments.

ate the available scientific evidence regarding the presence of noise-induced hearing loss and tinnitus in U.S. military personnel from World War II through 2002, when the legislation was enacted. Section 104 of this legislation is provided in Appendix A.

The National Academies assigned this work to the Medical Follow-up Agency of the Institute of Medicine (IOM). IOM staff worked with the VA to establish the following Statement of Task for the committee:

> An expert committee will provide recommendations to the Department of Veterans Affairs (VA) on the assessment of noise-induced hearing loss and tinnitus associated with military service in the Armed Forces. The committee will review staff-generated data on compliance with regulations regarding audiometric testing in the services at specific periods of time since World War II, review and assess available data on hearing loss in former service members, identify sources of potentially damaging noise during active duty, determine levels of noise exposure necessary to cause hearing loss or tinnitus, determine if the effects of noise exposure can be of delayed onset, identify risk factors for noise-induced hearing loss, and identify when hearing conservation measures were adequate to protect the hearing of service members. This study was mandated by Congress in Section 104 of Public Law 107-330. The committee will conduct its business through meetings over the course of the 24-month study and will issue a final report at the end of the study period.
>
> Staff of the Medical Follow-up Agency will identify veterans from each of the armed services (Army, Navy, Air Force, Marine Corps, and Coast Guard) and from each of the time periods from World War II to the present. A sample of the service medical records of these individuals will be obtained, examined for regulatory compliance regarding audiometric surveillance (including reference, periodic, and termination audiograms), abstracted, recorded, and tabulated.

The charge does not include consideration of effects of noise other than upon the auditory system, including hearing loss and tinnitus, nor of the issues surrounding assisted hearing through hearing aids or prosthetic devices. The study committee was selected to include members with expertise in audiology, bioacoustics, military preventive medicine, occupational medicine, industrial hygiene and hearing conservation programs, epidemiology, and otology.

It should be noted that Public Law 107-330 makes frequent reference to "acoustic trauma" in its charge to the committee (see Appendix A). At the committee's initial meeting in May 2004, discussion with congressional staff members clarified that the intent of the legislation was not the study of "acoustic trauma," which is a narrowly defined type of damage resulting from short-term, high-intensity noise exposure, but a study of the more broadly defined "noise-induced hearing loss," of which acoustic trauma is a subtype. It was also determined that the committee's charge did not include assessment of the disability or handicap resulting from noise-induced hearing loss or the means of assigning compensation to specific amounts or degrees of disability. The preceding Statement of Task incorporated these clarifications of the committee's charge.

The committee met five times from May 2004 through March 2005 and held numerous telephone conference calls through August 2005. During these meetings and conference calls, the committee reviewed and discussed the existing research literature on the topics central to its charge and received information during oral presentations made by representatives from various organizations, including several veterans and representatives of veterans' organizations, branches of the military, and consultants. In addition to these face-to-face meetings and telephone conference calls, the committee communicated frequently among themselves and with IOM staff via e-mail. This report represents the product of that information gathering and discussion. It has been divided into seven chapters. The primary purpose of this chapter, in addition to outlining the issues and the chronology of events subsequent to the passage of Public Law 107-330 as noted above, is to provide general background concerning the primary topics discussed in the ensuing chapters.

ACOUSTICS AND NOISE

Sound is produced by the propagation of pressure waves through a medium and originates from vibrating objects or from the rapid discharge or dissipation of energy, as in an explosive event. The pressure waves trigger responses in the auditory system of the listener. Noise generally refers to disagreeable or unwanted sound.

The magnitude or amplitude of a sound, including noise, can be measured in terms of sound pressure, in units of pascals, or sound intensity, in units of watts/m^2. More commonly, however, the level of the sound is expressed in terms of decibels (dB), which represent a logarithm of the ratio of two sound pressures or the two corresponding sound intensities. Specifically, the reference quantity in the denominator of the ratio is either a sound pressure of 20 micropascals or a sound intensity of 10^{-12} watts/m^2. The reference sound pressure level (SPL) for computation of decibels related to acoustic measurements was selected so that 0 dB SPL corresponds approximately to the lowest mid-frequency sound pressure that can be heard by the average normal-hearing young adult under ideal free-field listening conditions. At the other end of the scale, the maximum sound level that can be tolerated by most listeners is 120 dB SPL, with values exceeding 140 dB SPL, even for a brief instant, potentially resulting in permanent damage to the ear.

For application in the areas of the effects of noise on hearing, the sound levels are usually measured after being passed through a standardized filtering network, known as A-weighting, that attenuates the amplitude of the sound at frequencies below 500 Hz and above 10,000 Hz to roughly correspond to the perceived loudness of sound (see Appendix C). Sound levels measured with this filtering network are designated dBA.

For very brief impulse sounds, there are two common ways to express the level in dB. One is simply to use the fast-acting "peak" setting of a sound-level meter that is capable of measuring the true peaks of the sound wave. Such measures are often denoted as dBP. Another approach is to adjust the peak amplitude of the waveform for a steady-state sound (usually a 1000-Hz pure tone) so that it matches the peak amplitude of the waveform for the impulse. The level of the matching steady-state sound can then be measured with a sound level meter and, when doing so, the impulse is said to have the same "peak equivalent dB SPL" (pe dB SPL).

In addition to the overall level of the noise in dB, there are many other ways to characterize the relevant acoustic parameters of a noise. For the most part, however, these descriptions focus on characterizing the noise in either the time domain or the frequency domain. The frequency content of two noises, each with an overall level of 100 dBA, for example, can have a significant bearing on the resulting hearing loss measured (if any). Generally, all else being equal, sounds in the frequency range 2000–5000 Hz tend to be more damaging to human hearing than sounds with energy at lower or higher frequencies. With regard to the time domain, all else being equal, brief sounds are less damaging than longer sounds. For example, sounds with durations of less than a few milliseconds, frequently referred to in the present context as impulse noise, must exceed peak levels of 140 dBA to be considered hazardous, whereas a 15-minute steady-state sound is considered hazardous when its level exceeds 100 dBA (e.g., DoD, 2004). In the latter case, "hazardous" to hearing does not mean that hearing loss will occur following a single such exposure. With steady-state noise, the hazard occurs following repeated daily exposures for several years. This is the more common form of noise-induced hearing loss (NIHL), rather than that associated with a single extreme noise exposure, which is more appropriately referred to as "acoustic trauma."

Research over the past 60 to 70 years has shown that each of these acoustic parameters of noise—its sound pressure level, duration, type (impulse versus steady-state), and frequency content—can influence the hearing loss that is measured following the exposure to noise. The major influences of noise level and daily duration of exposure are captured in a single simplified metric, the noise *dose*. The noise dose represents the integration of noise level (more accurately, the underlying physical quantities) over the entire time of exposure. For a given exposure, the dose is of critical importance when evaluating the potential hazard to hearing of a particular noise.

The primary importance of the noise dose was recognized many years ago by the scientific community and has been incorporated into national and international standards designed to estimate the noise-induced hearing loss resulting from noise exposure (ISO-1999 [ISO, 1990]; ANSI S3.44 [ANSI, 1996]). Most often, the noise dose is specified in terms of the 8-hour equivalent continuous noise level in dBA and is derived from the time-weighted average (TWA) of the underlying physical quantities (e.g., sound pressure). When establishing a specific noise dose, a device known as a noise dosimeter is used. Parameters built into the noise dosimeter that can impact the measured noise dose include a dosimeter-specific threshold level, below which sound levels will not be measured, a criterion level, and an exchange rate. The latter two parameters are prescribed by various noise standards. Currently, a criterion level of 85 dBA and an exchange rate of either 3 dB or 5 dB are among the most widely implemented values. The exchange rate describes the trading relation between sound level and exposure duration that yields equivalent hazard for successive halvings of the exposure duration. To illustrate the tradeoff between sound level and duration of exposure that is built into noise dosimetry, assuming a criterion level of 85 dBA and a 3-dB exchange rate, an 8-hour continuous exposure to steady-state noise at 85 dBA would have the same noise dose as 88 dBA for 4 hours, 91 dBA for 2 hours, or 94 dBA for 1 hour.

For the specification of noise levels or dose to be relevant, the measures of noise dose must be made at the location of the individual being exposed and under actual or representative conditions. Ideally, it is the noise dose measured at the individual's ear at the time of exposure that is desired. Knowing, for example, that a jet engine produces a sound level of 140 dBA in the sound field at 1-meter distance is not particularly informative for an individual working daily (8 hours) about 30 meters from the engine (and while wearing a helmet or hearing protection). Assuming free-field conditions (and a point source), inverse-square-law behavior indicates that the increase in distance from the sound source will reduce the sound level to about 110 dBA at 32 meters. Furthermore, if a helmet or other hearing protection device is worn at the time of exposure, the noise level *at the ear* could be reduced by an additional 20–25 dB to safe levels.

It is known, however, that equivalent noise doses do not always yield equivalent noise-induced hearing loss. For example, consider three sequential noise exposures: one to a steady-state low-frequency noise, another to a steady-state high-frequency noise, and a third exposure to a series of impulses. Although the noise dose remains constant regardless of the sequence of these three noise exposures, the resulting hearing loss varies significantly with the order of their presentation (Mills, 1992; Ward, 1991).

THE MEASUREMENT OF HEARING AND TINNITUS

When studying the effects of noise on the auditory system, the two most common auditory complaints manifested in humans are hearing loss and tinnitus. In humans, hearing loss is typically measured by behaviorally determining the minimum sound pressure level that can be heard about 50 percent of the time, defined as the hearing "threshold," at each of several frequencies using pure tones ranging from 250 to 8000 Hz in octave steps. Pure tones at frequencies of 1500, 3000, and 6000 Hz are also frequently included, especially when sharp declines in high-frequency hearing are observed or anticipated. If an individual listener has a hearing threshold at a particular frequency that agrees perfectly with standardized values representing the average thresholds measured in a large group of young, normal-hearing adults for this same frequency, then the hearing threshold is said to be 0 dB "hearing level" or 0 dB HL. Hearing thresholds at each frequency from 250 to 8000 Hz do not need to be precisely equal to 0 dB HL, however, to be considered "normal." Rather, there is a range, generally accepted to be -10 to 25 dB HL at each frequency, that is considered to be representative of "normal" hearing in adults. The reliability of the clinical measurement of behavioral hearing thresholds in humans is such that a difference in thresholds must exceed 5 dB to be considered clinically significant (whether the threshold comparisons are across frequency at the same time or across time at the same frequency). As hearing thresholds increase beyond the maximum level of the normal range (> 25 dB HL), the degree of hearing loss in adults can be described as mild, moderate, moderately severe, severe, or profound. Table 1-1 shows threshold levels corresponding to each of these categories of hearing loss.

As described in more detail in Chapter 2, a hallmark of noise-induced hearing loss is the appearance of a hearing loss for high-frequency sounds, with the worse hearing thresholds typically occurring at frequencies of 3000–6000 Hz. Frequently, hearing is normal or near normal at lower frequencies (< 1000 Hz) and also returns toward the normal range at 8000 Hz. The result

TABLE 1-1 Categories of Hearing Loss and Corresponding Pure-Tone Thresholds for Adults

Category of Hearing Loss	Pure-Tone Threshold
Normal	< 25 dB HL
Mild	26–40 dB HL
Moderate	41–55 dB HL
Moderately severe	56–70 dB HL
Severe	71–90 dB HL
Profound	> 90 dB HL

is a pattern of hearing loss across frequency referred to as a "noise notch" (see Chapter 2). It is usually the noise-notch pattern of hearing loss across frequencies, together with supporting evidence from a detailed case history, that lead to the diagnosis of noise-induced hearing loss. Because noise-induced hearing loss is confined primarily to frequencies at or above 2000 Hz, the effects on auditory perception can be subtle. In addition to producing difficulty hearing high-frequency pure tones, the hearing loss may frequently have a negative impact on the perception of other high-frequency sounds, including several consonant sounds of speech and many environmental sounds. These difficulties may not be readily apparent in quiet listening conditions, but become prominent when there are competing sounds in the background, such as noise or other people talking. As the condition grows more severe, it can interfere with the ability to function socially and professionally (see NRC, 2005). Currently, damage to the ear as a result of noise exposure is not reversible in humans. The most common treatment is amplification of sound through the use of hearing aids.

Tinnitus is the other common complaint of those exposed to hazardous noise (see Chapter 4). Noise-induced tinnitus is a subjective, self-reported phenomenon that, unlike hearing loss, cannot be verified objectively, although certain perceptual attributes (e.g., loudness and pitch) can be established reliably under controlled conditions (psychoacoustic testing) (see Chapter 4). In addition, the mechanisms underlying tinnitus are less well understood than those underlying noise-induced hearing loss. Self-report questionnaires reveal a wide range of severity that is not directly correlated with the severity of any associated noise-induced hearing loss. Some individuals find the effects of tinnitus to be more debilitating than the effects of hearing loss. Although no current form of treatment can eliminate tinnitus, various treatment approaches are being used to reduce the adverse impact of tinnitus and potentially find a cure.

STUDYING THE EFFECTS OF NOISE ON HEARING AND TINNITUS

The modern era of research on the effects of noise on hearing began in the 1940s (e.g., Davis et al., 1949). Most of this research falls into one of three categories: (1) prospective studies of temporary hearing loss in humans; (2) retrospective analyses of permanent hearing loss in humans; and (3) laboratory animal studies of both temporary and permanent effects of noise on the auditory system. Research on tinnitus has used similar approaches, but the subjective nature of tinnitus has posed additional challenges as investigators have worked to develop and validate animal models of tinnitus.

Studies of Temporary Threshold Shift

Prospective studies of temporary hearing loss in humans have followed similar protocols. First, preexposure hearing thresholds are measured for pure tones at one or more frequencies. Next, the listener is exposed for minutes or hours to a sound of some type and level. Finally, the post-exposure hearing threshold is measured immediately following cessation of the exposure. Depending on the specific acoustic parameters for the noise exposure, the postexposure hearing thresholds may or may not be greater than the corresponding preexposure hearing thresholds. If the thresholds have worsened from preexposure to postexposure, this is designated as a temporary threshold shift (TTS). Its transient nature is confirmed through repeated postexposure measurements at different recovery times that reveal an eventual return to the preexposure hearing thresholds.

Some key advantages of this approach to the study of the effects of noise on hearing include the use of human subjects rather than laboratory animals, precise control of the acoustical parameters of the noise exposure, and careful measurement of hearing thresholds under optimal listening conditions. The primary shortcoming of this approach has been in generalizing the results of short-term experiments on TTS to the permanent hearing loss resulting from years of repeated exposures or exposure to high noise levels. That is, TTS may not be predictive of eventual permanent changes in hearing thresholds. There have been some suggestions that the mechanisms underlying the temporary and permanent changes in hearing following noise exposure may be different (Nordmann et al., 2000). In addition, ethical considerations restrict the range of exposure conditions that can be examined while still ensuring that the hearing loss is only temporary.

Studies of Permanent Threshold Shift

The laboratory studies of TTS in humans have been complemented by retrospective field studies of permanent threshold shifts (PTSs) in humans. The basic approach here has been to study the hearing loss measured in workers employed in industries with well-defined noise exposures for periods of many years. Of course, use of a valid and reliable means of measuring hearing thresholds is critical. Furthermore, these studies have almost always been cross-sectional rather than longitudinal.

The advantages inherent to retrospective field studies of PTS include the use of human subjects and more direct study of the permanent effects of noise on hearing than drawing inferences from the study of temporary effects. Disadvantages include the cross-sectional nature of the data, resulting in possible cohort bias, and limitations in the ability to generalize from the data because the study populations are often highly selected. With regard to cohort bias, the assumption in the cross-sectional approach is that the separate groups or cohorts differ only in terms of the independent variable under investigation. For field studies of PTS, the independent variable of interest is often the length of noise exposure. In this case, there may be other confounding factors associated with differences in experiences across generations, rather than with length of noise exposure. The hearing loss in a group of 40-year employees born in 1920, for example, might differ from the hearing loss in a group of 20-year employees born in 1940 for reasons other than their age and length of employment in a noisy industry.

Another important weakness of these studies is the lack of control over the noise exposure, which can vary from individual to individual or location to location, as well as the lack of control over noise exposures occurring outside of the workplace. In addition, a central issue in all such studies involves the interaction of the effects of noise and aging on hearing. Studies of noise-induced PTS, for example, usually "correct" the actual hearing thresholds measured by an amount corresponding to the hearing loss that is assumed to occur in individuals of the same age who were not exposed to noise. One such correction is simple decibel additivity in which the average age-associated hearing loss is subtracted from that measured in the individual, with the resulting amount being considered the "noise-induced permanent threshold shift," or NIPTS. For example, if a 60-year-old man has worked in a specific noise environment for 40 years and has a hearing loss of 70 dB HL at a particular frequency, and the typical hearing loss at this same frequency is 40 dB HL for a 60-year-old man who has not worked in noise, then the NIPTS is presumed to be 30 dB according to this simple dB-additivity rule. The implications of this rule are discussed in greater detail in Chapter 2, but it is apparent that both the amount of hearing loss assumed for a comparable age- and gender-matched non-noise-exposed cohort, as well as the manner in which the age-related and noise-related hearing loss combine, are critical to the derivation of the NIPTS and our understanding of it.

Laboratory Animal Studies

Animal studies of TTS and PTS, as well as other aspects of the effects of noise on hearing, offer an approach that eliminates several of the disadvantages inherent in human studies noted above. Specifically, noise exposures can be under strict control and both TTS and PTS can be measured in the same animals at various times during the animals' life spans. Furthermore, a variety of other measures beyond behavioral measurement of hearing thresholds, including a number of physiological and anatomical measurements, can also be obtained from the animals following completion of a particular noise exposure. The primary disadvantage to this approach, however, lies in the difficulty in generalizing the findings to humans, especially with regard to hazardous noise doses. Frequently, only qualitative comparisons can be made across species. The species most commonly used in laboratory studies of noise-induced hearing loss have been the guinea pig, chinchilla, gerbil, and cat.

In summary, each of the three fundamental approaches to the study of the effects of noise on hearing that have been used over the past 60–70 years has inherent advantages and disadvantages. The scientific community's understanding of the effects of noise on hearing has been enhanced through integration of findings making use of all of these approaches.

APPROACHES TO HEARING CONSERVATION

Three approaches can be taken to reduce the occurrence of noise-induced hearing loss and tinnitus, whether from industrial or military exposures to noise. First, through engineering, the equipment or devices producing the noise can be redesigned to reduce the sound levels generated at the source. Although there have been successful efforts to do so in many branches of the military (e.g., Yankaskas and Shaw, 1999), there are limitations to the effectiveness of this approach to hearing conservation. Many military operations, especially those on the battlefield or in the training for the battlefield, are inherently noisy. A second approach is to identify those individuals who are susceptible to noise-induced hearing loss or tinnitus prior to exposure to high-intensity sound and isolate or protect those with greater vulnerability to the damaging effects of noise. Identification of individual differences in susceptibility to noise-induced hearing loss, however, has proven to be an elusive goal (see Chapter 2). A third approach is to design and implement a hearing conservation program, which can also contribute to protection against tinnitus. Such programs educate noise-exposed populations about the hazards of high-intensity noise, measure the hearing thresholds of personnel on a regular basis, and instruct individuals in the use of personal hearing protection devices. In this approach, the goal is to attenuate the noise to safe levels *at the ears* of at-risk individuals. Hearing conservation programs focus on the prevention of damage to hearing and do not typically include work with hearing aids or other devices to assist individuals with hearing impairments. The implementation of hearing conservation programs has been the most viable approach in the majority of industrial and military settings and is frequently the method of choice, but the effectiveness of the programs varies. In the military, the types of hearing protection devices available include earplugs, earmuffs, and helmets. The attenuation characteristics of these devices, as well as the pros and cons of each type, are reviewed in Chapter 5.

The use of hearing protection devices as the primary means of hearing conservation, however, has some limitations, especially in the military context. For example, there are many unexpected exposures to high-intensity sounds in the military, especially under combat conditions or training for such conditions (see Chapter 3). Depending on the specific circumstances of the exposure (see Chapter 2), it is possible for a single such exposure to result in significant hearing loss and tinnitus (e.g., Mrena et al., 2004). Furthermore, the most commonly used hearing protection devices have been conventional passive devices that provide the same amount of attenuation regardless of sound level. As a result, the device designed to protect the wearer's hearing from high-level noise also makes it difficult to hear lower level sounds, such as the voice of a commander, a fellow soldier, or an approaching enemy. The recent widespread introduction (in 2004) into the military of level-dependent hearing protection devices designed to provide increasing attenuation for higher level impulse sounds, leaving low- and moderate-level sounds unaffected, is a potentially important development.

Research is also being done to explore pharmacological approaches to reducing susceptibility to noise-induced hearing loss. For example, studies with laboratory animals have found beneficial effects from the administration of antioxidants (e.g., Henderson et al., 1999; Kopke et al., 2005; McFadden et al., 2005). A clinical trial is testing an antioxidant compound in Marine Corps recruits (Boswell, 2004), but results had not been reported at the time the committee completed its work. Studies in animals and humans have also investigated protective effects of supplemental oral magnesium (e.g., Attias et al., 1994; Scheibe et al., 2000; Attias et al., 2004).

Various means have been used to evaluate the effectiveness of hearing conservation programs. These approaches to evaluation are described in Chapter 5. One metric with widespread use in the military is the measurement of significant threshold shift (STS). Although the precise definition of STS has changed in the military over time and across branches of the military (see Chapters 3 and 5), the basic approach has been to try to identify individuals *as soon as* they show any signs of possible noise-induced hearing loss. It is important to note, however, that STS is not a measure of hearing loss in dB HL. Rather, it is a relative shift in threshold between the current hearing threshold and a previously established reference threshold for that same individual. If 10 dB is used to define an STS, for example, then this could represent a change in hearing from 0 to 10 dB HL (both still within "normal" hearing) between the two measurements or from 20 to 30 dB HL (from "normal" hearing to "mild" hearing loss).

Regular measurement of hearing is critical to evaluating programs. The participants in hearing conservation programs of the military are currently required to have hearing thresholds measured annually. Obviously, if this is not taking place, then an STS-based approach to hearing conservation will not be appropriate. As a result, one measure of program effectiveness can simply be the percentage of individuals in the program who receive the required annual measurement of hearing thresholds. Another measure of program effectiveness is the incidence of STS among those individuals in the hearing conservation program. Ideally, STS-based approaches include steps to verify that observed STS values are not the result of TTS, usually through follow-up measures of hearing thresholds obtained after prescribed periods of quiet. STS cases that remain unchanged at follow-up, or those STS cases that do not receive follow-up, are considered to be permanent threshold shifts. The incidence of PTS cases, therefore, is another possible metric of hearing conservation program effectiveness. Military hearing conservation programs currently mandate follow-up testing, but it is not always completed. These metrics are examined in greater detail in Chapters 3 and 5.

EVALUATING THE STRENGTH OF EVIDENCE

To address the questions posed to the committee by the statement of task, efforts were made to identify a relevant body of evidence through searches of the indexed medical literature and catalogues of reports prepared by or for the military services. Studies and reports were also identified from the reference lists of other documents, and some documents were provided by the military services at the committee's request.

Published peer-reviewed reports generally carried the most weight in drawing conclusions because the methods and findings of those reports could be assessed. Reports that had not undergone peer review and some unpublished data were also considered by the committee and evaluated in the context of the available body of published literature.

Ideally, in addressing the charge to the committee, the committee would have preferred to draw on data from reports of longitudinal, population-based studies of noise-induced hearing loss or tinnitus in humans in military settings. Clearly, such studies would offer the greatest strength of evidence to support the committee's findings and recommendations. Unfortunately, there are few such studies. Therefore, the committee was compelled to turn to other sources of evidence to address its charge.

The sources of evidence considered by the committee included epidemiological, laboratory, and clinical studies directly addressing the question at hand. Epidemiological studies generally carry the most weight in evaluating evidence for or against an association between an exposure (noise) and the resulting health outcome (hearing loss or tinnitus) in humans. These studies measure health-related exposures and outcomes in a defined set of human subjects and use that information to make inferences about the nature and strength of associations between such exposures and outcomes in the population from which the study sample was drawn. Epidemiological studies can be categorized as experimental (clinical trial) or observational and as controlled (analytic) or uncontrolled (descriptive).

The primary outcome of interest in epidemiological studies is usually the incidence or prevalence of the health condition under investigation. The incidence of a particular condition refers to the number of newly occurring cases of that condition that develop over a specific period of time in a particular population and is expressed as either a risk (a probability) or a rate. A condition's prevalence is the proportion of individuals in a sample who have that condition at a single point in time or during an interval of time. Risk, in the epidemiological sense, is defined as the probability of developing a particular health condition. The term "relative risk" refers to the ratio of the incidence of the condition in a population exposed to some potential hazard of interest, such as occupational noise, to the corresponding incidence in a similar but nonexposed group. Cross-sectional studies do not directly measure the risk associated with an exposure for two important reasons: (1) these studies do not automatically define whether the exposure or the condition came first; and (2) cross-sectional samples usually contain old as well as new cases (i.e., incident and prevalent cases), further obscuring the temporal sequence of exposure and condition.

Among the various epidemiological designs, experimental studies generally have the advantage of random assignment to exposures and, therefore, have the potential to be the most influential in assessing the strength and direction of an association, although they are subject to a potential selection bias. Experimental studies of noise exposure and hearing loss or tinnitus in humans must be designed to prevent permanent harm. As a result, such studies can be conducted to study only those exposures resulting in *temporary* hearing loss or tinnitus.

Most of the epidemiological studies considered by the committee were observational. There were few prospective observational studies relevant to the committee's charge. Most were cross-sectional rather than longitudinal. Observational studies that compare exposed subjects and unexposed controls are more definitive than uncontrolled studies, but uncontrolled studies are also important for showing the presence of an outcome in an exposed population. Most of the epidemiological studies considered by the committee were studies without control groups. Furthermore, the biggest drawback with cross-sectional epidemiological studies is that the outcome of interest (e.g., a hearing threshold > 25 dB HL) is measured only once—at the time of the study or some other specified point—making it impossible to demonstrate that the outcome occurred *after* the exposure of interest (e.g., noise), a temporal relationship necessary to establish a causal link between the two.

Among epidemiological research designs, case reports and case series are generally weakest. They are inadequate by themselves to establish an association, but they can be valuable in drawing the attention of the scientific community to the problem and in generating testable hypotheses. The committee did not rely on case reports in reaching its conclusions.

The vast majority of data available on noise-induced hearing loss and tinnitus in military personnel is not epidemiological. The data came from a variety of clinical, descriptive, cross-sectional studies of variously defined groups of military personnel. The data were reported in ways that gave little or no indication of the prevalence or incidence of either hearing loss or tinnitus. Instead, the dependent measures were generally hearing thresholds at various pure-tone frequencies, which were reported as average thresholds for groups defined by age or length of service in the military. In the absence of control groups in most of these studies, the committee turned to standardized compilations of "control data" on hearing thresholds for groups of screened or unscreened individuals of various ages for comparison purposes. These data and the limitations associated with this approach are described in more detail in Chapter 3.

Some of the questions posed in the charge to the committee could be addressed by existing data on noise-induced hearing loss, much of which is based on laboratory studies of humans and animals. As noted, the laboratory studies in humans are necessarily restricted to the investigation of temporary effects, either TTS or temporary tinnitus. Central to this experimental approach is the assumption that the TTS observed at 2 minutes postexposure has a defined relationship with the PTS that occurs in humans following 10–20 years of exposure to industrial noise (CHABA, 1968). To the extent that this association is valid, laboratory studies of TTS in humans can provide insights into exposure parameters affecting PTS. In nearly all cases of laboratory studies in humans, the dependent variable has been some measure of hearing threshold. These studies cannot provide precise estimates of the risk of experiencing hearing loss or tinnitus from noise exposure.

As noted previously, studies of noise-induced hearing loss in laboratory animals make it possible to examine the associations among TTS, PTS, and underlying cochlear damage in the same set of subjects and under strict laboratory control of the exposure. Such data represent a powerful tool for understanding the mechanisms underlying noise-induced hearing loss in a variety of mammalian species, including variables impacting the development of and recovery from noise-induced hearing loss, as well as establishing the relationships among TTS, PTS, and underlying pathology. The dependent measures in these laboratory studies typically include some measure of hearing thresholds, measured behaviorally or physiologically, and measures of anatomical damage. Where the results from laboratory animal studies are consistent with findings from human studies, they add assurance that the human results are biologically reasonable.

Some aspects of the committee's charge were best addressed with data from well-designed and carefully executed human epidemiological studies.

When such data were not available, the committee turned to alternate data with the resulting caveats to its findings noted. Other aspects of the committee's charge were best addressed with data from well-designed and carefully executed laboratory studies with humans and animals. Both forms of evidence are considered valid, depending on the issue or questions being addressed, and have been weighed by the committee in evaluating the strength of evidence supporting its findings.

With the foregoing in mind, the committee adopted the following scale for the strength of evidence. As will be seen, the strength of evidence in this scale is tied to the presence and number of "strong studies" supporting a particular committee finding. In general, observational epidemiological studies cannot by themselves establish causal associations. Strong epidemiological studies in support of a statistical association between an exposure and a condition, whether causal or not, could include well-designed cross-sectional studies where the likelihood of chance findings has been minimized, known confounding factors have been considered in the analysis, and known or potential biases have been eliminated. However, in support of a causal association, "strong studies" are generally well-designed, prospective observational human population studies or randomized controlled trials in which chance, bias, and confounding are similarly treated. With respect to laboratory studies, "strong studies" are well-designed and carefully executed and interpreted human *or animal* studies in which chance, bias, and confounding have also been treated in a similar way.

Sufficient evidence of a causal relationship: Consistent evidence from *many* strong longitudinal studies.

Sufficient evidence (of an association): Evidence from *several* strong longitudinal or cross-sectional studies.

Limited or suggestive evidence (of an association): No evidence from strong studies, but some evidence from other studies of sufficient quality.

Not sufficient evidence to determine whether an association exists: *Few or no* studies of sufficient quality.

Sufficient evidence that no association exists: Several strong studies that find no association.

However, when applying the foregoing scale for strength of evidence, the context of the specific question being addressed must be kept in mind. For example, if the specific question posed or the issue addressed pertains to the effect of noise *on humans* and the *only* evidence available is from studies of laboratory animals, this evidence is considered not to be sufficient regardless of the number of "strong" studies available from laboratory animals.

THE COMMITTEE'S REPORT

The remainder of the report summarizes the evidence regarding the questions put to the committee concerning military service and noise-induced hearing loss and tinnitus and presents the committee's findings. Chapter 2 reviews the mechanisms of noise-induced hearing loss and evidence regarding the impact of various risk factors. Chapter 3 reviews noise and noise hazards associated with military service. Chapter 4 focuses on tinnitus, especially its association with noise exposure and hearing loss. Chapter 5 turns to the nature and effectiveness of hearing conservation programs in the armed services. Chapter 6 presents the results of an audit of the service medical records of military personnel sampled from various periods of service from World War II to 2002. Finally, Chapter 7 provides a summary that draws on the information presented in preceding chapters to address the specific questions and issues posed in the Statement of Task and in Public Law 107-330.

REFERENCES

ANSI (American National Standards Institute). 1996. *ANSI S3.44 Determination of Occupational Noise Exposure and Estimation of Noise-Induced Hearing Impairment.* New York: Acoustical Society of America.

Attias J, Weisz G, Almog S, Shahar A, Wiener M, Joachims Z, Netzer A, Ising H, Rebentisch E, Guenther T. 1994. Oral magnesium intake reduces permanent hearing loss induced by noise exposure. *American Journal of Otolaryngology* 15(1):26–32.

Attias J, Sapir S, Bresloff I, Reshef-Haran I, Ising H. 2004. Reduction in noise-induced temporary threshold shift in humans following oral magnesium intake. *Clinical Otolaryngology and Allied Sciences* 29(6):635–641.

Boswell S. 2004, February 17. New weapon against hearing loss? Marines participating in clinical trial to prevent hearing loss. *ASHA Leader*. Pp. 1, 20–21.

CHABA (National Academy of Sciences–National Research Council Committee on Hearing, Bioacoustics, and Biomechanics). 1968. *Proposed Damage-Risk Criterion for Impulse Noise (Gunfire).* Report of Working Group 57 (Ward WD, ed.). Washington, DC: National Academy of Sciences.

Clark W. 1991. Noise exposure from leisure activities: A review. *Journal of the Acoustical Society of America* 90(1):175–181.

Davis H, Parrack H, Eldredge D. 1949. Hazards of intense sound and ultrasound. *Annals of Otology, Rhinology, and Laryngology* 58(3):732–738.

Department of Veterans Affairs. 2005a. Service-Connected Compensation—Veterans with Hearing Impairment as of December 21, 2004. Data provided to the Institute of Medicine Committee on Noise-Induced Hearing Loss and Tinnitus Associated with Military Service from World War II to the Present, Washington, DC.

Department of Veterans Affairs. 2005b. Veterans Receiving VA Disability Compensation with Tinnitus as Their Major Disability and Their Monthly Dollar Awards by Period of Service. Data provided to the Institute of Medicine Committee on Noise-Induced Hearing Loss and Tinnitus Associated with Military Service from World War II to the Present, Washington, DC.

DoD (Department of Defense). 2004. *Department of Defense Instruction 6055.12: DoD Hearing Conservation Program*. Washington, DC: DoD.

Donahue AM, Ohlin DW. 1993. Noise and the impairment of hearing. In: Deeter DP, Gaydos JC, eds. *Occupational Health: The Soldier and the Industrial Base*. Falls Church, VA: Office of the Surgeon General.

Edwards RR, Price DR. 1989. *Descriptive Analysis of Medical Attrition in U.S. Army Aviation*. Fort Rucker, AL: U.S. Army Aeromedical Research Laboratory.

Henderson D, McFadden SL, Liu CC, Hight N, Zheng XY. 1999. The role of antioxidants in protection from impulse noise. *Annals of the New York Academy of Sciences* 884:368–380.

ISO (International Organization for Standardization). 1990. *ISO 1999: Acoustics—Determination of Occupational Noise Exposure and Estimation of Noise-Induced Hearing Impairment*. Geneva, Switzerland: ISO.

Johnson KO. 1957. Problems in military audiometry: A CHABA symposium. Veteran's compensations for hearing loss. *Journal of Speech and Hearing Disorders* 22(5):731–733.

Kopke R, Bielefeld E, Liu J, Zheng J, Jackson R, Henderson D, Coleman JK. 2005. Prevention of impulse noise-induced hearing loss with antioxidants. *Acta Otolaryngologica* 125(3):235–243.

McFadden SL, Woo JM, Michalak N, Ding D. 2005. Dietary vitamin C supplementation reduces noise-induced hearing loss in guinea pigs. *Hearing Research* 202(1-2):200–208.

Mills JH. 1992. Noise-induced hearing loss: Effects of age and existing hearing loss. In: Dancer A, Henderson D, Salvi RJ, Hamernik RP, eds. *Noise-Induced Hearing Loss*. St. Louis, MO: Mosby Year Book. Pp. 237–245.

Mrena R, Savolainen S, Pirvola U, Ylikoski J. 2004. Characteristics of acute acoustical trauma in the Finnish Defence Forces. *International Journal of Audiology* 43:177–181.

Nordmann AS, Bohne BA, Harding GW. 2000. Histopathological differences between temporary and permanent threshold shift. *Hearing Research* 139(1-2):13–30.

NRC (National Research Council). 2005. *Hearing Loss: Determining Eligibility for Social Security Benefits*. Dobie RA, Van Hemel S, eds. Washington, DC: The National Academies Press.

Rench ME, Johnson S, Sanders T. 2001. *Cost Benefit Analysis for Human Effectiveness Research: Bioacoustic Protection*. Wright-Patterson Air Force Base, OH: Air Force Research Laboratory.

Scheibe F, Haupt H, Ising H. 2000. Preventive effect of magnesium supplement on noise-induced hearing loss in the guinea pig. *European Archives of Oto-Rhino-Laryngology* 257(1):10–16.

Veterans Benefits Administration. 2004. *Veterans Benefits Administration Annual Benefits Report for FY 2003*. Washington, DC: Department of Veterans Affairs.

Walden BE, Worthington DW, McCurdy HW. 1971. *The Extent of Hearing Loss in the Army: A Survey Report*. Washington, DC: Walter Reed Army Medical Center.

Ward WD. 1991. The role of intermittence in PTS. *Journal of the Acoustical Society of America* 48:561–574.

Yankaskas KD, Shaw MF. 1999. Landing on the roof: CVN noise. *Naval Engineers Journal* 111(4):23–34.

Yarington CT. 1968. Military noise induced hearing loss: Problems in conservation programs. *Laryngoscope* 78(4):685–692.

2

Noise-Induced Hearing Loss

The purpose of this chapter is to provide background material on noise-induced hearing loss to facilitate understanding of the evidence on noise-induced hearing loss in military personnel presented in Chapter 3. The chapter begins with a general discussion of the structure and function of the auditory system, with particular emphasis on the periphery, and the impact of noise on the peripheral auditory system. The effects of noise on hearing thresholds are reviewed next, followed by a review of the time course for the development of hearing loss from noise exposure. Next, exogenous and endogenous risk factors that may alter an individual's susceptibility to noise-induced hearing loss are reviewed. This is followed by a discussion of national and international standards that have been developed to estimate the amount of noise-induced hearing loss to be expected from a given noise exposure and to separate the effects of noise from age-related changes in hearing.

MECHANISMS AND MODELS OF NOISE-INDUCED HEARING LOSS

Structure and Function of the Hearing Apparatus

In humans and other mammals, the auditory system consists of the external, middle, and inner ears (Figure 2-1), as well as the central auditory pathways in the brain. Sound waves enter the external ear through the pinna, travel through the external ear canal, and strike the eardrum. The external ear boosts high-frequency (2000–5000 Hz in humans) sound en-

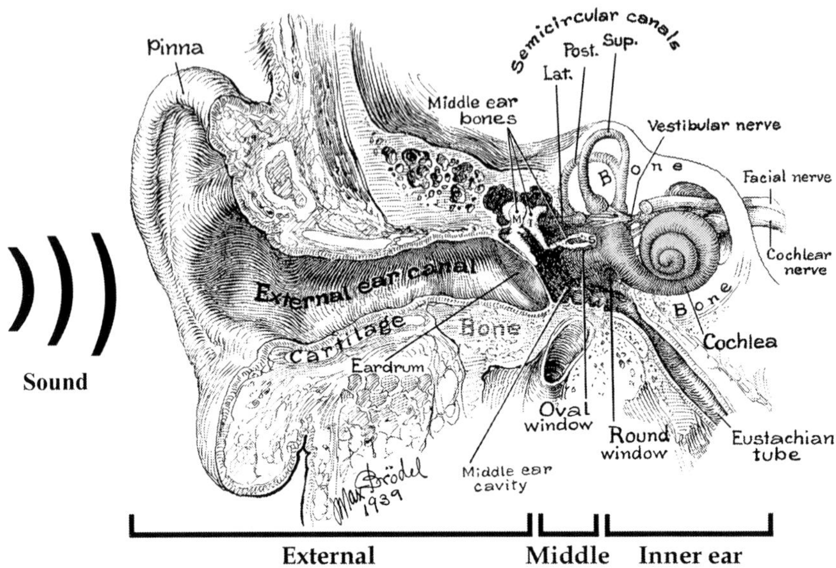

FIGURE 2-1 Semi-schematic drawing of the human ear. Sound waves enter the pinna, travel through the external ear canal, and strike the eardrum, setting it in motion. Motion of the eardrum sets the middle ear bones (malleus [M], incus [I], and stapes [S]) in motion and ultimately generates pressure waves in the fluids of the inner ear. Sensory cells in the hearing portion of the inner ear (i.e., cochlea) are then stimulated. When the fibers of the cochlear nerve are stimulated by the sensory cells, auditory information is transmitted to the brain.
SOURCE: Modified from Brödel (1946).

ergy by about 20 dB before it strikes the eardrum (Shaw, 1974). The eardrum vibrates when sound waves strike it, setting the middle-ear bones (malleus, incus, stapes) (Figure 2-1) in the air-filled middle-ear cavity in motion. The base of the stapes is fitted into the oval window of the hearing portion of the fluid-filled inner ear, the cochlea. Movement of the stapes sets up pressure waves in the fluids inside the cochlea, which contains the organ of Corti, the sensory organ for hearing, spiraling from base to apex. The primary sensory receptors for hearing, the inner hair cells, are found within the organ of Corti as are the outer hair cells, which primarily facilitate the sensory response of the inner hair cells. The pressure waves within the cochlea vibrate the basilar membrane and the attached organ of Corti (Figure 2-2). Specific sound frequencies vibrate specific places along the length of the cochlea, with high-frequency sound causing maximum vibration in the base of the cochlea and low-frequency sound causing maximum vibration in the apex. In addition, as the intensity of sound increases, the

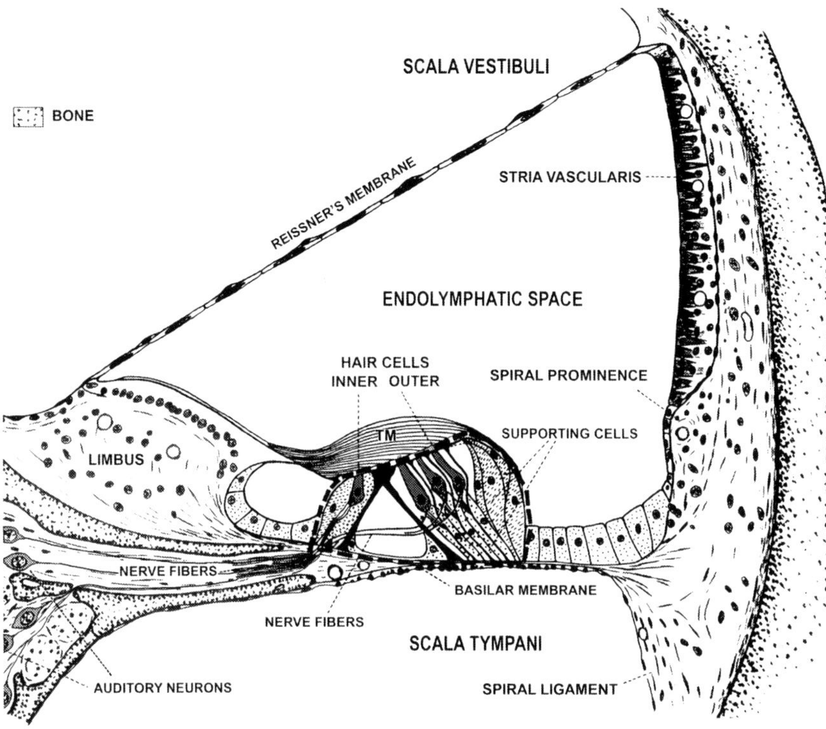

FIGURE 2-2 Cross-section of one turn of the spiral-shaped cochlea. The organ of Corti (outlined by the black dashed line) is attached to the flexible basilar membrane and is surrounded by large fluid spaces (i.e., scala vestibuli, endolymphatic space, scala tympani). The organ of Corti contains sensory cells (i.e., inner and outer hair cells) that respond to pressure waves in the fluid spaces by releasing neurotransmitter from their bases. The nerve fibers that terminate on the hair cell bases are extensions of the auditory neurons. The nerve fibers conduct auditory information to the brain when the hair cells release neurotransmitter. TM = tectorial membrane.
SOURCE: Modified from Davis and Associates (1953).

amplitude of basilar membrane vibration also increases, although in a nonlinear, compressive manner over much of its operating range.

The mechanical activity of the basilar membrane leads to mechanical stimulation of the inner and outer hair cells. From the surface of each hair cell, thin hair-like processes (stereocilia) project into the overlying gelatinous tectorial membrane (Figure 2-2). Movement of the basilar membrane and organ of Corti relative to the tectorial membrane deflects the stereocilia and opens ion channels in the hair cells. Channel opening depolarizes the hair cells so they release a neurotransmitter from their bases. This conversion of mechanical energy from basilar membrane vibration to neuroelectrical energy by the sensory cells in the organ of Corti is a process involving high levels of metabolic activity. The nerve fibers connected to the hair cells, primarily the inner hair cells, are excited by the neurotransmitter and transfer the auditory information to the brain.

Effects of Noise on Hearing

The magnitude of hearing loss that results from excessive exposure to noise depends on factors associated with the exposure (e.g., sound pressure level [SPL], duration, type of noise, and frequency), as well as the characteristics of the individual being exposed (e.g., susceptibility to noise damage, age, prior history of hearing/ear damage). In the next section, we examine the influence of the type of noise in greater detail.

Impulse/Impact Noise

High-level, short-duration noise can arbitrarily be categorized as impulse noise, which is the product of explosive devices (e.g., gunfire), or impact noise, which is generated by the forceful meeting of two hard surfaces (e.g., a hammer to a nail, impact wrenches). The typical measures of impulse noise are the initial peak level and the duration of the first overpressure. This is the A-duration and is less than 1 millisecond (msec) for handguns and several msec for large cannons. For impact noise, the two principal descriptors are the highest peak in a series of successive peaks (reverberations) and the so-called B-duration, the duration from the highest peak level to a point in time when the reverberations have decayed either 10 or 20 dB. B-durations range from 50 to 300+ msec. The distinction between impulse and impact noise becomes blurred in many real-life situations because impulse noise can reflect off the ground, or other surfaces, and the reflections add to the initial impulse noise, creating a large, more complicated waveform that is best described using the B-duration (Hamernik and Hsueh, 1991).

Impulse noise creates several special hazards to the auditory system. First, the high peak levels associated with gunfire (140–190 dB pe SPL)[1] may damage the cochlea by causing rapid mechanical failure and injury (Henderson and Hamernik, 1986). A series of rapidly occurring impulses

[1]As noted in Chapter 1, various metrics have been used in the literature to quantify the sound levels associated with impulse and impact noise, including dBP and dB pe SPL. Sound levels for steady-state noise, on the other hand, are more commonly expressed as dBA. Since simple conversions among these various metrics are not possible, the committee chose to report sound levels using the specific metric employed in the studies reviewed.

can be partially attenuated by the acoustic reflex, a reflexive contraction of the middle-ear muscles, while isolated impulses reach the cochlea before the activation of the acoustic reflex. Thus, intense explosions may result in large cochlear lesions and significant hearing losses. This damage is termed "acoustic trauma," and hearing at most frequencies may be affected (e.g., Ward and Glorig, 1961). Additional symptoms include a sense of fullness in the ears, speech sounding muffled, and a ringing in the ears (i.e., tinnitus) (Kraus, 1959; Ward and Glorig, 1961). Although some recovery of hearing takes place after an acoustic trauma episode, the individual is often left with a severe, permanent hearing loss (Ward and Glorig, 1961; Van Campen et al., 1999). Exposure to impulse noise can result in acoustic trauma from a limited number of exposures, including a single exposure, but can also result in conventional noise-induced hearing loss from extended periods of exposure to impulse noise over many weeks, months, or years.

The relationship between noise-induced hearing loss and the peak amplitude of an impulse or impact noise is complicated. Systematic research with the chinchilla has shown that at the lower range of exposure to impulse noise (< 140 dB pe SPL) or impact noise (< 115 dB pe SPL), the chinchilla develops a hearing loss that is proportional to the total energy of the exposure (peak level × number of impulses). However, above these peak sound pressure levels, the auditory system is damaged primarily by the large displacements caused by high peak levels. The dividing line between the "energy" and "peak-level" behavior is referred to as the "critical level." It should be noted that the critical levels of about 140 dB SPL for impulse noise and 115 dB SPL for impact noise are general approximations for the chinchilla. The actual critical level is dependent on the specific waveform of the impulse and impact noise (Henderson and Hamernik, 1986). Based on across-species comparisons from chinchillas to humans, the critical levels for humans are likely to be approximately 10 dB higher than those observed in chinchillas. However, because of the high risk of hearing loss from high-level impulses and the variability in subsequent noise-induced hearing loss, a more conservative criterion of 140 dB SPL has been adopted for humans.

Below the critical level, hearing loss grows by the rate of approximately 1–3 dB of hearing loss for each dB of increase in peak level. However, above the critical level, hearing loss grows 3–7 dB for each dB increase in the level of the impulse or impact noise. This accelerated growth of hearing loss with increase in peak sound pressure level above the critical level is one of the factors that make high-level impulse and impact noise particularly dangerous (Henderson and Hamernik, 1986).

Impulse and impact noise also present a heightened risk when either occurs with other steady-state background noise (approximately 85–95 dBA). Experimental studies with laboratory animals have shown that exposure to combinations of relatively benign impact and steady-state noise can lead to multiplicative interactions with hearing loss and cochlear damage, with the effects of the combined exposure being greater than the simple additive effects of impulse or continuous noise (Hamernik et al., 1981). Lei et al. (1994) have developed a metric, based on the distribution of noise levels during exposure, that captures the extra hazard to hearing associated with such combined exposures in laboratory animals.

Intermittent and Continuous Exposures to Steady-State Noise

Exposure to less intense noise (i.e., < 90 dBA) for short durations (i.e., ≤ 24 hrs) may result in a mild (< 30 dB) temporary hearing loss (Mills et al., 1970; Melnick, 1976). A noise-induced temporary hearing loss, or temporary threshold shift (TTS), is characterized by an increase in the hearing thresholds at some frequencies immediately following exposure, depending on the frequencies comprising the noise and its intensity. The threshold shift generally disappears within 24–48 hours after the exposure terminates (Mills et al., 1970; Melnick, 1976). Typically, exposure to more intense noise (> 90 dBA) or moderate noise for longer durations (> 24 hours) results in a larger amount of TTS (i.e., > 40 dB). In these cases, postexposure improvement of thresholds may continue for 30 days or longer, but in general, thresholds will not return to preexposure values. The individual likely will be left with a residual permanent threshold shift (PTS) (Taylor et al., 1965; Mills and Talo, 1972; Mills, 1973; Henderson et al., 1974a; Henderson and Hamernik, 1982).

Hearing loss that results from exposure to sound with energy spread across a wide range of frequencies, such as many broad-band noises and impulses common to most industrial and military settings, is typically characterized by a gradual increase in threshold as frequency increases. Typically, the hearing loss abruptly reaches a maximum between 3000 and 6000 Hz, followed by a return toward normal hearing at still higher frequencies. This particular pattern of hearing loss, as illustrated in Figure 2-3, is typically referred to as the "noise-notch" audiogram. It is a clinical hallmark often used to distinguish noise-related high-frequency hearing loss from that associated with other etiologies, such as ototoxic medications or aging. Several mechanisms have been offered to explain the extra vulnerability of the higher frequencies to the damaging effects of a broad-band noise, including better transmission of the higher frequencies through the outer and middle ears to the inner ear (e.g., Saunders and Tilney, 1982; Rodriguez and Gerhardt, 1991) and specific vascular (e.g., Axelsson and Vertes, 1982) or metabolic (e.g., Thalmann, 1976) vulnerabilities of this region of the cochlea. However, none of these mechanisms can fully explain

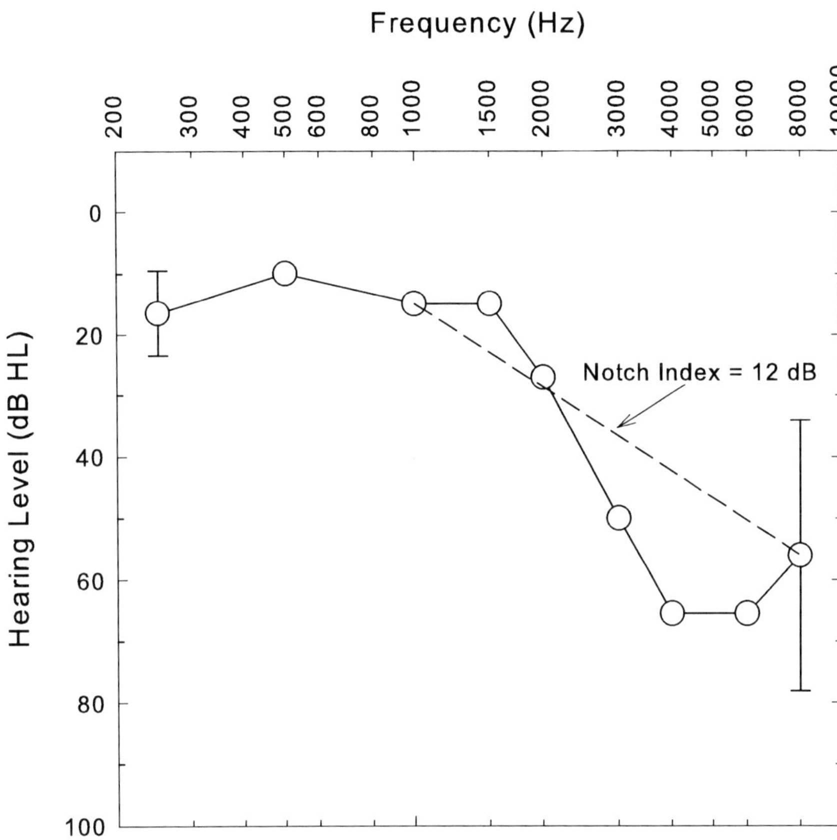

FIGURE 2-3 Illustration of a typical noise-notch audiogram. Average audiogram (*n* = 450 ears) from Cooper and Owen (1976) shown here. Error bars at 250 and 8000 Hz represent ±1 standard deviation and were the only standard deviations reported by the authors of this study for the average pure-tone thresholds at individual frequencies. The dashed line connecting thresholds at 1000 and 8000 Hz provides a visual representation of the Notch Index (NI) metric.

all of the features of the increased vulnerability of the 3000–6000 Hz region of the cochlea to noise damage.

Although the group data from Cooper and Owen (1976) in Figure 2-3 reveal a clear decrease in hearing from 1000 Hz to 4000 Hz followed by a return toward better hearing at still higher frequencies (8000 Hz), a pattern that typifies a noise notch, this is not always readily apparent for individual data. Discerning a noise notch in the pattern of hearing loss may be especially challenging in older adults for whom age-related hearing loss is superimposed on a preexisting noise notch (see pp. 62–63). As a result, there have been attempts to define the presence or absence of a noise notch more objectively than by simply relying on visual inspection of the pattern of hearing loss in the high frequencies, the latter approach not being particularly reliable (e.g., McBride and Williams, 2001a,b). One such approach to objectively define the presence or absence of a noise notch was advocated initially by Coles et al. (2000) and further refined by Dobie and Rabinowitz (2002). A graphic demonstration is provided by drawing a line to connect the hearing thresholds at 1000 and 8000 Hz, as illustrated by the dashed line in Figure 2-3. Having thresholds between 1000 and 8000 Hz (especially those at 2000, 3000, and 4000 Hz) that fall at or below the dashed line is thought to indicate the presence of a high-frequency notch in the hearing loss. Dobie and Rabinowitz (2002) describe a corresponding metric, referred to as the notch index (NI), that is simply the mean of the hearing thresholds at 1000 and 8000 Hz subtracted from the mean of the hearing thresholds at 2000, 3000, and 4000 Hz. Values of NI greater than 0 dB are thought to indicate the presence of a notch, whereas those less than 0 dB do not. For the hearing thresholds displayed in Figure 2-3, the notch index is 12 dB and is consistent with poorer hearing thresholds at 2000–4000 Hz than at 1000 and 8000 Hz. Other approaches to objective determination of the presence or absence of a noise notch have been described previously (e.g., Gates et al., 2000). The simplicity of the notch index and similar metrics is appealing, although additional research is needed to establish its reliability, as well as sensitivity and specificity in the identification of noise-induced hearing loss in the general population.

In summary, there are four key acoustic parameters of a given noise exposure that determine the type and amount of the resulting hearing loss. These are the sound pressure level of the noise, the duration and temporal pattern of the exposure (hours/day, impulses/day, number of years), the type of noise (steady-state, impulse/impact, blast), and the spectral content of the noise. Knowledge of values for each of these four parameters is necessary, but not sufficient, to fully assess the hazard of a given exposure to hearing. Although there can be some variation in the audiometric pattern of hearing loss for pure-tone thresholds following exposure to noise, the hallmark of noise-induced hearing loss is a characteristic noise notch in the audiogram that typically occurs between 3000 and 6000 Hz.

Effects of Noise on the Structure of the Hearing Apparatus

Acoustic trauma can occur following exposure to very intense noise, typically blasts > 150 dBA. Humans experiencing blasts at very high sound levels (~ 180 dB SPL) may suffer damage to the middle ear, including hemorrhage in or perforation of the eardrum and fracture of the malleus (Davis et al., 1949; Hirsch, 1968; Ward, 1973; Henderson et al., 1974b;

Roberto et al., 1989). If the eardrum does not rupture during such an intense exposure, the organ of Corti is likely to rupture off the basilar membrane (Ward, 1973; Henderson et al., 1974a,b; Roberto et al., 1989). When a portion of the organ of Corti ruptures, it does not reattach to the basilar membrane. Rather, it eventually degenerates. As noted, the hearing loss associated with acoustic trauma often is severe and spans a wide range of frequencies, much broader than that represented by the high-frequency, noise-notch pattern of hearing loss associated with other types of noise exposures.

Because sound levels in areas free from reflective surfaces, known as free fields, decay 6 dB per doubling of distance from the sound source, a key factor in such exposures may be the proximity of the individual to the blast. For example, consider two individuals, A and B, such that A is located 1 meter from a blast and the sound level recorded at that location was 160 dBA. Individual B, on the other hand, is located 32 meters from this same blast. Assuming that the blast can be modeled as a point source in a free field, the sound level at the location occupied by B will be 130 dBA. Whereas A may experience acoustic trauma from such a blast, including the development of a severe hearing loss affecting a wide range of frequencies, this will be much less likely for B, who might, through repeated such exposures over time, develop the more common noise-notch pattern of hearing loss. Importantly, individuals A and B may each subsequently report having heard a loud blast in a case history, but the hazard to hearing will be much greater for individual A, who is closer to the blast. Observations from Operation Iraqi Freedom suggest that even when personnel are close enough to suffer a blast injury that results in medical evacuation, four out of ten such individuals escape without permanent hearing loss, although many do experience acoustic trauma resulting in a severe or profound hearing loss (Chandler, 2005).

Individuals with mild or moderate permanent noise-induced hearing losses typically have some structural damage in their cochleas. The damage may initially involve scattered loss of sensory cells, primarily outer hair cells, in the organ of Corti (undamaged sensory cells shown in Figures 2-2 and 2-4, part A). Permanent noise-induced hearing loss may also result in damage to or destruction of other important structures in the cochlea, including fibrocytes in the spiral ligament and limbus and cells of the stria vascularis (Liberman and Mulroy, 1982; Hirose and Liberman, 2003) (Figure 2-2). In humans and other mammals, outer hair cells are usually the first type of sensory cell to be damaged or destroyed by excessive noise (Bredberg, 1968; McGill and Schuknecht, 1976) (Figure 2-4, part B). With larger permanent hearing losses, the degeneration involves both outer and inner hair cells (Bredberg, 1968; Liberman and Mulroy, 1982; Bohne and Harding, 2000) (Figure 2-4, part C). With severe permanent hearing losses,

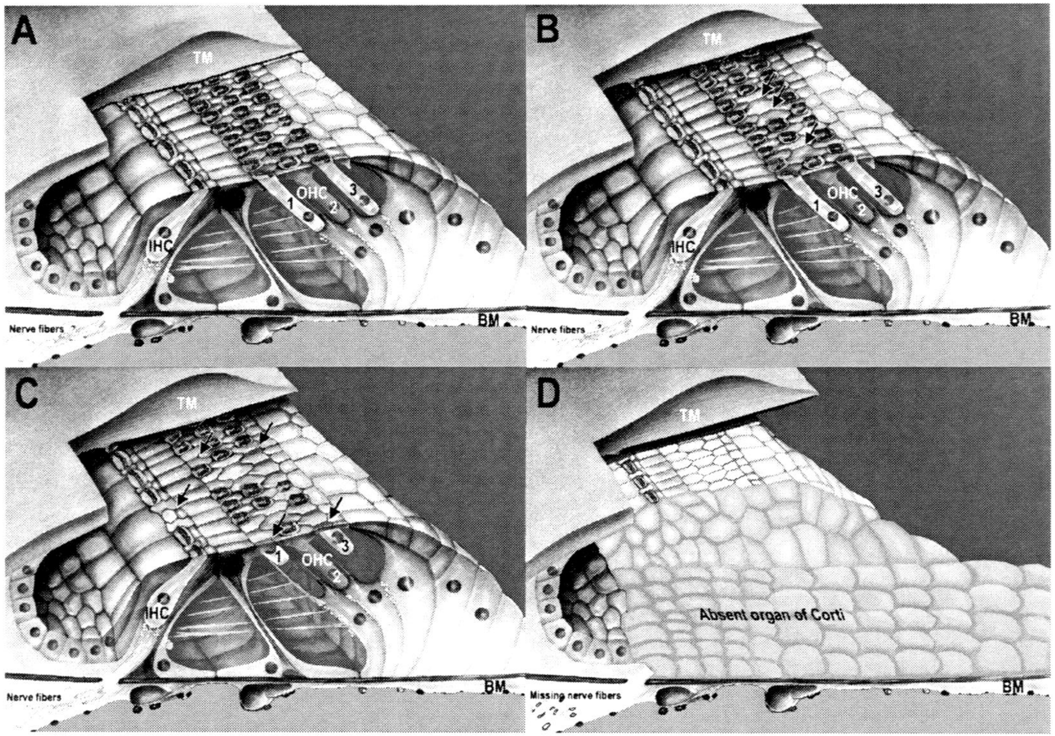

FIGURE 2-4 Drawings of the organ of Corti showing: (A) undamaged organ of Corti where all inner hair cells (IHC) and outer hair cells (OHC 1, 2, 3) are present; (B) beginning noise damage where 3 outer hair cells are missing (arrows); (C) moderate noise damage where 11 outer hair cells and 1 inner hair cell are missing (extent of loss indicated by arrows); and (D) severe noise damage where an entire portion of the organ of Corti is absent and is replaced on the basilar membrane (BM) by an undifferentiated, squamous epithelium. Nerve fibers to the area are also missing. TM = tectorial membrane.

SOURCE: Modified from an original painting by David Bellucci.

a variable amount of the organ of Corti (i.e., both sensory and supporting cells) is missing. In these cases, an undifferentiated layer of squamous epithelium covers the basilar membrane where the organ of Corti degenerated, and the nerve fibers that originally innervated the missing sensory cells also disappear (Johnsson and Hawkins, 1976; McGill and Schuknecht, 1976; Bohne and Harding, 2000) (Figure 2-4, part D).

The time course of cell degeneration and scar formation in the cochlea following a damaging noise exposure can be determined from animal studies only. A number of studies have shown that outer hair cells often begin to degenerate during an exposure. Additional outer hair cells, as well as inner hair cells and various supporting cells, may degenerate for days to a few weeks following termination of the exposure. While the various cells are degenerating, scars are forming in the organ of Corti. Phalangeal scars, formed from supporting-cell processes, replace missing hair cells, and squamous epithelial scars, formed from supporting cells on the basilar membrane, replace degenerated portions of the organ of Corti. Nearly all scar formation is completed by 1 month postexposure (e.g., Stockwell et al., 1969; Bohne, 1976; Fredelius, 1988; Wang et al., 2002).

Although there are some exceptions, especially for high-intensity, low-frequency sounds (e.g., Jerger et al., 1966; Burdick et al., 1978; Mills et al., 1983), good consistency has been observed in human and animal studies between the frequency content of the exposure stimulus and the location in the cochlea experiencing the greatest damage or injury (e.g., Johnsson and Hawkins, 1976; Moody et al., 1976). For narrow-band stimuli, the maximum cochlear insult is often one-half to one octave higher in frequency than the exposure stimulus (Ward, 1973). For broad-band noises and impulses, much more commonly encountered in military and industrial settings, the damage is greatest in the high-frequency (i.e., basal) portion of the cochlea (e.g., Gravendeel and Plomp, 1959; Ward, 1973; Ylikoski and Ylikoski, 1994). Furthermore, these differences in location of the greatest cochlear damage are accurately reflected in the pattern of hearing loss. For example, the noise-notch pattern of hearing loss (Figure 2-3) is associated with underlying damage to the sensory cells in the basal portion of the cochlea; that is, the portion of the cochlea tuned to those frequencies. In addition, as suggested by the sequence of illustrations shown in Figure 2-4, there is also a positive correlation between the amount of damage at a particular location in the cochlea and the severity of the hearing loss measured for a frequency associated with that location, although this correlation is believed to be weaker for low-frequency sounds (e.g., Bredberg, 1968). The pattern of hearing loss measured following noise exposure provides valuable information about the extent and severity of the underlying damage, especially in the middle and high frequencies following exposures to broad-band sounds.

In summary, although the specific site of lesion varies with the type of noise to which one is exposed (steady-state, impulse/impact, blast), the sensory receptors within the cochlea are the most common site of permanent destruction from noise exposure. Behavioral pure-tone thresholds, although not perfect indicators of the underlying cochlear pathology, represent the best available and most widely used measure of the underlying damage in humans.

TIME RELATION BETWEEN EXPOSURE TO NOISE AND THE DEVELOPMENT OF HEARING LOSS AND COCHLEAR DAMAGE

Consider exposure to an intense sound for a specified duration, with the measurement of hearing thresholds performed at periodic intervals during the exposure and at several times following the exposure. Thresholds measured in brief intervals during the exposure represent the growth or development of hearing loss, whereas those measured after the exposure is terminated represent the recovery at specific postexposure time intervals. For humans, the research conducted on the growth and recovery of hearing loss associated with noise exposure has primarily used changes in behavioral measures of hearing threshold, either TTS or PTS. Thus, conclusions about cochlear damage resulting from noise exposure in humans are limited by the imperfect association between behavioral measures of hearing threshold and underlying cochlear damage noted previously (e.g., Bredberg, 1968).

This less than perfect correlation has also been noted in the specific context of the study of recovery from noise exposure. For example, Davis et al. (1950) repeatedly exposed humans to a series of intense sounds with the exposures spaced so that there was complete recovery of TTS from one exposure before beginning the next exposure. In only 3 of more than over 60 intensity-duration exposure combinations, the individuals' hearing thresholds failed to fully recover to baseline levels established before the series of exposures; that is, some PTS occurred in only 3 of 60 cases (5 percent). On this basis, it was assumed that if an individual completely recovered from a TTS, the inner ear had not sustained permanent structural damage. However, because the inner ears of these individuals were not examined microscopically, the occurrence of permanent structural damage could not be ruled out. Subsequently, laboratory animal studies in the 1960s and 1970s using cats and chinchillas, two common animal models of human hearing, showed that animals who had completely recovered from TTS of at least 50 dB always sustained some permanent cochlear damage (e.g., Miller et al., 1963; Carder and Miller, 1972). Although it is likely that the same phenomenon is true for humans, there is insufficient evidence to support this generalization.

With regard to the effects of the duration of the exposure, when the duration is defined in terms of hours of continuous exposure to intense noise, hearing thresholds in humans begin to deteriorate after about 1–4 hours of exposure to moderately intense noise (i.e., ~85 dBA) and reach a maximum threshold shift of 10–18 dB after 8–12 hours (Mills et al., 1970; Melnick, 1976; Mills et al., 1979). With exposure to more intense noise (i.e., 92–120 dBA), hearing thresholds begin to deteriorate less than 30 minutes after the start of the exposure in both humans (Davis et al., 1950; Mills et al., 1970; Mills et al., 1979) and other mammals (Miller et al., 1963; Carder and Miller, 1972; Mills, 1973). With regard to recovery following such continuous exposures to noise, generally, threshold shifts that are less than 20 dB fully recover by 48 hours after termination of the exposure (Mills et al., 1970; Carder and Miller, 1972; Melnick, 1976). Threshold shifts of about 30 dB may require 3–6 days for the temporary component of the threshold shift to disappear. With larger maximum thresholds shifts (e.g., > 40 dB), recovery may be complete or incomplete. In animals, threshold shifts that require a week or more for recovery have been shown to be associated with permanent cochlear damage, primarily the loss of outer hair cells (Eldredge et al., 1973a; Eldredge et al., 1973b).

We speculate that delayed recovery of thresholds in humans also signifies the occurrence of cochlear damage. Mills (1976) exposed four chinchillas to a 4-kHz octave-band noise at 80 dB SPL for 90 days and measured thresholds and threshold shifts behaviorally before, during, and for 150 days after the exposure. Threshold shifts grew to an asymptote of 50 dB by 24 hours of exposure. After termination of the exposure, thresholds recovered to a 30-dB shift over 8 days, reached a plateau, and then recovered another 10 dB between 60 and 150 days.

It is important to note, however, that with the exception of a very short-term "bounce" in the recovery process that occurs typically within the first 2 minutes following cessation of the sound exposure (Hirsh and Ward, 1954), thresholds measured following these exposures are all consistent with a process of recovery. That is, thresholds *always improve or remain stable* as postexposure time increases.

Limited data on PTS in humans appear to be consistent with these data from laboratory animals on recovery following noise exposure. Segal et al. (1988), for example, using a classification scheme based on the severity of hearing loss (mild, moderate, or severe), reported that approximately 91 percent of the 1,514 hearing losses observed following acoustic trauma in military personnel remained in the same severity classification for follow-up periods of 4 months to 4 years, 7 percent improved, and 2 percent worsened by one category. Similar findings (4 percent classified as having worse hearing at follow-up) were reported by Melinek et al. (1976) for a smaller sample of 626 ears from military personnel with acoustic trauma. In both studies, additional noise exposure between the initial and final measurement of hearing thresholds was possible because these individuals remained in the military for the period between these measurements. Given the reliability of clinical threshold measurements and the use of a severity classification scheme based on those measurements, these data provide evidence in humans of the stability of the postexposure hearing thresholds over an extended period of time.

Laboratory studies of TTS and PTS in humans and laboratory animals that have focused on *recovery* of hearing thresholds have typically been terminated soon after the thresholds have fully recovered (TTS) or appear to have stabilized for a period of days or weeks (PTS). Although at least some studies have followed participants of various ages with hearing loss for up to 15 years (e.g., Macrae, 1991), no life-span studies of humans have followed the same subjects longitudinally for their remaining lifetime after a noise exposure that produced a significant TTS or PTS early in life. In laboratory animals, Mills et al. (1997) exposed one ear of six 18-month-old ("middle-aged") gerbils to high-intensity sound that produced PTS of 10–15 dB at 6 weeks postexposure. The animals were no longer exposed to noise and thresholds were assessed again at age 36 months, which is near the end of the gerbil's life. During the period from 18 to 36 months of age, hearing losses of 0–50 dB can occur in gerbils that are non-noise exposed at a young age and reared entirely in a quiet environment. At age 36 months, hearing thresholds in the "noise-exposed and aged" ears had increased further by about 3 dB (or, a total shift of 13–18 dB compared with the preexposure baseline). Thresholds in the "non-noise exposed and aged" ears (of the same animals), on the other hand, had increased by about 10–20 dB over this same time period. In other words, at an age of 36 months, hearing thresholds in the ears subject to both noise exposure and aging were nearly equivalent to the hearing levels in the control ears subject only to aging (and substantially lower than the thresholds predicted using ISO/ANSI rules for the combined effects of noise and aging; see below). Clearly, the noise exposed animals in this study did not have unusual changes in hearing levels that were manifested long after the noise exposure was terminated. Other than this study, however, there are no other data from studies of laboratory animals in which hearing thresholds were measured throughout the animals' life spans after recovery appeared to be complete. Thus, few data are available from laboratory animals or humans with which one can address the issue of "delayed effects" in later life of a noise exposure experienced much earlier in life.

It is noteworthy, however, that the modeling of noise-induced hearing loss and age-related hearing loss in ISO-1999 suggests that noise-induced hearing loss is almost asymptotic beginning at about 10–20 years of exposure and continuing through 40 years of exposure, especially for the milder amounts of hearing loss (ISO, 1990). (Also see Figure 2-5 in the subsequent discussion of estimating noise-induced hearing loss.) Thus, if continued exposure to noise for an additional two decades (from 20 to 40 years of exposure) results in little additional noise-related hearing loss following an initial 10–20 years of exposure, it hard to imagine that removal from the noise after 10–20 years of exposure would result in further declines in hearing.

Finally, for the study of PTS in humans, "duration" may also be defined by *years* of daily exposure (typically, 5 days/week) to intense sounds of various types. Here, cross-sectional data from humans indicate that PTS increases with years of exposure, typically growing most rapidly during the first 10–15 years of exposure at 4000 Hz and the first 10–20 years at 2000 Hz (e.g., Nixon and Glorig, 1961; Gallo and Glorig, 1964; Taylor et al., 1965).

FINDING: The evidence from laboratory studies in humans and animals is sufficient to conclude that the most pronounced effects of a given noise exposure on pure-tone thresholds are measurable immediately following the exposure, with the length of recovery, whether partial or complete, related to the level, duration, and type of noise exposure. Most recovery to stable hearing thresholds occurs within 30 days.

FINDING: There is not sufficient evidence from longitudinal studies in laboratory animals or humans to determine whether permanent noise-induced hearing loss can develop much later in one's lifetime, long after the cessation of that noise exposure. Although the definitive studies to address this issue have not been performed, based on the anatomical and physiological data available on the recovery process following noise exposure, it is unlikely that such delayed effects occur.

RISK FACTORS FOR NOISE-INDUCED HEARING LOSS: INDIVIDUAL DIFFERENCES AND NONACOUSTIC FACTORS

Thus far, the focus in the preceding pages has been placed on the effects of noise on the hearing of the "typical" or "average" person. For at least 175 years, however, it has been known that one of the hallmarks of noise-induced hearing loss is the wide range of individual differences in hearing loss that can result from seemingly identical noise exposures (Temkin, 1933; Ward, 1965, 1968, 1995). Various factors, some inherent to the individual (endogenous) and some external to the individual (exogenous), have been examined by researchers to assess their role in susceptibility to noise-induced hearing loss (e.g., Humes, 1984; Boettcher et al., 1987; Boettcher et al., 1992; Henderson et al. 1993; Ward, 1995). In some cases, such as exposures to solvents or use of aminoglycoside antibiotics, the exogenous factors not only interact with noise exposure but also produce hearing loss themselves. Here these factors are examined with regard to whether they interact with noise exposure to increase the hearing loss beyond that resulting from exposure to either agent alone. The details of earlier reviews will not be repeated here. Rather, the emphasis in this section is placed on their conclusions and recent findings (since 1990) that serve to supplement these earlier reviews.

Although an effort was made to focus on recent human studies, for some factors, the only studies have been in laboratory animals. In many of the animal studies, the exposures were exploratory and often not relevant to the occupational setting. In addition, in the human studies, details of the exposures to noise and other agents were not always available, especially for an individual subject. Dose-response information was rarely available. Finally, the studies reviewed included a wide range of methods to assess hearing loss. With these caveats in mind, we proceed to a review of evidence on the effects of certain exogenous and endogenous factors on the amount of noise-induced hearing loss measured following noise exposure.

Exogenous Factors

As noted in reviews by Humes (1984), Boettcher et al. (1987), and Boettcher et al. (1992), interactions of the effects of noise exposure and various drugs and chemical agents have received considerable attention. When noise, diuretics, or aminoglycosides (common antibiotics, such as gentamicin) are used in combination, for example, synergistic interactions occur such that the hearing loss from the combination of agents is greater than the hearing loss from either agent alone. This result was supported by Aran et al. (1992), who found, using guinea pigs, that exposure to sound at moderate to high levels increased the ototoxic effects of the drugs. Gratton et al. (1990) studied the interaction of a drug commonly used in chemotherapy, cisplatin, with concurrent noise exposure. Using chinchillas, administration of cisplatin during noise exposure resulted in greater hearing loss and hair cell loss than occurred for either agent alone, but the interaction was dependent on the noise level. For sodium salicylate, a compound related to aspirin, conclusions regarding the interaction with effects of noise exposure have been somewhat equivocal, with most studies suggesting that the addition of salicylates does not make the ear more susceptible to the damaging effects of noise (e.g., Spongr et al., 1992, in chinchillas).

Although hearing loss in the industrial or military environment is most often attributed to noise exposure, there is increasing attention to whether co-exposure to chemical agents present in these environments may potentiate noise-induced hearing loss. Carbon disulfide, for example, is a chemical common in the clothing industry and is often neurotoxic; an interaction between noise exposure and exposure to carbon disulfide has been observed in some human studies (Franks and Morata, 1996).

Vapors from organic solvents, such as toluene, are often found in occupational environments involved in printing and painting, as well as in shipyards where paint spraying is common. Effects of toluene have been shown in laboratory animal studies to combine in an additive or synergistic manner with noise (e.g., Cary et al., 1997; Fechter, 2004; see also Table D-1 in Appendix D). Styrene, which is widely used in manufacturing and in the chemical and petroleum industries, is another organic solvent whose effects have been shown in animal studies to combine synergistically with noise (Morata and Campo, 2001). Overall, results from animal studies generally suggest that although exposure to noise and solvents individually may be at safe levels, simultaneous or successive exposures to both may increase susceptibility to noise-induced hearing loss. However, inconsistencies in results of animal studies have been observed, and the extent of the interaction is highly species-specific. Some of the inconsistencies may relate to differences in the sites of damage of the ototoxins and their mode of transmission. For example, the damaging effects of noise affect primarily the inner ear and occur via acoustic transmission, whereas chemical exposures may reach the inner ear and the central nervous system by being inhaled or absorbed through the skin and circulated through the bloodstream.

Results of laboratory studies with animals showing increased risk for noise-induced hearing loss with exposure to solvents led to efforts to confirm these results in humans, given the widespread occupational exposures to these chemicals in noisy environments. Many observational and epidemiological studies of noise-induced hearing loss have been conducted in humans exposed to solvents (e.g., Morata et al., 1993; reviews in Morata et al., 1994; Morata and LeMasters 1995; Morata, 1998, 2003). Some studies show an increased prevalence of hearing loss in workers exposed to noise and solvents relative to workers exposed to either agent alone (e.g., Morata et al., 1993). An increased risk of hearing loss was also observed for aircraft maintenance personnel exposed to jet fuel and noise on an Air Force base (Kaufman et al., 2005). However, taken together, results of human studies are equivocal, and the design of these occupational studies is often confounded by a lack of control groups and poor quantification of the exposures. Typically, the hearing of a group of unexposed workers is compared to the hearing of groups of workers exposed to noise only, to solvent only, and to both noise and solvent. However, the noise or solvent exposure within the "noise and solvent" group may not be equivalent to the exposures within the noise-only or solvent-only groups. Moreover, noise and chemical exposure records for individual study participants often are not available. Therefore, exposures must be estimated from area surveys rather than from dosimetry or biological monitoring. Nonoccupational noise exposures are also difficult to estimate and control for in analyses. In addition, workers of different ages or with different durations of employment may be subject to uncontrolled cohort differences in their histories of noise or solvent exposure and in other factors that could affect hearing. Given these design limitations, there is insufficient evidence in humans to reach conclusions regarding the interactive effects of solvents and noise on noise-induced hearing loss.

Exposure to chemical asphyxiants, such as hydrogen cyanide and carbon monoxide, is common in some workplaces, and the interaction of these asphyxiants with noise exposure has also received attention recently. Cyanides are used in electroplating and metal leaching. Carbon monoxide is among the most common workplace air pollutants, especially for individuals working around gas-combustion engines, such as mechanics and other engine workers. Recent reviews of research in animal models indicate that low to moderate levels of carbon monoxide and hydrogen cyanide potentiate noise-induced hearing loss, especially at high noise levels (Cary et al., 1997; Fechter et al., 2000; also see Table D-2 in Appendix D). This potentiating effect, moreover, has been observed whether the exposures to each agent were simultaneous or successive. A recent review by Fechter (2004) notes that solvents (toluene, ethyl benzene, styrene) are likely to result in an additive effect to noise whereas asphyxiants (carbon monoxide, hydrogen cyanide) appear to result in synergistic effects to noise. There are no known human studies on the interaction of the effects of chemical asphyxiants and noise on noise-induced hearing loss.

Cigarette smoking may have a negative effect on hearing (e.g., Cruickshanks et al., 1998), but it is unclear whether it interacts with noise exposure to increase noise-induced hearing loss (see Table D-3 in Appendix D). Recent studies have yielded mixed results, with some finding no synergistic effects between smoking and noise exposure (Starck et al., 1999; Toppila et al., 2000), one observing a small interaction (Palmer et al., 2004), and one showing a significant increase in the prevalence rate for noise-induced hearing loss among factory workers who smoked (Mizoue et al., 2003). The design limitations described previously for human studies of the interactive effects of noise and solvents are also evident in these cross-sectional studies of the effects of smoking on noise-induced hearing loss.

Finally, whole-body vibration increases temporary hearing loss when noise is present (see review by Humes, 1984), but only when body temperature is elevated (Manninen, 1988). An increase in body temperature is known to increase the effects of noise on hearing (e.g., Drescher, 1976). Exercise during noise exposure and cardiovascular fitness have been shown to decrease (Manson et al., 1994), increase, or have no effect on temporary hearing loss (Wilson and Herbstein, 2003). The effect of exercise on noise-induced hearing loss may also relate to increases in body temperature (Pekkarinen, 1995). Many of these studies included relatively small ($n < 20$) convenience samples, with no randomized designs or control groups. Results from studies of the interactive effects of noise and electromagnetic fields in magnetic resonance imaging devices have been equivocal, and only limited results are available (Pekkarinen, 1995).

FINDING: *Nonacoustic factors may interact with the effects of noise to increase the measured noise-induced hearing loss. For many exogenous factors, evidence in animal models reveals that the effects of drugs or chemical agents may combine in an additive or synergistic manner with the effects of noise to increase noise-induced hearing loss. In particular, aminoglycosides, cisplatin, and solvents (toluene and styrene) interact in laboratory animals with noise presented simultaneously or sequentially to increase the amount of noise-induced hearing loss. However, there is not sufficient evidence to confirm this finding in humans. In particular, the evidence is not conclusive in humans with regard to additive or synergistic effects of noise and the following exogenous factors on hearing: aminoglycosides, cisplatin, diuretics, salicylates, solvents, carbon disulfide, carbon monoxide, cigarette smoking, whole-body vibration, body temperature, exercise, and electromagnetic fields.*

Endogenous Factors

Previous reviews of endogenous factors affecting noise-induced hearing loss have been performed (Humes, 1984; Henderson et al., 1993; Ward, 1995). Factors covered in these review papers include age (both young and old), gender, race, eye color, and prior hearing loss. With regard to age, the focus here, given the committee's charge, will be only on the older end of the age continuum. In each of the reviews, the basic thesis is that there have been inconsistent trends reported in the literature for each of these endogenous factors, but in general, the research points to a lack of effect of each of these factors on noise-induced hearing loss. It is certainly not the case that one could adjust a general prediction of noise-induced hearing loss for the "average person" to be more appropriate for a given individual with knowledge of any of these endogenous factors.

With regard to gender differences in susceptibility to noise-induced hearing loss, audiometric surveys of employees in noisy industries have shown that women, on average, have better hearing than men, even when gender-specific age corrections have been applied (e.g., Berger et al., 1978). It is not clear, however, that the source of these differences can be attributed to gender alone (Ward et al., 2000).

Most of the more recent publications that have appeared since the latest of these general reviews by Ward (1995) have been concerned with the effects of old age on susceptibility to noise-induced hearing loss. These recent publications are reviewed first, followed by several citations with regard to the effects of race and prior hearing loss on noise-induced hearing loss.

The general conclusion from several recent studies (Sun et al., 1994; Erway et al., 1996; Mills et al., 1997; McFadden and Campo, 1998; Ohlemiller et al., 2000; Boettcher, 2002; Fraenkel et al., 2003), all of which were conducted with laboratory animals, is that aging per se does not enhance susceptibility to noise-induced hearing loss or cochlear damage. However, there is a suggestion that genetic factors that lead to hearing loss, including a predisposition to age-related hearing loss, may render an individual more susceptible to noise-induced hearing loss (Erway et al., 1996; Ohlemiller et al., 2000). This is also consistent with recent work by Holme and Steel (2004), who have identified a gene in mice, Cdh23, that is linked to increased susceptibility to noise-induced hearing loss. Other studies in mice indicate that a genetic basis for resistance to noise-induced hearing loss may exist as well (Yoshida et al., 2000).

The basic approach pursued in most of the foregoing studies on aging and susceptibility was to examine the effects of specific noise exposures on two or more groups of subjects differing in age. As noted, in general, older adult animals were neither more nor less susceptible than younger adults to the effects of noise on hearing. A related, but separate, question regarding interactions of noise and aging is whether prior noise-induced hearing loss impacts the subsequent progression of age-related hearing loss. Two recent studies with older humans (Gates et al., 2000; Rosenhall, 2003) suggest that evidence of prior noise exposure results in a more rapid deterioration of hearing with aging, at least for some frequencies (e.g., 2000 Hz). On the other hand, two other recent studies (Cruickshanks et al., 2003; Lee et al., 2005), also with humans, failed to find evidence supporting differences in the amount of age-related hearing loss or the progression of age-related hearing loss based on a history of prior noise exposure. (Table D-4 in Appendix D summarizes the key features of these four studies.) The reasons for these differences in findings are unclear, and this is an area that deserves further attention and research.

Race is another endogenous factor that has received some attention recently by researchers studying noise-induced hearing loss. Henselman et al. (1995), for example, conducted a retrospective study among soldiers in the U.S. Army and noted that, for similar noise exposures and length of service exceeding 10 years, black soldiers demonstrated less age-corrected noise-induced hearing loss than white soldiers. It is unclear, however, whether this is a valid indication of racial differences in noise susceptibility. The study used the same (primarily white) unscreened database to generate the age corrections to the audiograms. African Americans have been observed to have less age-related hearing loss than whites (Royster et al., 1978). Thus, the reduced noise-induced hearing loss reported for African Americans in this study could be due to the subtraction of an inappropriately high age-related threshold derived from data for white subjects to generate the age-corrected noise-induced hearing loss for African Americans.

Ishii and Talbott (1998) conducted another retrospective study of noise-induced hearing loss in industry and found that white workers in a metal fabricating plant had significantly higher high-frequency (3000–8000 Hz) hearing levels than nonwhites (86 percent African American, 14 percent Hispanic) working in the same plant. The differences in hearing levels in the high frequencies, however, are similar to the race differences in age-related hearing loss noted previously. This suggests that appropriate race-specific age correction of these hearing levels, rather than the use of the same age correction for all racial/ethnic groups, would have resulted in little difference between racial/ethnic groups in the actual *noise-induced* hearing loss.

Although the effects of prior hearing loss, including previous noise-induced hearing loss, on the susceptibility to subsequent noise-induced hearing loss have not received much attention in recent research, several earlier studies in laboratory animals and humans have addressed this topic (e.g., Glorig et al., 1961; Resnick et al., 1962; Ward, 1968, 1973; Mills, 1973; Man et al., 1975; Howell, 1978; Botte and Variot, 1979; Humes, 1984; Humes and Jesteadt, 1991). The subjects in two of these earlier studies were military personnel (Resnick et al., 1962; Man et al., 1975). In most of the studies cited, the existing hearing loss resulted from prior noise exposure. When expressed in terms of relative threshold shift (TTS or PTS), the observed amount of postexposure threshold shift always decreased as the amount of prior hearing loss increased, giving the impression that individuals with prior hearing loss are less susceptible to noise-induced hearing loss. When examined in terms of the hearing threshold to which the individuals have been shifted (so-called "shifted threshold" in dB HL or dB SPL), however, it is apparent that individuals with and without prior hearing loss are equally susceptible to noise-induced hearing loss (Mills, 1973; Humes, 1980; Humes and Koval, 1981; Humes and Jesteadt, 1991).

Consider a noise exposure that is known to produce, on average, a noise-induced hearing loss of 20 dB HL at 4000 Hz. An individual who enters this noise exposure with a hearing threshold of 0 dB HL at 4000 Hz will demonstrate a threshold shift of 20 dB, whereas another individual who enters this noise exposure with a hearing threshold of 40 dB HL at 4000 Hz will show a threshold shift of 0 dB. On the other hand, if a given noise exposure is known to produce a noise-induced hearing loss of 40 dB HL at 4000 Hz, individuals with preexposure thresholds of 0 and 20 dB HL at 4000 Hz will demonstrate, on average, threshold shifts of 40 and 20 dB, respectively. Thus, the relative threshold shift in individuals with preexisting noise-induced hearing loss is always less than that observed in individuals with normal hearing prior to exposure, but the shifted thresholds (in dB

HL or dB SPL) are frequently similar, depending on the particular combination of thresholds and exposure conditions involved. Basically, the nonlinear additivity rules used to combine hearing thresholds from age-related hearing loss and noise-induced hearing loss, reviewed in detail below, also apply to the addition of previous and current noise-induced hearing loss. The basic conclusion from review of the research in this area is that individuals with previous noise-induced hearing loss are neither more nor less susceptible to subsequent noise-induced hearing loss than individuals without such pre-existing hearing loss.

FINDING: *Several endogenous factors have been examined, including (old) age, gender, race, eye color, and prior hearing loss, but there is not sufficient evidence in humans to conclude that any of these factors predicts susceptibility to noise-induced hearing loss.*

ESTIMATING NOISE-INDUCED HEARING LOSS

Large-scale, systematic investigations of the temporary and permanent effects of noise on hearing, in both human and laboratory animal subjects, began in the 1940s and early 1950s at several research laboratories, largely because of the extreme sound levels to which military personnel were exposed during World War II. Soon, a need developed for the scientific community to coordinate, evaluate, and guide the research in this area, and the National Academy of Sciences, with support of various branches of the military, created the Committee on Hearing and Bioacoustics (CHABA; subsequently appending "and Biomechanics" to its name while leaving the acronym, CHABA, intact). Over the ensuing decades, CHABA and subsidiary working groups generated several critical reviews and summaries of the state of our knowledge with regard to noise-induced hearing loss (e.g., Kryter et al., 1966; CHABA, 1968; NRC, 1992, 1993). These reports frequently led to additional research that, in turn, led to updated reports. Through this iterative process, much came to be known about the effects of noise on hearing.

There was particular interest in being able to estimate the average permanent hearing loss, as well as its distribution, expected from years of exposure to various noise types and levels. The greatest progress was made in the understanding of noise-induced hearing loss resulting from years of continuous or intermittent daily exposures to steady-state noise. The work in this area resulted in the adoption of an international standard (ISO-1999 [ISO, 1990]) and an American standard (ANSI S3.44 [ANSI, 1996]) designed to estimate the median noise-induced hearing loss for a given exposure, along with some measures of the statistical distribution of the resulting noise-induced hearing loss (e.g., 10th and 90th percentiles). Although both standards state that they can be applied to a wide variety of noise exposure conditions, including intermittent and time-varying exposures to steady-state noise and exposures to impulse noise, the formulas included in each standard were derived primarily from data on noise-induced hearing loss among industrial workers with 10–40 years of exposure to broadband, steady-state noise for 8 hours per day (5 days/week). In addition, both standards restrict their application to equivalent continuous sound levels of 75–100 dBA for a normal 8-hour working day. As a result, the ISO/ANSI predictions are considered to be most valid for similar exposure conditions. The standards, therefore, are less likely to provide valid estimates of permanent hearing loss in humans for other types of noise (e.g., impulse), noise exposures (e.g., daily exposures that differ significantly from 8 hours), or exposure durations (less than 10 or more than 40 years). These limitations of the standards should be kept in mind when they are applied to estimate noise-induced hearing loss, including their use to estimate noise-induced hearing loss in the military.

These national and international standards represent the synthesis of the best available data on industrial noise-induced hearing loss. The available data, however, are not without limitations. For example, sample sizes were often small, the studies were cross-sectional rather than longitudinal and subject to cohort effects, and specification of the noise exposure was by group or area of the industry, rather than for each individual.

Figure 2-5 provides an illustration of the age-corrected "noise-induced permanent threshold shift," NIPTS, estimated with ISO-1999 (1990).[2] The top panel demonstrates the development of NIPTS at one frequency, 4000 Hz, with increasing years of exposure to 8-hour equivalent continuous noise levels of 85, 90, 95, or 100 dBA. These estimates reveal two key features of NIPTS: (1) that NIPTS increases with noise level and, for an 8-hour equivalent continuous noise level of 85 dBA (or less), is negligible at 4000 Hz, the frequency showing the greatest amount of NIPTS; and (2) that NIPTS grows most rapidly during the first 10–15 years, with only slight increases beyond that. It is important to emphasize here, however, that this is the hearing loss associated *with noise exposure only* (NIPTS). Also, the values displayed are the predicted median values from ISO-1999. The same standard provides for the generation of other percentiles or quartiles. In general, the interquartile ranges (the difference between

---

[2]Although the relative NIPTS in dB is calculated in ISO-1999, the committee chose to plot this both as NIPTS in dB and noise-only hearing loss in dB HL in Figures 2-5 and 2-6 for consistency with the other figures in this report. Since it is assumed in both figures that the age-related component is 0 dB HL throughout, X dB NIPTS is also X dB HL of "pure" noise-induced hearing loss. Such "pure" noise-induced hearing loss, in the absence of concomitant age-related hearing loss, can only occur hypothetically and within such a model framework.

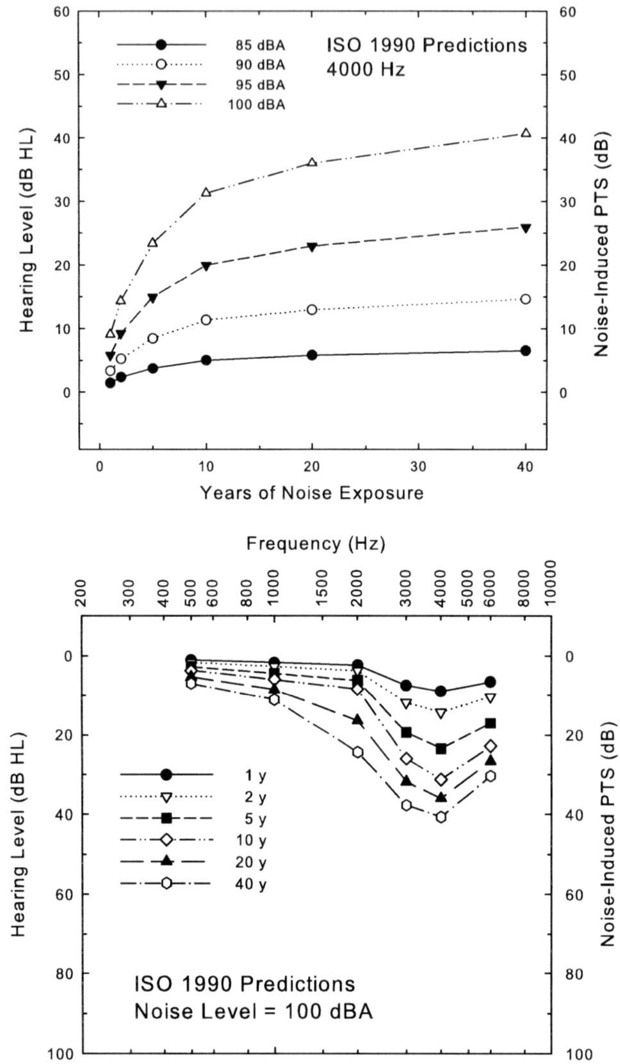

FIGURE 2-5 Depiction of the development of the noise-induced permanent threshold shift (NIPTS) in dB, or the noise-only noise-induced hearing loss in dB HL, at 4000 Hz (top) and at several frequencies (bottom) with increasing years of noise exposure as estimated by ISO-1999 (1990). The 8-hour equivalent continuous noise level is the parameter in the top panel, whereas the noise exposure is fixed at the highest level (100 dBA) included in the standards in the bottom panel, with length of exposure as the parameter. In both panels, age-related hearing loss is fixed at 0 dB HL throughout to illustrate that portion of the hearing loss associated with noise exposure only. This enables the hypothetical NIPTS in dB (left ordinate) to be plotted as hearing loss in dB HL (right ordinate).

thresholds at the 25th to 75th percentiles) around the median values in Figure 2-5 are typically 10–20 dB. Of course, while the individual is being exposed to such noise for 40 years, the individual also ages, and aging by itself can result in elevated hearing thresholds. This issue is addressed in more detail below. Here it is important to note that because aging effects are not included, it is unlikely that these would be the actual thresholds measured, especially for more than 10–20 years of noise exposure (assuming that the exposure begins in early adulthood).

The bottom panel of Figure 2-5 shows ISO-1999 (1990) predictions for NIPTS for an 8-hour equivalent continuous noise level of 100 dBA, and for progressively longer durations of exposure up to the maximum of 40 years. Here, however, NIPTS in dB (left ordinate), as well as the equivalent hearing thresholds in dB HL (right ordinate), for the noise-only portion of the hearing loss have been illustrated in conventional clinical format for frequencies from 500 through 6000 Hz to demonstrate the typical pattern of hearing loss at specific ages. Note the emergence of a noise-notch pattern over time, with the maximum hearing loss occurring at 4000 Hz. It is noteworthy that, consistent with the data depicted in the top panel, hearing loss at 4000 Hz increases most with durations of exposure up to 10 years, but that this is not the case at 2000 Hz, where increases in hearing loss continue *after* the first 10 years of exposure. Nonetheless, the noise-notch pattern of hearing loss is maintained and the average hearing loss *associated with noise exposure only* is about 40 dB HL in the high frequencies after 40 years of exposure to the highest noise level included in the ISO/ANSI standards.

One of the key issues covered in the ISO/ANSI standards is the way each addresses the combined effects of noise-induced and age-related hearing loss. Consider, for example, a man who either begins his job in industry or enlists in the military at 20 years of age. Thirty years later, this same individual retires. Estimates of the resulting noise-induced hearing loss using either standard would take into consideration this individual's 30 years of noise exposure. However, while this individual was exposed to noise, he also aged and was 50 years old when the daily noise exposure ended. It is well known that hearing loss also increases with age and that the average 50-year-old man will have worse hearing, especially in the high frequencies, than the average 20-year-old man. Thus, both the 30 years of noise exposure and advancing age can have an effect on high-frequency hearing, and it is the combined effects of both factors (as well as others) that will contribute to the hearing loss measured in this 50-year-old man.

How should the separate effects of noise and aging on hearing be disentangled? Continuing with this hypothetical example, let's assume that the noise exposure alone is expected to produce a permanent hearing loss corresponding to a hearing level of 25 dB HL at 4000 Hz after 10 years of exposure, with no further changes for another 20 years until retirement at age 50. Let us also assume that the median hearing level for a 50-year-old man in the United States is 20 dB HL at 4000 Hz. What hearing level would one expect to measure in a 50-year-old man who has worked in this noise for 30 years? One model that has been considered is simple decibel additivity of the hearing levels associated with aging and noise exposure, or 25 dB HL (NIPTS) + 20 dB HL (aging) = 45 dB HL. Another model suggests simple additivity of the corresponding sound powers, expressed as $10 \log_{10} [10^{(25/10)} + 10^{(20/10)}]$ or 26 dB HL. Based on empirical fits to the existing data, the ISO and ANSI standards both adopted the same compressive nonlinear model that yields values between these two extremes. The model used in the standards also is similar to a more general nonlinear compressive model that has proven successful when adding or combining elevated thresholds of various types (Humes et al., 1987; Humes and Jesteadt, 1989), including those arising from noise exposures (Humes and Jesteadt, 1991).

The specific form of the model used in the ISO and ANSI standards is as follows:

$$\text{HTLAN} = \text{HTLA} + \text{HTLN} - [(\text{HTLA} \times \text{HTLN}) / 120],$$

where HTLAN is the hearing threshold level from the combined effects of age (A) and noise (N), HTLA is the hearing threshold level associated with age alone, and HTLN is the hearing threshold level associated with the noise exposure alone. In the hypothetical example considered previously, according to this model, the 50-year-old man would be expected to have a hearing threshold at 4000 Hz of about 41 dB HL after 30 years of noise exposure [i.e., 20 + 25 = 45; (20 × 25) / 120 = 4.2; 45 – 4.2 = 40.8 dB HL]. In this case, the result is much closer to the prediction made according to simple decibel additivity (45 dB HL) than with power additivity (26 dB HL), but this is less true as the amount of hearing loss expected from each factor increases. For example, assume that both noise and age result in expected hearing threshold levels of 50 dB HL at 4000 Hz, then the ISO/ANSI additivity model predicts a threshold of about 79 dB HL, which is 21 dB less than simple decibel additivity (100 dB HL). In many situations, the ISO/ANSI additivity model is applied in the reverse direction. That is, the thresholds for a group or individual following X years of continuous exposure to noise at an 8-hour equivalent continuous level of Y dBA, as well as the age and gender of the group or individual are known, and the task is to determine how much of the hearing loss is associated with the noise exposure and how much is associated with age-related changes.

Regardless of the way such models are used, many assumptions are involved. One key assumption has to do with the time course of the development of hearing loss associated with each factor. For example, consider two men, A and B, both 20 years old when they began work in the same noisy industrial or military environment. Individual A happens to be susceptible to the damaging effects of noise and develops a 40-dB hearing loss within the first few years of employment in this noisy environment, then experiences no additional decline in hearing over the next several decades (not susceptible to age effects). Individual B, on the other hand, is not susceptible to the damaging effects of noise and shows typical or average age-related hearing loss throughout adulthood. Let us assume further that, at age 70, the median hearing loss at this same frequency for males is 40 dB. So, at age 70, both A and B have a 40-dB hearing loss. Given the foregoing, the ISO/ANSI additivity model would attribute all of the hearing loss for each 70-year-old male to aging, an accurate estimate for individual B. Yet, for individual A, who had noise-induced hearing loss most of his adult life, attributing the hearing loss entirely to age would be inappropriate.

The foregoing examples demonstrate a key point regarding models designed to estimate noise-induced hearing loss and allocate the hearing loss to either noise-related or age-related components: These models were built from and are designed for *group* data, not individual data. Given the same 40-dB HL hearing loss of individuals A and B at age 70 in the foregoing example, it is virtually impossible without other data to determine whether the scenario represented by A or that represented by B is the actual underlying scenario for a given individual. This serves to underscore the critical importance of periodic measurement of hearing thresholds for those exposed to high levels of noise. Given the wide range of individual differences in susceptibility, the possible effects of exogenous factors, and our inability to predict these effects from other measures, regular measurement of hearing thresholds is the only way to determine if a change in hearing has occurred in an individual during the period of a particular noise exposure. In the absence of such information, estimates of noise-induced hearing loss must be confined to statistical descriptions of population values (i.e., medians, percentiles) expected from the specific noise exposure, but there is no way to determine where among the population a particular individual would be found (e.g., "average," "most susceptible 10 percent," etc.).

ISO-1999 and ANSI S3.44 both allocate a portion of the observed hearing loss to age-related hearing loss and a portion to noise-induced hearing loss. We have noted several concerns with regard to the noise-induced component in the preceding paragraphs. Concerns have also been expressed, however, regarding the validity of the component representing age-related hearing loss. Specifically, the validity of the screened database (database A) on age-related hearing loss has been called into question, due, in part, to its derivation from data that are now several decades old and may be subject to cohort bias (Wiley et al., 2001). This serves to underscore the critical importance of annual measurement of hearing thresholds for individuals not exposed to noise, who can serve as a control group for the noise-exposed workers to document the effects of aging.

Thus far in our discussion of the estimation of noise-induced hearing loss from models such as ISO-1999 and ANSI S3.44, hearing thresholds at only one particular frequency have been considered. Although both exposures to industrial or military noise and advancing age result in high-frequency hearing loss, the pattern of hearing loss across frequencies differs for these two etiologies. As noted previously, it is the noise-notch pattern of hearing loss, together with detailed case-history information, that is used clinically to distinguish noise-induced hearing loss from other forms of high-frequency hearing loss, such as age-related hearing loss. This difference in pattern of hearing loss for noise and age is illustrated in Figure 2-6. The top panel in this figure shows the average hearing loss as a function of age in males when data from several cross-sectional studies of age-related hearing loss were synthesized (Robinson and Sutton, 1979). These data form the basis of one of the databases (database A) of age-related hearing loss included in the ISO-1999 and ANSI S3.44 standards. They represent composite values from studies that took some care in screening out subjects with prior exposures to noise or previous hearing loss attributable to other etiologies (based on self-report)—so-called "highly screened" samples. From the functions depicted in the top panel of Figure 2-6, it is apparent that hearing loss is greatest in the highest frequencies initially, and at older ages, the hearing loss is still worse at the highest frequencies, but lower frequencies are also affected. The ISO and ANSI standards both make provision for the use of other databases representing age-related hearing loss. Database B in both standards represents the hearing thresholds for a large unscreened sample from a U.S. Public Health Service survey completed in 1962 (Glorig and Roberts, 1965). Although the specific values for hearing thresholds at each frequency differ between databases A and B, with better hearing thresholds found in database A, greater high-frequency hearing loss at older ages is common to both databases.

In contrast, the patterns of NIPTS, or noise-only hearing loss in dB HL, in the lower panel of Figure 2-6 illustrate hypothetical data on the progression of noise-induced hearing loss patterned after actual data (Robertson et al., 1978). Of course, the actual amount of noise-induced hearing loss will depend on a variety of factors, as noted previously, such as the type of noise, the level of the noise, the noise spectrum, and the duration and pattern of exposure. The lower panel of Figure 2-6 illustrates the increase in hearing loss with longer durations (in years) of continuous exposure to broad-band, steady-state noise at a specific noise level, similar to that depicted in Figure 2-5. Unlike the patterns of hearing loss associated with age in the top panel of Figure 2-6, the hearing loss in the bottom panel of this figure reaches a maximum at 6000 Hz and then returns toward milder

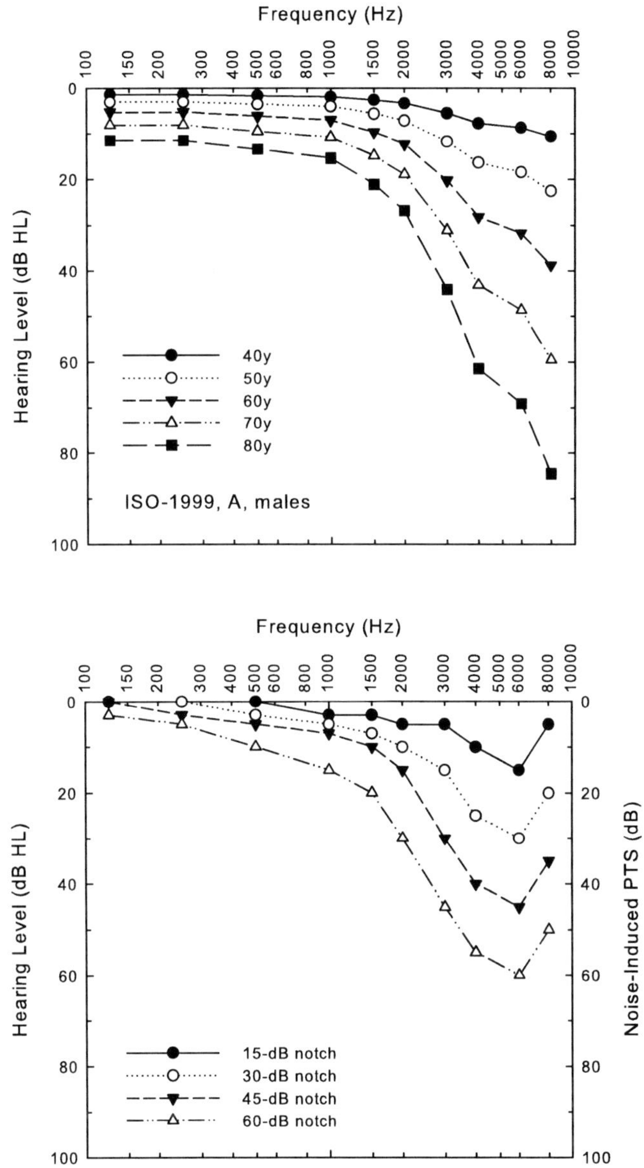

FIGURE 2-6 Age-related hearing loss for men (ISO-1999, database A, top panel) and hypothetical progression of noise-induced hearing loss with increased length of exposure in years (bottom panel). The bottom panel displays hypothetical NIPTS in dB, as well as the noise-only hearing loss data plotted in dB HL.

hearing loss at 8000 Hz. This demonstrates the characteristic noise-notch pattern of hearing loss in which the notch is located at 6000 Hz. As noted previously, the notch location varies with the noise exposure and across individuals experiencing the same noise exposure, but it is generally located at 3000, 4000, or 6000 Hz. A noise notch located at 6000 Hz was chosen for illustration purposes in the bottom panel of Figure 2-6 because of the frequent appearance of notches at this frequency in the data on military noise-induced hearing loss reviewed subsequently in Chapter 3.

Perhaps, therefore, the *pattern* of hearing loss across frequency can assist in determining how much, if any, of an older adult's hearing loss can be attributed to prior noise exposure. That is, rather than just considering the hearing threshold at one frequency for the two 70-year-old individuals, A and B, in our previous example, one of whom had no prior noise-induced hearing loss and the other who had sizable noise-induced hearing loss, perhaps the presence or absence of a noise notch will assist in sorting this out. To examine the impact of aging on the pattern of hearing loss across frequency, the additivity model of ISO-1999 and ANSI S3.44 was applied to the two sets of data in the top and bottom panels of Figure 2-6. The top panel was assumed to represent a "pure" age-related hearing loss for each age decade, and the bottom panel was interpreted as four different degrees of noise notch developed in four young men during the first few years of noise exposure (rather than the progression of noise-induced hearing loss over time, as originally indicated). The case represented by the "45-dB notch" has a threshold at 4000 Hz that is about 40 dB HL and is representative of individual A in our previous examples. Individual B, on the other hand, was assumed to have no noise-induced hearing loss, and thresholds for this individual would be best represented by the age-only curves ("no notch") depicted in the top panel of Figure 2-6.

Figure 2-7 illustrates the combined effects of noise (noise notches of various depths) and age (50-, 60-, 70-, and 80-year-olds) that result from using the ISO/ANSI additivity model to combine the sets of hearing thresholds from the two panels of Figure 2-6. When examining the predictions for each age, clear notching is visible in patterns of hearing loss for those individuals with initial noise-notch patterns at ages 50 and 60 years, but appears to be absent at ages 70 and 80 years. The other clear trend with age is the convergence of all the hearing losses by the age of 80 years. Hearing losses that differed by about 50 dB in the high frequencies for 50-year-olds differ by about half that much for 80-year-olds. Thus, there is less difference in *pattern* of hearing loss by the time these individuals reach their 70s and 80s, and the *severity* of the loss no longer differs as much among these individuals. As a result, two individuals who have similar hearing thresholds when measured at 70 or 80 years of age may have had entirely different patterns of hearing loss as young adults and throughout much of their

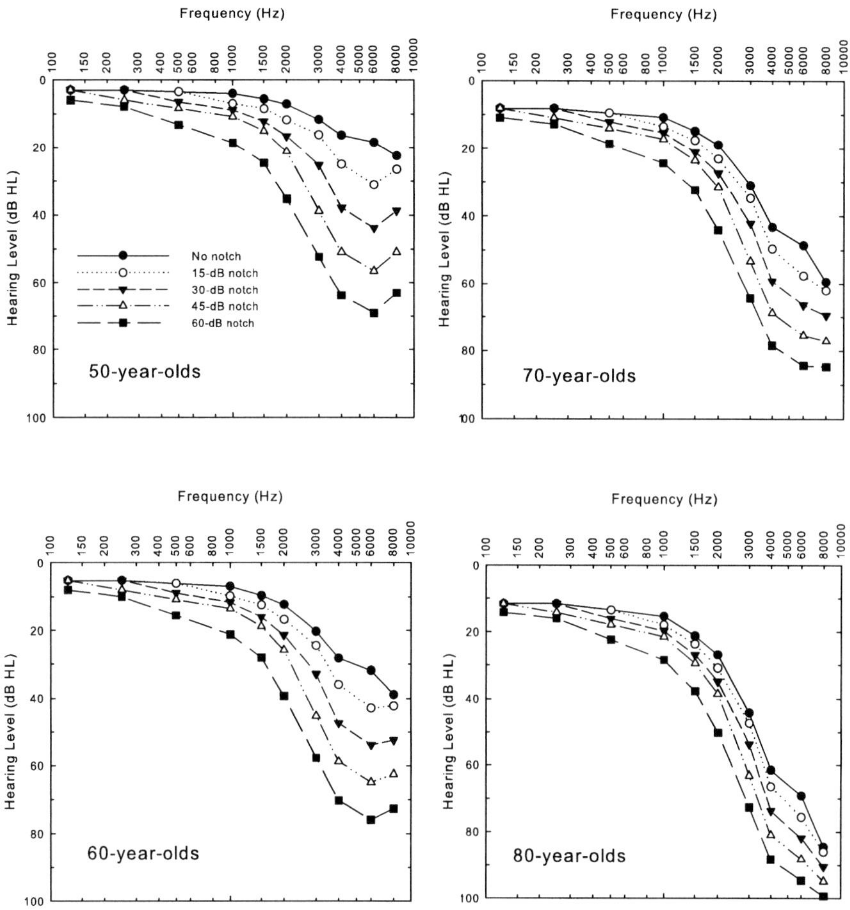

FIGURE 2-7 Illustrations of the combined effects of aging (top panel from Figure 2-6) and noise exposure (bottom panel from Figure 2-6) using the ISO-1999/ANSI S3.44 model for additivity. Each panel depicts the combined hearing loss for a separate decade (50-, 60-, 70-, or 80-year-old men).

adult lives. Once again, this underscores the critical importance of measuring hearing thresholds periodically (preferably annually) for individuals exposed to noise and, ideally, before and after employment or military service. With only the hearing thresholds from a much later stage in life, it is virtually impossible to discern how much, if any, of an individual's hearing loss can be attributed to noise exposure or for how long this hearing loss might have been present.

With regard to the estimation of noise-induced hearing loss, the following represents a summary of the main points of this section of the chapter:

- Without measurement of pure-tone thresholds prior to and following a given exposure to noise, it is impossible to document the effects of that exposure on hearing or to know what portion of the hearing loss in an older individual is due to earlier noise exposure.
- ISO-1999 and ANSI S3.44 provide estimates of median values and range of variation in noise-induced hearing loss for a given noise exposure. Predictions are best for the noise-induced hearing loss that results from continuous or intermittent exposures to steady-state noise at levels between 75 and 100 dBA for 8 hours per day (for an assumed 5-day work week) for periods of 10 to 40 years.
- Age-related hearing loss occurs at several of the same frequencies for which noise-related hearing loss occurs, and the measured thresholds are presumed to be a combination of these two forms of hearing loss.
- Combined effects of noise and age on hearing thresholds range from energy summation to decibel summation. The combined effects of noise and age included in ISO-1999 and ANSI S3.44 lie somewhere between these two extremes.

FINDING: The evidence from cross-sectional studies of noise-induced hearing loss in humans is sufficient to conclude that daily time-weighted average noise exposures greater than approximately 85 dBA for 8 hours for periods of many years pose a hazard to human hearing and that the hazard increases as the time-weighted average exposure exceeds this value.

FINDING: The evidence is not sufficient to determine the probability of acquiring a noise-induced hearing loss, or to estimate the magnitude of the noise-induced hearing loss, that a specific individual is likely to experience from a given noise exposure.

REFERENCES

ANSI (American National Standards Institute). 1996. *ANSI S3.44 Determination of Occupational Noise Exposure and Estimation of Noise-Induced Hearing Impairment.* New York: Acoustical Society of America.

Aran JM, Hiel H, Hayashida T, Erre JP, Aurousseau C, Gulhaume A, Dulon D. 1992. Noise, aminoglycosides, and diuretics. In: Dancer AL, Henderson D, Salvi RJ, Hamernik RP, eds. *Noise-Induced Hearing Loss.* St. Louis, MO: Mosby Year Book. Pp. 188–195.

Axelsson A, Vertes D. 1982. Histological findings in cochlear vessels after noise. In: Hamernik RP, Henderson D, Salvi R, eds. *New Perspectives on Noise-Induced Hearing Loss.* New York: Raven Press. Pp. 49–68.

Berger EH, Royster LH, Thomas WG. 1978. Presumed noise-induced permanent threshold shift resulting from exposure to an A-weighted Leq of 89 dB. *Journal of the Acoustical Society of America* 64:192–197.

Boettcher FA. 2002. Susceptibility to acoustic trauma in young and aged gerbils. *Journal of the Acoustical Society of America* 112(6):2948–2955.

Boettcher FA, Henderson D, Gratton MA, Danielson RW, Byrne CD. 1987. Synergistic interactions of noise and other ototraumatic agents. *Ear and Hearing* 8(4):192–212.

Boettcher FA, Gratton MA, Bancroft BR, Spongr V. 1992. Interaction of noise and other agents: Recent advances. In: Dancer AL, Henderson D, Salvi RJ, Hamernik RP, eds. *Noise-Induced Hearing Loss*. St. Louis, MO: Mosby Year Book. Pp. 175–187.

Bohne BA. 1976. Healing of the noise-damaged inner ear. In: Hirsh SK, Eldredge DH, Hirsh IJ, Silverman SR, eds. *Hearing and Davis: Essays Honoring Hallowell Davis*. St. Louis, MO: Washington University Press. Pp. 85–96.

Bohne BA, Harding GW. 2000. Degeneration in the cochlea after noise damage: Primary versus secondary events. *American Journal of Otology* 21(4):505–509.

Botte MC, Variot MH. 1979. [Auditory fatigue in individuals having sustained an acoustic trauma (author's transl)]. *Annales d'Oto-Laryngologie et de Chirurgie Cervico-Faciale* 96(12):827–833.

Bredberg G. 1968. Cellular pattern and nerve supply of the human organ of Corti. *Acta Oto-Laryngologica Supplement* 236:1–135.

Brödel M. 1946. *Three Unpublished Drawings of the Anatomy of the Human Ear*. Assisted by Malone PD, Guild SR, Crowe SJ. Philadelphia, PA: W.B. Saunders Co.

Burdick CK, Patterson JH, Mozo BT, Camp RT. 1978. Threshold shifts in chinchillas exposed to octave bands of noise centered at 63 and 1000 Hz for three days(a). *Journal of the Acoustical Society of America* 64(2):458–466.

Carder HM, Miller JD. 1972. Temporary threshold shifts from prolonged exposure to noise. *Speech and Hearing Research* 15(3):603–623.

Cary R, Clarke S, Delic J. 1997. Effects of combined exposure to noise and toxic substances—critical review of the literature. *Annals of Occupational Hygiene* 41(4):455–465.

CHABA (National Academy of Sciences–National Research Council Committee on Hearing, Bioacoustics, and Biomechanics). 1968. *Proposed Damage–Risk Criterion for Impulse Noise (Gunfire)*. Report of Working Group 57 (Ward WD, ed.). Washington, DC: National Academy of Sciences.

Chandler D. 2005. *Current Scope of Problem*. Washington, DC: Walter Reed Army Medical Center.

Coles RR, Lutman ME, Buffin JT. 2000. Guidelines on the diagnosis of noise-induced hearing loss for medicolegal purposes. *Clinical Otolaryngology* 25(4):264–273.

Cooper JC, Owen JH. 1976. Audiologic profile of noise-induced hearing loss. *Archives of Otolaryngology* 102(3):148–150.

Cruickshanks KJ, Klein R, Klein BEK, Wiley TL, Nondahl DM, Tweed TS. 1998. Cigarette smoking and hearing loss: The Epidemiology of Hearing Loss Study. *Journal of the American Medical Association* 279(21):1715–1719.

Cruickshanks KJ, Tweed TS, Wiley TL, Klein BEK, Klein R, Chappell R, Nondahl DM, Dalton DS. 2003. The 5-year incidence and progression of hearing loss: The Epidemiology of Hearing Loss Study. *Archives of Otolaryngology and Head and Neck Surgery* 129:1041–1046.

Davis H and Associates. 1953. Acoustic trauma in the guinea pig. *Journal of the Acoustical Society of America* 25(6):1180–1189.

Davis H, Parrack H, Eldredge D. 1949. Hazards of intense sound and ultrasound. *Annals of Otology, Rhinology and Laryngology* 58(3):732–738.

Davis H, Morgan CT, Hawkins JE, Galambos R, Smith FW. 1950. Temporary deafness following exposure to loud tones and noise. *Acta Oto-Laryngologica Supplement* 88: 1–56.

Dobie RA, Rabinowitz PM. 2002. Change in audiometric configuration helps to determine whether a standard threshold shift is work-related. *Spectrum* 19(Suppl 1):17.

Drescher DG. 1976. Effect of temperature on cochlear responses during and after exposure to noise. *Journal of the Acoustical Society of America* 59(2):401–407.

Eldredge DH, Mills JH, Bohne BA. 1973a. Anatomical, behavioral, and electrophysiological observations on chinchillas after long exposures to noise. *Advances in Oto-Rhino-Laryngology* 20:64–81.

Eldredge DH, Miller JD, Mills JH, Bohne BA. 1973b. Behavioral, physiological and anatomical studies of threshold shifts in animals. In: *Proceedings of the International Congress on Noise as a Public Health Problem.* Washington, DC: U.S. Environmental Protection Agency.

Erway LC, Shiau YW, Davis RR, Krieg EF. 1996. Genetics of age-related hearing loss in mice. III. Susceptibility of inbred and F1 hybrid strains to noise-induced hearing loss. *Hearing Research* 93(1-2):181–187.

Fechter LD. 2004. Promotion of noise-induced hearing loss by chemical contaminants. *Journal of Toxicology and Environmental Health* 67(8-10):727–740.

Fechter LD, Chen GD, Rao D, Larabee J. 2000. Predicting exposure conditions that facilitate the potentiation of noise-induced hearing loss by carbon monoxide. *Toxicological Sciences* 58(2):315–323.

Fraenkel R, Freeman S, Sohmer H. 2003. Susceptibility of young adult and old rats to noise-induced hearing loss. *Audiology Neurootology* 8(3):129–139.

Franks JR, Morata TC. 1996. Ototoxic effects of chemicals alone or in concert with noise: A review of human studies. In: Axelsson A, Borchgrevink HM, Hamernik RP, Hellstrom PA, Henderson D, Salvi RJ, eds. *Scientific Basis of Noise-Induced Hearing Loss.* New York: Thieme. Pp. 437–446.

Fredelius L. 1988. *Degeneration Pattern in the Guinea Pig Organ of Corti After Pure Tone Acoustic Overstimulation. A Morphological and Electrophysiological Investigation.* Ph.D. dissertation. Stockholm, Sweden.

Gallo R, Glorig A. 1964. Permanent threshold shift changes produced by noise exposure and aging. *American Industrial Hygiene Association Journal* 25:237–245.

Gates GA, Schmid P, Kujawa SG, Nam B, D'Agostino R. 2000. Longitudinal threshold changes in older men with audiometric notches. *Hearing Research* 141(1-2):220–228.

Glorig A, Roberts J. 1965. Hearing levels of adults by age and sex, United States, 1960–1962. *Vital and Health Statistics.* Series 11, No. 11. Washington, DC: Department of Health, Education, and Welfare.

Glorig A, Ward WD, Nixon J. 1961. Damage risk criteria and noise-induced hearing loss. *Archives of Oto-Rhino-Laryngology* 74:413–423.

Gratton MA, Salvi RJ, Kamen BA, Saunders SS. 1990. Interaction of cisplatin and noise on the peripheral auditory system. *Hearing Research* 50(1-2):211–223.

Gravendeel DW, Plomp R. 1959. The relation between permanent and temporary noise dips. *American Medical Association Archives of Otolaryngology* 69(6):714–719.

Hamernik RP, Hsueh KD. 1991. Impulse noise: Some definitions, physical acoustics and other considerations. *Journal of the Acoustical Society of America* 90(1):189–196.

Hamernik RP, Henderson D, Coling D, Salvi R. 1981. Influence of vibration on asymptotic threshold shift produced by impulse noise. *Audiology* 20(3):259–269.

Henderson D, Hamernik RP. 1982. Asymptotic threshold shift from impulse noise. In: Hamernik RP, Henderson D, Salvi R, eds. *New Perspectives on Noise-Induced Hearing Loss.* New York: Raven Press. Pp. 265–281.

Henderson D, Hamernik RP. 1986. Impulse noise: Critical review. *Journal of the Acoustical Society of America* 80(2):569–584.

Henderson D, Hamernik RP, Sitler R. 1974a. Audiometric and anatomical correlates of impulse noise exposure. *Archives of Otolaryngology* 99:62–66.

Henderson D, Hamernik RP, Sitler RW. 1974b. Audiometric and histological correlates of exposure to 1-msec noise impulses in the chinchilla. *Journal of the Acoustical Society of America* 56(4):1210–1221.

Henderson D, Subramaniam M, Boettcher FA. 1993. Individual susceptibility to noise-induced hearing loss: An old topic revisited. *Ear and Hearing* 14(3):152–168.

Henselman LW, Henderson D, Shadoan J, Subramaniam M, Saunders S, Ohlin D. 1995. Effects of noise exposure, race, and years of service on hearing in U.S. Army soldiers. *Ear and Hearing* 16(4):382–391.

Hirose K, Liberman MC. 2003. Lateral wall histopathology and endocochlear potential in the noise-damaged mouse cochlea. *Journal of the Association for Research in Otolaryngology* 4(3):339–352.

Hirsch FG. 1968. Effects of overpressure on the ear—a review. *Annals of the New York Academy of Science* 152(1):147–162.

Hirsh IJ, Ward WD. 1954. Recovery of the auditory threshold after strong acoustic stimulation. *Journal of the Acoustical Society of America* 24:131–141.

Holme RH, Steel KP. 2004. Progressive hearing loss and increased susceptibility to noise-induced hearing loss in mice carrying a Cdh23 but not a Myo7a mutation. *Journal of the Association for Research in Otolaryngology* 5(1):66–79.

Howell RW. 1978. A seven-year review of measured hearing levels in male manual steelworkers with high initial thresholds. *British Journal of Industrial Medicine* 35(1):27–31.

Humes LE. 1980. Temporary threshold shifts for masked pure tones. *Audiology* 19:335–345.

Humes LE. 1984. Noise-induced hearing loss as influenced by other agents and by some physical characteristics of the individual. *Journal of the Acoustical Society of America* 76(5):1318–1329.

Humes LE, Jesteadt W. 1989. Models of the additivity of masking. *Journal of the Acoustical Society of America* 85(3):1285–1294.

Humes LE, Jesteadt W. 1991. Modeling the interactions between noise exposure and other variables. *Journal of the Acoustical Society of America* 90(1):182–188.

Humes LE, Koval CB. 1981. Temporary threshold shifts for masked pure tones. II. Broadband masker. *Audiology* 20(4):325–335.

Humes LE, Dirks DD, Bell TS, Kincaid GE. 1987. Recognition of nonsense syllables by hearing-impaired listeners and by noise-masked normal hearers. *Journal of the Acoustical Society of America* 81(3):765–773.

Ishii EK, Talbott E. 1998. Race/ethnicity differences in the prevalence of noise-induced hearing loss in a group of metal fabricating workers. *Journal of Occupational and Environmental Medicine* 40(8):661–666.

ISO (International Organization for Standardization). 1990. *ISO 1999: Acoustics—Determination of Occupational Noise Exposure and Estimation of Noise-Induced Hearing Impairment.* Geneva, Switzerland: ISO.

Jerger J, Alford B, Coats A. 1966. Effects of very low frequency tones on auditory thresholds. *Journal of Speech and Hearing Research* 9(1):150–160.

Johnsson LG, Hawkins JE. 1976. Degeneration patterns in human ears exposed to noise. *Annals of Otology, Rhinology and Laryngology* 85(6 Pt 1):725–739.

Kaufman LR, LeMasters GK, Olsen DM, Succop P. 2005. Effects of concurrent noise and jet fuel exposure on hearing loss. *Journal of Occupational and Environmental Medicine* 47(3):212–218.

Kraus RN. 1959. An evaluation of patients suspected of having noise induced hearing loss. *Laryngoscope* 69(11):1429–1442.

Kryter KD, Ward WD, Miller JD, Eldredge DH. 1966. Hazardous exposure to intermittent and steady-state noise. *Journal of the Acoustical Society of America* 39(3):451–464.

Lee FS, Matthews LJ, Dubno JR, Mills JH. 2005. Longitudinal study of pure-tone thresholds in older persons. *Ear and Hearing* 26:1–11.

Lei SF, Ahroon WA, Hamernik RP. 1994. The application of frequency and time domain kurtosis to the assessment of hazardous noise exposures. *Journal of the Acoustical Society of America* 96(3):1435–1444.

Liberman CM, Mulroy MJ. 1982. Acute and chronic effects of acoustic trauma: Cochlear pathology and auditory nerve pathophysiology. In: Hamernik RP, Henderson D, Salvi R, eds. *New Perspectives on Noise-Induced Hearing Loss.* New York: Raven Press. Pp. 106–135.

Macrae JH. 1991. Presbycusis and noise-induced permanent threshold shift. *Journal of the Acoustical Society of America* 90(5):2513–2516.

Man AL, Naggan L, Swet RT. 1975. Bilateral hearing loss as a sequel to unilateral acoustic trauma. *Israel Journal of Medical Sciences* 11:5–9.

Manninen O. 1988. Changes in hearing, cardiovascular functions, haemodynamics, upright body sway, urinary catecholamines and their correlates after prolonged successive exposures to complex environmental conditions. *International Archives of Occupational and Environmental Health* 60(4):249–272.

Manson J, Alessio HM, Cristell M, Hutchinson KM. 1994. Does cardiovascular health mediate hearing ability? *Medicine and Science in Sports and Exercise* 26(7):866–871.

McBride DI, Williams S. 2001a. Audiometric notch as a sign of noise induced hearing loss. *Occupational and Environmental Medicine* 58(1):46–51.

McBride DI, Williams S. 2001b. Characteristics of the audiometric notch as a clinical sign of noise exposure. *Scandinavian Audiology* 30(2):106–111.

McFadden SL, Campo P. 1998. Cubic distortion product otoacoustic emissions in young and aged chinchillas exposed to low-frequency noise. *Journal of the Acoustical Society of America* 104(4):2290–2297.

McGill TJI, Schuknecht HF. 1976. Human cochlear changes in noise induced hearing loss. *Laryngoscope* 86(9):1293–1302.

Melinek M, Naggan L, Altman M. 1976. Acute acoustic trauma—a clinical investigation and prognosis in 433 symptomatic soldiers. *Israel Journal of Medical Sciences* 12(6):560–569.

Melnick W. 1976. Human asymptotic threshold shift. In: Henderson D, Hamernik RP, Dosanjh DS, Mills JH, eds. *Effects of Noise on Hearing.* New York: Raven Press. Pp. 277–289.

Miller JD, Watson CS, Covell WP. 1963. Deafening effects of noise on the cat. *Acta Oto-Laryngologica Supplement* 176:1–91.

Mills JH. 1973. Temporary and permanent threshold shifts produced by nine-day exposures to noise. *Speech and Hearing Research* 16(3):426–438.

Mills JH. 1976. Threshold shifts produced by 90-day exposure to noise. In: Henderson D, Hamernik RP, Dosanjh DS, Mills JH, eds. *Effects of Noise on Hearing.* New York: Raven Press. Pp. 265–275.

Mills JH, Talo SA. 1972. Temporary threshold shifts produced by exposure to high-frequency noise. *Speech and Hearing Research* 15(3):624–631.

Mills JH, Gengel RW, Watson CS, Miller JD. 1970. Temporary changes of the auditory system due to exposure to noise for one or two days. *Journal of the Acoustical Society of America* 48(2, Pt 2):524–530.

Mills JH, Gilbert RM, Adkins WY. 1979. Temporary threshold shifts in humans exposed to octave bands of noise for 16 to 24 hours. *Journal of the Acoustical Society of America* 65(5):1238–1248.

Mills JH, Osguthorpe JD, Burdick CK, Patterson JH, Mozo B. 1983. Temporary threshold shifts produced by exposure to low-frequency noises. *Journal of the Acoustical Society of America* 73(3):918–923.

Mills JH, Boettcher FA, Dubno JR. 1997. Interaction of noise-induced permanent threshold shift and age-related threshold shift. *Journal of the Acoustical Society of America* 101(3): 1681–1686.

Mizoue T, Miyamoto T, Shimizu T. 2003. Combined effect of smoking and occupational exposure to noise on hearing loss in steel factory workers. *Occupational and Environmental Medicine* 60:56–59.

Moody DB, Stebbins WC, Johnsson LG, Hawkins JE. 1976. Noise-induced hearing loss in the monkey. In: Henderson D, Hamernik RP, Dosanjh DS, Mills JH, eds. *Effects of Noise on Hearing*. New York: Raven Press. Pp. 309–322.

Morata TC. 1998. Assessing occupational hearing loss: Beyond noise exposures. *Scandinavian Audiology Supplement* 48:111–116.

Morata TC. 2003. Chemical exposure as a risk factor for hearing loss. *Journal of Occupational Environmental Medicine* 45(7):676–682.

Morata TC, Campo P. 2001. Auditory function after single of combined exposure to styrene: A review. In: Henderson D, Prasher D, Kopke R, Salvi R, Hamernik R, eds. *Noise Induced Hearing Loss: Basic Mechanisms, Prevention and Control.* London: Noise Research Network Publications. Pp. 293–304.

Morata TC, Lemasters GK. 1995. Epidemiologic considerations in the evaluation of occupational hearing loss. *Occupational Medicine* 10(3):641–656.

Morata TC, Dunn DE, Kretschmer LW, Lemasters GK, Keith RW. 1993. Effects of occupational exposure to organic solvents and noise on hearing. *Scandinavian Journal of Work, Environment & Health* 19(4):245–254.

Morata TC, Dunn DE, Sieber WK. 1994. Occupational exposure to noise and ototoxic organic solvents. *Archives of Environmental Health* 49(5):359–365.

Nixon JC, Glorig A. 1961. Noise-induced permanent threshold shift at 2000 cps and 4000 cps. *Journal of the Acoustical Society of America* 33:904–908.

NRC (National Research Council). 1992. *Hazardous Exposure to Impulse Noise.* Committee on Hearing, Bioacoustics, and Biomechanics. Washington, DC: National Academy Press.

NRC. 1993. *Hazardous Exposure to Steady-State and Intermittent Noise.* Committee on Hearing, Bioacoustics, and Biomechanics. Washington, DC: National Academy Press.

Ohlemiller KK, Wright JS, Heidbreder AF. 2000. Vulnerability to noise-induced hearing loss in 'middle-aged' and young adult mice: A dose-response approach in CBA, C57BL, and BALB inbred strains. *Hearing Research* 149(1-2):239–247.

Palmer KT, Griffin MJ, Syddall HE, Coggon D. 2004. Cigarette smoking, occupational exposure to noise, and self reported hearing difficulties. *Occupational and Environmental Medicine* 61:340–344.

Pekkarinen J. 1995. Noise, impulse noise, and other physical factors: Combined effects on hearing. *Occupational Medicine* 10(3):545–559.

Resnick DM, Albrite JP, Shutts RE. 1962. Noise-induced hearing loss among army helicopter trainees. *Laryngoscope* 72:262–273.

Roberto M, Hamernik RP, Turrentine GA. 1989. Damage of the auditory system associated with acute blast trauma. *Annals of Otology, Rhinology and Laryngology* (Suppl 140):23–34.

Robertson RM, Page JC, Williams CE. 1978. *The Prevalence of Hearing Loss Among Selected Navy Enlisted Personnel*. NAMRL-1251. Pensacola, FL: Naval Aerospace Medical Research Laboratory.

Robinson DW, Sutton GJ. 1979. Age effect in hearing—a comparative analysis of published threshold data. *Audiology* 18(4):320–334.

Rodriguez GP, Gerhardt KJ. 1991. Influence of outer ear resonant frequency on patterns of temporary threshold shift. *Ear and Hearing* 12(2):110–114.

Rosenhall U. 2003. The influence of ageing on noise-induced hearing loss. *Noise Health* 5(20):47–53.

Royster LH, Thomas WG, Royster JD, Lilley D. 1978. Potential hearing compensation cost by race and sex. *Journal of Occupational Medicine* 20(12):801–806.

Saunders JC, Tilney LG. 1982. Species difference in susceptibility to noise exposure. In: Hamernik RP, Henderson D, Salvi R, eds. *New Perspectives on Noise-Induced Hearing Loss*. New York: Raven Press. Pp. 229–248.

Segal S, Harell M, Shahar A, Englender M. 1988. Acute acoustic trauma: Dynamics of hearing loss following cessation of exposure. *American Journal of Otology* 9(4):293–298.

Shaw EA. 1974. Transformation of sound pressure level from the free field to the eardrum in the horizontal plane. *Journal of the Acoustical Society of America* 56(6):1848–1861.

Spongr VP, Boettcher FA, Saunders SS, Salvi RJ. 1992. Effects of noise and salicylate on hair cell loss in the chinchilla cochlea. *Archives of Otolaryngology–Head and Neck Surgery* 118(2):157–164.

Starck J, Toppila E, Pyykko I. 1999. Smoking as a risk factor in sensory neural hearing loss among workers exposed to occupational noise. *Acta Oto-Laryngologica* 119(3):302–305.

Stockwell CW, Ades HW, Engström H. 1969. Patterns of hair cell damage after intense auditory stimulation. *Annals of Otology, Rhinology and Laryngology* 78:1144–1168.

Sun JC, Bohne BA, Harding GW. 1994. Is the older ear more susceptible to noise damage? *Laryngoscope* 104(10): 1251–1258.

Taylor W, Pearson J, Mair A, Burns W. 1965. Study of noise and hearing in jute weaving. *Journal of the Acoustical Society of America* 38:113–120.

Temkin J. 1933. Die Schädigung des Ohres durch Lärm und Erschütterung. *Monatsschrift für Ohrenheilkunde und Laryngo-Rhinologie* 67:257–299.

Thalmann RR. 1976. Quantitative biochemical techniques for studying normal and noise-damaged ears. In: Henderson D, Hamernik RP, Dosanjh DS, Mills JH, eds. *Effects of Noise on Hearing*. New York: Raven Press. Pp. 129–154.

Toppila E, Pyykko II, Starck J, Kaksonen R, Ishizaki H. 2000. Individual risk factors in the development of noise-induced hearing loss. *Noise Health* 2(8):59–70.

Van Campen LE, Dennis JM, Hanlin RC, King SB, Velderman AM. 1999. One-year audiologic monitoring of individuals exposed to the 1995 Oklahoma City bombing. *Journal of the American Academy of Audiology* 10(5):231–247.

Wang Y, Hirose K, Liberman MC. 2002. Dynamics of noise-induced cellular injury and repair in the mouse cochlea. *Journal of the Association for Research in Otolaryngology* 3:248–268.

Ward WD. 1965. The concept of susceptibility to hearing loss. *Journal of Occupational Medicine* 7(12):595–607.

Ward WD. 1968. Susceptibility to auditory fatigue. In: Neff WD, ed. *Contributions to Sensory Physiology*. Vol. 3. New York: Academic Press. Pp. 191–226.

Ward WD. 1973. Noise-induced hearing damage. *Otolaryngology* 2:377–390.

Ward WD. 1995. Endogenous factors related to susceptibility to damage from noise. *Occupational Medicine* 10(3):561–575.

Ward WD, Glorig A. 1961. A case of firecracker-induced hearing loss. *Laryngoscope* 71:1590–1596.

Ward WD, Royster JD, Royster LH. 2000. Auditory and nonauditory effects of noise. In: Berger EH, Royster LH, Royster JD, Driscoll DP, Layne M, eds. *The Noise Manual.* 5th ed. Fairfax, VA: American Industrial Hygiene Association. Pp. 101–147.

Wiley TL, Torre P, Cruickshanks KJ, Nondahl DM, Tweed TS. 2001. Hearing sensitivity in adults screened for selected risk factors. *Journal of American Academy of Audiology* 12(7):337–347.

Wilson WJ, Herbstein N. 2003. The role of music intensity in aerobics: Implications for hearing conservation. *Journal of the American Academy of Audiology* 14(1):29–38.

Ylikoski ME, Ylikoski JS. 1994. Hearing loss and handicap of professional soldiers exposed to gunfire noise. *Scandinavian Journal of Work, Environment & Health* 20(2):93–100.

Yoshida N, Hequembourg SJ, Atencio CA, Rosowski JJ, Liberman MC. 2000. Acoustic injury in mice: 129/SvEv is exceptionally resistant to noise-induced hearing loss. *Hearing Research* 141(1-2):97–106.

3

Noise and Noise-Induced Hearing Loss in the Military

The focus of this chapter is on noise and noise-induced hearing loss in the U.S. military. The committee was asked to identify sources of potentially damaging noise in the military setting and to review and assess available evidence on hearing loss incurred by members of the armed services as a result of noise exposure during military service since World War II. Concern about noise exposure and hearing loss among military personnel has been evident throughout this period (e.g., Glorig, 1952; Carmichael, 1955; CHABA, 1968; Yarington, 1968; Walden et al., 1971; Yankaskas and Shaw, 1999).

The first part of the chapter briefly reviews the services' policies and programs to collect data on noise levels generated by equipment used by military personnel and the noise doses received by military personnel working in certain settings. Examples of the kinds of data collected through these efforts are provided. The remainder of the chapter focuses on the committee's assessment of data on hearing thresholds and hearing loss among military service members since World War II.

NOISE IN THE MILITARY ENVIRONMENT

The sources of noise in the military are as varied as the activities carried out by the members of the Army, Navy, Air Force, Marine Corps, and Coast Guard. Obvious sources of potentially hazardous noise are weapons systems and jet engines, but vehicles, other aircraft, watercraft, communication systems, and industrial-type activities also serve as sources of potentially damaging noise.

Collection of Data on Noise Levels and Estimated Noise Exposures

Since World War II, numerous measurements of the sound pressure levels in proximity to various weapon systems and other military equipment have been collected. In addition, information has been collected on estimated noise doses for personnel working in steady-state noise. In the late 1970s, the Department of Defense (DoD) established, as part of an overall hearing conservation program, a department-wide requirement for periodic surveys of noise-hazardous environments and, subsequently, requirements for noise dosimetry. Each military service was responsible for collecting and maintaining information about hazardous noise environments and noise exposures. Many military sites had been collecting such information well before the DoD requirements were put in place. This section briefly reviews DoD-level requirements concerning measurement of noise levels and noise exposure. It also reviews the services' data collection activities and the availability of these data. Noise-exposure limits are discussed in Chapter 5.

Department of Defense Requirements

In 1978, DoD established a requirement that each of the military services conduct sound surveys to identify and periodically monitor noise-hazardous environments (DoD, 1978). By 1987, the requirements included provisions for measuring noise exposures for workers exposed to noise levels of 85 dBA or more (DoD, 1987). Also included were separate specifications for the measurement of impulse noise and performance criteria for the measurement devices to be used. The current requirements, contained in DoD Instruction 6055.12, DoD Hearing Conservation Program, specify that sound pressure levels (SPLs) are to be measured in all potentially hazardous noise work areas at least once and within 30 days of any change in operations affecting noise levels (DoD, 2004).

Noise exposure (i.e., dose) is to be measured as time-weighted average (TWA) noise levels for military personnel working in industrial-type operations with hazardous noise levels.[1] The surveys must be conducted by trained personnel using sound-level meters or dosimeters meeting or exceeding relevant standards established by the American National Standards Institute (ANSI). The DoD instruction does not require measurement of noise doses associated with military activities, whether actual operations or training exercises. Noise exposure during such activities can be highly variable, and typical dosimeters are not designed to capture the rapid rise to briefly sustained peak sound pressure levels in excess of 140 dB that occur

[1]The DoD instructions also apply to noise-exposed civilian workers.

with weapons fire and other impulse noise (e.g., Kardous and Willson, 2004).

Requirements for maintaining data from noise surveys and exposure assessments have changed over time. The initial requirement in 1978 was that data be maintained for 5 years. By 1987, the period had been extended to 30 years. In 1996, DoD specified that noise exposure data were to be maintained "for the duration of employment plus 40 years" (DoD, 1996).

DoD also has established design standards for noise levels of new materiel designed or purchased for the military services. In the most recent version of these standards (DoD, 1997), the stated purpose is to lead to equipment that minimizes noise-induced hearing loss, permits acceptable speech communication in a noisy environment, minimizes aural detection by an enemy, minimizes community annoyance, and provides acceptable habitability of personnel quarters (DoD, 1997). The design standards include limits for steady-state and impulse noise in occupied areas and noise from shipboard equipment and aircraft, including rotary-wing aircraft. However, for both new and older equipment used in "military-unique" settings,[2] DoD regulations give priority to maintaining combat readiness and allow for tradeoffs between noise reduction and weight, speed, cost, or other factors crucial to the effectiveness of the equipment (DoD, 2004).

Air Force

The Air Force began requiring noise measurements in 1948 with its first regulation regarding hazardous noise: AFR 160-3, Precautionary Measures Against Noise Hazards (Gasaway, 1988). By 1956, regulations required the use of either direct measurement or published data to plot master plans of bases to indicate where exposure to hazardous noise might occur. Data on noise levels of aircraft and other power machinery were published periodically by the Wright Air Development Center at Wright-Patterson Air Force Base. One notable compendium contains measurements from within cockpit areas of hundreds of types and models of aircraft (Gasaway, 2002; also see Gasaway, 1986).

Dosimetry measurements began at selected airbases in the late 1970s (Fairman and Johnson, 1979). A 1982 Air Force regulation required evaluation of individual noise exposures for personnel whose exposure exceeded

[2]The term "military-unique" settings refers to DoD military and civilian operations and workplaces that are unique to the national defense mission (DoD, 1998). They include combat, combat training, and operation, testing, and maintenance of military equipment and systems, among which are weapons, aircraft, ships, submarines, missiles, ordnance, and tactical vehicles. The designation applies to such operations as peacekeeping missions, field maneuvers, naval operations, and military flight.

the equivalent of an 8-hour time-weighted average of 84 dBA (Department of the Air Force, 1982). For a group of employees doing similar work with similar noise exposures (referred to as a "similar exposure group"), it was and remains permissible to monitor the most highly exposed individual and to assign the resulting measurement to each group member, essentially a "worst case" assumption. Current noise evaluation procedures (Department of the Air Force, 1994) require dosimetry for a minimum of 3 worker-days (defined as one worker for 3 days, or three workers for 1 day) to identify the average daily exposure.

Starting in the late 1980s, various Air Force installations began automating their recordkeeping for sound pressure levels and dosimetry. In 2005, each installation continues to maintain its own database. The dosimetry data are used primarily for local shop or worksite decisions, or occasional installation-level uses. Compilation of such data across the entire Air Force is possible but is not done for routine analysis (Weisman, 2005).

Navy and Marine Corps

Navy requirements for the collection of noise survey data date back at least to a 1983 regulation requiring noise measurements and personal dosimetry with appropriate equipment and calibration (Department of the Navy, 1983). The Navy is also responsible for noise surveys for Marine Corps facilities. To date, data on sound pressure levels are routinely collected at Navy and Marine Corps facilities but are not routinely transferred to a central database.

Noise dosimetry data are routinely collected by local Navy medical units to perform exposure assessments and to make recommendations for placement of personnel into the hearing conservation program. Under current procedures, such noise exposure data, in the form of time-weighted average sound levels, must be provided to the exposed individuals, the command, and the entity providing medical surveillance (Navy Environmental Health Center, 2004b). Starting in 2002, noise dosimetry data collected for a variety of industrial, shipboard, and other naval operations, including Marine Corps activities, have been added annually to the Navy Occupational Exposure Database (Crowder, 2005). Some of these data date back to 1980. The data are not routinely used in the Navy's overall hearing conservation program, but they are used on a case-by-case basis to respond to inquiries (Crowder, 2005).

Army

The Army has both a centralized program to evaluate the sound pressure levels of new weapons systems and equipment and a distributed program for noise measurement and analysis at worksites at all Army installations. Through the centralized Health Hazard Assessment Program, begun in 1980, new equipment is tested to assess various potential hazards, including noise, chemicals, radiation, and vibration. The measurements of sound pressure levels are used to estimate likely time-weighted averages during use of the equipment, but noise dosimetry is not carried out as part of this program. The test information is used to make recommendations regarding the need for personal hearing protection as well as possible restrictions on training time with the systems (personal communication, F. Sachs, U.S. Army Center for Health Promotion and Preventive Medicine, August 18, 2004). The Army also has comprehensive data on sound pressure levels from weapons and equipment beginning from the 1970s and a more limited set of data going back to the 1960s.

In addition, each Army installation evaluates work environments for potential noise hazards from steady-state noise in industrial-type operations. Since 1988, the sound pressure level and noise dosimetry measurements have been collected in the Health Hazard Information Module database of the Army's Occupational Health Management Information System. Dosimetry measurements are not routinely attempted for military-unique activities in the Army, in part because the impulse noise components are not readily measured by current instrumentation (U.S. Army Center for Health Promotion and Preventive Medicine, 1999; personal communication, D. Ohlin, U.S. Army Center for Health Promotion and Preventive Medicine, 2005).

Coast Guard

Coast Guard noise surveys were part of the Coast Guard hearing conservation program by the late 1960s and early 1970s (McConnell, 2004). Sound pressure level and noise dosimetry measurements made by the Coast Guard are provided to units in the form of written reports (McConnell, 2005).

Defense Occupational and Environmental Health Readiness System–Industrial Hygiene

In 2005, as this report was being written, all the services were still using their own databases on sound pressure levels and noise dosimetry. However, development of a DoD-wide database for recording, storing, and retrieving sound pressure level and noise dosimetry data, as well as information related to other occupational exposures, is in advanced stages. Introduction of the Defense Occupational and Environmental Health Readiness System–Industrial Hygiene (DOEHRS-IH) is planned for fall 2005 (personal communication, K. Wisniewski, U.S. Army Center for Health

Promotion and Preventive Medicine, February 2005). This database is intended to provide a longitudinal record of noise and other occupational exposures for DoD personnel.

Sound Levels and Noise Doses in the Military Environment

Information on noise sources and noise levels in the military environment is plentiful and detailed but not complete and not easily summarized. Sound levels vary depending on the distance from the sound source and the conditions under which the sound is being generated. Important characteristics of impulse noise include not only the peak sound pressure level, but the time pattern of the impulses and the frequency spectrum.

Table 3-1 provides examples of some of the measurements made since the 1950s of average sound levels found in ground vehicles and aircraft and peak sound pressure levels generated by certain weapons.[3] On aircraft carriers, flight operations create an environment with combinations of aircraft noise, mechanical noise, and impact noise (Yankaskas and Shaw, 1999). Below the flight deck, sound levels have been measured at 106 dBA during aircraft launches. Exposure to high sound levels has also been reported for military personnel in positions such as radio operators (Robertson et al., 1990) and sonar technicians (Marshall and Carpenter, 1988) in the Navy and cryptolinguists in the Air Force (Ritter and Perkins, 2001). In addition, military personnel may encounter potentially damaging noise from equipment and activities comparable to those found in industrial settings, such as the operation of heavy equipment (Chandler and Fletcher, 1983). Data are also available on the acoustic spectra of some types of noise in the military environment (e.g., Johnson and Nixon, 1974; Gasaway, 2002), but they are not illustrated in Table 3-1.

The examples of noise levels associated with equipment and weaponry in the military included in Table 3-1 clearly demonstrate that there are many sources of high sound pressure levels in the military environment that exceed criteria for safe exposure. Data on sound pressure levels, however, are not sufficient by themselves to determine the noise dose received by an individual. As described above, dosimetry data have also been collected, but the committee found little published dosimetry data that could be used to draw conclusions about typical exposures (e.g., Fairman and Johnson, 1979; Jordan and Jones, 1983).

[3]Appendix F provides an illustrative list of documents, most of which are available in the published literature or in electronic form from government sources, that report sound levels generated by a variety of military aircraft, vehicles, equipment, and weapons systems.

TABLE 3-1 Examples of Sound Levels Associated with Military Equipment

a. *Aircraft in Use in the 1950s*

Name	Model	Location	Condition	Sound Level dBC/F	Sound Level dBA
Douglas Skyraider (USN; in use 1950s–1960s)	A-1J (AD-7)	Cockpit	Takeoff	132	128
			Climb	118	117
			Level	121	120
			Cruise	113	109
Douglas Skyraider (USN)	A-1J (AD-7)	Passenger area	Taxi	87	81
			Takeoff	111	111
			Normal cruise	106	103
			High cruise	109	108

NOTES: dBC/F, C-weighted or flat-weighted levels; dBA, A-weighted levels.
SOURCE: Gasaway (2002).

b. *Military Equipment in Use in the 1960s*

Name	Model	Location	Condition	Sound Level dB
Tanks		Interior		115 (±10)
Personnel carrier (APC)		Interior		120 (±10)
NATO rifle	M-14	Operator's right ear	20 rounds, full automatic	159
3.7-in rocket launcher (bazooka)		Operator's right ear		163
105mm howitzer				190 (impulse)
Sergeant missile		100 ft from launch site		145
Helicopter (Shawnee)	H-21C	Crew chief, 15-20 ft from aircraft	Rotor engaged	110

SOURCE: Adapted from Yarington (1968).

c. *Flight Operations at Landing Signal Officer Platform, U.S. Navy Aircraft Carrier, 1970s*

Event (n)	Aircraft	Average Noise Level dB(SPL)	Noise Level dBA	Average Duration at ≥ 90 dBA (seconds per event)
Trap (35)	F4	128	122	5
Trap (1)	A7	123	114	4
Trap (2)	C1	114	108	5
Bolter (6)	F4	120	116	5
Wave-off (8)	F4	125	124	5
Touch and go (10)	F4	129	121	4
Deck launch (2)	C1	127	123	20

SOURCE: Robertson et al. (1978a).

d. *Military Equipment in Use in 2005*

Army Vehicles	Model	Condition	Location	Speed km/hr (mph)	Sound Level dBA
High-mobility multi-purpose wheeled vehicle (HMMWV), non-heavy variants	M966, also: M996, M997, M998, M1037, and others	2/3 payload	Crew positions	0(idle)	78
				48(30)	84
				88(55)	94
HMMWV, heavy variant	M1097, M1097A2, M1113, M1114	2/3 payload	Crew positions	Up to 50(31)	< 85
				64(40)	88
				80(50)	92
				96(60)	98
HMMWV heavy variant	M1097	Full payload	Crew positions	Up to 40(25)	< 85
				96(60)	100

continued

TABLE 3-1 continued

Army Vehicles	Model	Condition	Location	Speed km/hr (mph)	Sound Level dBA
Commercial utility cargo vehicle	M1008, M1009, M1010, M1028		In cab	Below 88(55)	< 85
				88(55)	85–91
Ambulance	M1010		Patient areas	All speeds	< 85
Armored personnel carrier, A3 version	M113A3 family, including M106A2, M1064A3, M1059A3, M58A3, M730A2, M901A3, M981A3			Idle	85–92
				16(10)	106
				32(20)	109
				48(30)	114
				63(40)	118
Abrams tank	M1A2, M1, M1A1		In vehicle	Idle	93
				Tactical idle	103
				16(10)	108
				48(30)	114
				63(40)	117

Army Helicopters	Model	Sound Level Location	dBA
Chinook	CH-47D	Cockpit	103
Blackhawk	UH-60A	Pilot	106
		Copilot	106
Apache	YAH-64	Pilot	104
		Copilot	101

Weapons (Impulse Noise)	Model	Condition	Location	Sound Level dBP
9mm pistol	M9		Shooter	157
5.56mm rifle	M16A2		Shooter	157
5.56mm squad automatic weapon fired from HMMWV	M249		Gunner	160
7.62mm machine gun	M60	Fired from HMMWV	Gunner	155
0.50 caliber machine gun	M2	Fired from HMMWV	Gunner	153
Machine gun	MK 19, Mod 3	Fired from HMMWV	Gunner	145
Grenade	M26		At 50 ft	164
MAAWS recoilless rifle	M3		Gunner	190
Light antitank weapon	M72A3		Gunner	182
Javelin antitank missile		Open position	Gunner	160
		Enclosed position	Gunner	166
		Fighting position	Gunner	172
105mm towed howitzer	Ml 19	At charge 8	Gunner	183
155mm towed howitzer	M198	Firing M203 propellant	Gunner	178

NOTES: In flight, helicopter crews wear helmets with integral hearing protectors. Passengers must rely on their own hearing protectors (e.g., earplugs) or ones supplied by air operations. HMMWV, high-mobility multipurpose wheeled vehicle; MAAWS, multi-role anti-armor anti-personnel weapon system.
SOURCE: Adapted from U.S. Army Center for Health Promotion and Preventive Medicine (2004d).

Estimating Cumulative Noise Exposures

Despite the existence of data on sound pressure levels generated by weapons and equipment, and dosimetry estimates of noise exposure for certain personnel, arriving at an estimate of the cumulative noise exposure of any service member or group of service members is nearly impossible. To an even greater extent than civilian workers, military personnel are not likely to experience homogeneous noise exposures over the course of their military service.

The impulsive and intermittent nature of many military noise exposures (e.g., gunfire, plane launches and landings, tank operation, 6-month shipboard deployments) adds an element of uncertainty to considerations of exposure and effect. For example, "high noise levels" for about 3 hours of a 14-hour period have been described as typical of high-tempo flight operations on an aircraft carrier (Yankaskas and Shaw, 1999). As discussed in Chapter 2, intermittent noise exposure may permit recuperation, thus ameliorating to some extent the hazardous effect of noise exposures. However, military personnel may also have noise exposures that are prolonged compared to those of civilians. At sea, for example, sailors are exposed to ambient shipboard noise continuously and may encounter potentially hazardous noise levels even in their sleeping quarters, giving their auditory systems no opportunity for short-term recovery (Yankaskas and Shaw, 1999; Yankaskas, 2001, 2004).

Even for personnel assigned to a specific occupational specialty, it is reasonable to assume that the typical activities for an individual vary over time and that the activities at any given time vary among personnel at different military installations. In addition, members of the military frequently change assignments and may be exposed to different degrees of noise hazard in different work settings. Among a sample of Navy enlisted personnel who served during the period 1982–2004 and were still serving after 2001, for example, the average length of service was 80 months, and time on shore duty averaged 40 months (Shaw and Trost, 2005). Finally, as described in detail in Chapter 5, wide variations in the effective use of hearing protection devices among military personnel can dramatically affect an individual's noise exposure.

In sum, despite the availability of data on sound pressure levels and some dosimetry data, the complexity of military noise exposures precludes ready estimates of service members' cumulative noise doses.

FINDING: The evidence is sufficient to conclude that hazardous noise levels are and have been present in many military settings.

FINDING: Extensive collections of data on sound pressure levels produced by equipment and activities in military settings are available from World War II to the present. Many estimates of noise exposures (doses) from specific activities also are available from more restricted time periods. However, because of the changing nature of assignments in the military, the unpredictable aspects of military training and combat, the intermittent nature of many military noise exposures, and the sporadic use of hearing protection while in the military, these data do not provide a sufficient basis for estimating cumulative noise exposures over the course of military service for individuals or for subgroups (e.g., occupational specialties, branches, or eras).

EVIDENCE REGARDING THE EFFECTS OF NOISE ON HEARING AMONG U.S. MILITARY PERSONNEL

The committee was asked to review the evidence that hearing loss has been incurred by members of the armed services as a result of noise exposures during military service since World War II. To investigate this subject, the committee examined information from various sources, including studies reported in the published literature, reports prepared for the military services, and data from the military services' hearing conservation databases, which were provided at the request of the committee. The committee also undertook additional analyses of some of the data.

The available information proved to offer an incomplete picture of changes in hearing thresholds over the course of military service and virtually no direct measures of the noise exposure or noise dose for individuals or groups. In the remainder of the chapter, the nature of the available data and their limitations are discussed, followed by a review of the data examined by the committee and presentation of the committee's conclusions.

Studying Noise-Induced Hearing Loss Among Military Personnel

Potentially damaging noise exists in the military environment, but assessing its effects on the hearing of military personnel is not straightforward. Predictive formulas have been adopted as a basis for estimating the amount of hearing loss to be expected in an industrial population exposed to specified levels and durations of occupational noise (i.e., ISO-1999 [ISO, 1990]; ANSI S3.44 [ANSI, 1996]), but they are based on assumptions of exposures to noise of up to 8 hours per day, 5 days per week, and for periods of 10 to 40 years. Frequently, noise exposures in military settings do not conform to these conditions.

An alternative would be to have longitudinal data obtained using consistent measurement tools to track noise doses and hearing thresholds for individual military service members, or at least defined subgroups, over the course of their military service. Other factors that might affect hearing would also have to be taken into account, including nonoccupational noise exposures. Such analyses have not been conducted and are impossible with the existing data.

In the studies reviewed by the committee, two approaches primarily have been used to assess risk of noise-induced hearing loss in the military. They are: (1) measurement of noise doses and estimation of the hearing loss expected from the measured dosages; and (2) measurement of pure-tone hearing thresholds in groups of military personnel. In the latter case, the data may consist of the actual thresholds measured at several frequencies or the relative shift in thresholds over some period of time.

Noise Doses

Comprehensive data on cumulative noise doses received by individual service members are not available. Furthermore, as discussed above, the available dosimetry data are selective and not in a form that would support reasonable efforts to incorporate them in analyses of noise-induced hearing loss.

As an alternative, some studies use data on short-term noise doses or sound levels in given settings. Other indicators, such as military occupational categories, are also used. If the categories are broad, however, it may be difficult to know if there are significant variations in actual noise exposures within each group. A range of noise exposures within groups would tend to blur distinctions among groups.

In addition, differences in age or length of service cannot be assumed to correspond to differences in the amount of noise exposure. Personnel showing signs of threshold shifts may be moved to jobs with lower noise levels. Those who remain in military service for longer periods may move into lower noise assignments. Use of hearing protection can vary over time and among groups. Many recent analyses have been based on data for personnel who are enrolled in the services' hearing conservation programs and, by definition, considered to be at risk for damaging noise exposures. However, their effective noise doses are not necessarily greater than those of other personnel because of the protective measures that are part of the hearing conservation program. There is also no information from which to assess whether noise exposures at a given age or length of service are comparable in personnel who have served in different eras.

Measurement of Hearing Thresholds

Threshold data from periodic hearing tests have been recorded in service members' individual medical records since the 1950s. Service members' medical records may also include the results of annual tests conducted in conjunction with the services' hearing conservation programs. Access to these data is often difficult or impractical, and only limited use has been made of them for analytic studies.[4] Some of the studies reviewed by the committee used actual databases from the hearing conservation programs, but others relied on test data collected specifically for the study or for some other purpose. Most of the reports reviewed were cross-sectional analyses.

The committee focused on reports from the 1970s and later, although a few earlier studies are noted. The emphasis on the more recent data reflects, in part, the evolution of audiometric testing standards and equipment between World War II and 1970. Data collected since the 1970s are considered more reliable and consistent than earlier data, although changes since then, such as the shift from self-recording audiometers to microprocessor-based audiometers, could affect the comparability of measurements. Differences may also exist among data sources in the range of values recorded during hearing testing, with the military seldom using levels below 0 dB HL.

No specific metric has been developed to establish unequivocal evidence for the presence of noise-induced hearing loss. A primary indicator of the presence of noise-induced hearing loss is the pattern of hearing loss in the high frequencies, referred as the "noise notch," in which hearing thresholds at 3000, 4000, or 6000 Hz are worse than those at lower frequencies or at 8000 Hz. Chapter 2 described work being done to try to define a "notch index" that may characterize such notches quantitatively, but currently, a variety of approaches are used to present data in studies of noise-induced hearing loss.

Frequently, data reviewed by the committee were presented in terms of group averages, such as mean or median values for thresholds at individual frequencies or groups of frequencies (e.g., 3000, 4000, and 6000 Hz). These averages obscure the variation within the group. A few reports provided data on the distribution of pure-tone thresholds at various frequencies, making it possible to determine the percentage of the group whose thresholds exceeded specific values. Some reports for Army and Air Force personnel also included the percentage of personnel whose thresholds met criteria for hearing-loss classification systems used in medical fitness profiles.

In many cases, threshold data available to the committee were not measured pure-tone thresholds, but the percentages of personnel in the services' hearing conservation programs who experienced a significant threshold shift (STS). STS provides an indication of the relative changes in

---

[4]The present Institute of Medicine study included a review of audiometric reports contained in the medical records of a sample of veterans. That review focused on procedural aspects of the testing and reporting, not the audiometric results. The findings from the record review are discussed in Chapter 6.

hearing thresholds from baseline measurements following some unspecified period of time (usually, at least 1 year). Follow-up testing is required to determine whether the threshold shift is permanent.

An STS indicates only that a change in hearing of at least a certain magnitude has occurred. It does not capture the amount of change or the new hearing thresholds. The committee received annual STS data from the services' hearing conservation programs and also reviewed published reports on STS in the military. Although the data are potentially useful, several administrative problems make their validity suspect at present. Those problems are described below.

Comparisons between groups are also used to explore the effects of differences in exposures. The committee undertook age-specific comparisons of average high-frequency thresholds among specific groups of military personnel and those in a reference population. The group data available to the committee permit only a limited analysis. The committee identified a small number of other studies that were able to go further and explore the likelihood of finding hearing loss among military veterans on the basis of data for individuals and that used statistical methods to account for other relevant differences among the members of the comparison groups. One study examined the risk of hearing loss among veterans with and without service in Vietnam. The other study compared veterans and nonveterans.

Most of the reports of hearing thresholds in the military reviewed by the committee, including reports on STS, are based on a single measurement at a specific point in time and are not appropriate for drawing conclusions about longitudinal trends. Comparisons of age groups within studies or across studies conducted at different times are subject to the cohort effects of differences over time in the characteristics and exposures of the members of the groups, as well as differences in data definitions, measurement tools, and other aspects of data collection. In studying noise-induced hearing loss in the military population during the 60 years since World War II, a few important considerations include the irregular timing of armed conflicts and changes in hearing conservation programs, audiometer technology, and the types of hearing protection available (see Table 3-2).

In summary, there are no data that permit a comprehensive review of noise-induced hearing loss experienced by military personnel in each of the armed services during the period since World War II. Most of the data identified by the committee came from cross-sectional reports of average (a mix of mean and median) hearing levels among various categories of service members. The reports typically provided data stratified by age or length of service. The committee reviewed at least 12 such reports. (See Table D-5 in Appendix D for summary information on the features of these reports.) No data on Coast Guard personnel were available to the committee. Only very

TABLE 3-2 Time Line of Major Conflicts, Milestones in Hearing Conservation Programs, and Hearing Protection Devices

Year	Major Conflicts	Hearing Conservation Milestones	Hearing Protection Available
1941	World War II begins		Cotton, fingers, or no protection
1945	World War II ends		Some early hearing protection available: V-51R, early earmuffs
1948		Air Force issues first regulation concerning hazardous noise	
1950	Korean War begins	1950s: Introduction and increasing use of pure-tone audiometry	
1953	Korean War ends		
1956		Air Force introduces first of services' hearing conservation programs	
1960		Pure-tone audiometry replaces whisper test for screening military applicants	1960s: V-51R, triple-flange earplugs, improved earmuffs
1964	Vietnam War begins		
1970		First Navy comprehensive hearing conservation program Air Force begins automated handling and storage of audiometric monitoring data (late 1974)	1970s: Same earplugs and earmuffs, new foam earplugs, helmets with noise attenuation introduced
1975	Vietnam War ends		
1978		DoD Instruction establishing hearing conservation programs	
1980		First comprehensive Army hearing conservation program Army begins data repository for audiometric data	1980s: Conventional earplugs and muffs same as 1960s and 1970s; some tanker helmets introduced with active noise reduction (ANR) included
1983		OSHA Final Noise Standard	
1990	Gulf War deployments begin: Operation Desert Shield		1990s: Performance of existing earplugs and muffs essentially the same; new use of earphone in foam earplugs for tank and helicopter communication under helmet

continued

TABLE 3-2 continued

Year	Major Conflicts	Hearing Conservation Milestones	Hearing Protection Available
1991	Gulf War: Operation Desert Storm Gulf War ends		Widespread use of ANR in tanker helmets, some in aircraft flight helmets
1999		Introduction of DOEHRS-HC to standardize audiometric data repositories across services	
2001	War on Terrorism: Operation Enduring Freedom (Afghanistan) begins		2000s: Introduction of noise-level-dependent earplugs (Combat Arms Earplug)
2003	War on Terrorism: Operation Iraqi Freedom begins		
2004		Services adopt OSHA definition for STS	

NOTES: Additional information on developments in hearing protection can be found in Table 5-2. DoD, Department of Defense; DOEHRS-HC, Defense Occupational and Environmental Health Readiness System-Hearing Conservation; OSHA, Occupational Safety and Health Administration; STS, significant threshold shift.
SOURCES: Gasaway (1988); Nixon (1998); personal communication, E. Berger, E-A-R/Aearo Company, March 2005; Department of Veterans Affairs (2005); Ohlin (2005).

limited data were available on women in the services, and many of the committee's analyses were based on data for men alone. In some cases, including the reports from DOEHRS-HC on STS, data for men and women were not reported separately. Women have been a small proportion of the military population, but their numbers are increasing. It will be important to understand how their hearing loss experience compares with that of men to interpret changes in hearing thresholds over time.

Although the committee's focus is on data since the 1970s, this is not to imply that there were no attempts to examine noise-induced hearing loss in the military by the scientific community prior to this era (e.g., Glorig et al., 1957; Solomon and Fletcher, 1958). Rather, the 1970s represent the initial period for which substantial amounts of hearing-threshold data, obtained with appropriate equipment and procedures, became available from reasonable samples of personnel and from multiple branches of the military.

The next section presents the data reported since 1970 on the average hearing thresholds measured in military personnel from various branches of the military. This is followed by presentation of data on the variations in hearing thresholds across personnel over this same time period.

EVIDENCE BASED ON AVERAGE HEARING THRESHOLDS

Following a brief presentation of results for the special case of acoustic trauma, average pure-tone hearing thresholds across frequency are presented for groups of military personnel. This section concludes with the presentation of several summaries of average hearing thresholds for several branches of the military.

Acoustic Trauma

Acoustic trauma presents, perhaps, the clearest cause-and-effect link between noise exposure and hearing loss. Even in those cases, however, the amount of change in hearing thresholds as a result of that event cannot be determined conclusively without information about preexposure thresholds. The committee found no systematic data on the number of events that occur among U.S. military personnel, and acoustic trauma has not been a reportable injury for military personnel (AMSA, 1998; Ohlin, 2004a). Reports of hearing loss among 62 percent of personnel with blast injuries who were treated at Walter Reed Army Medical Center from March 2003 through May 2005 (n = 414) (Chandler, 2005) provide an indication that acoustic trauma is occurring, and likely occurred in the past, but does not provide sufficient evidence to estimate its overall incidence. It is likely that cases of hearing loss resulting from acoustic trauma are included in the remainder of the data discussed.

Average Thresholds

Examples of Noise-Notch Patterns of Hearing Loss Among Military Personnel

The earliest data examined in detail were from studies of Army personnel in combat arms branches—infantry, armor, and artillery (Walden et al., 1971; Walden et al.,1975). An initial pilot study in 1971 was followed in 1974 by a more systematic study based on a random sample of Army enlisted personnel in these three branches at 10 posts. The mean hearing levels for the three groups were similar, with those for armor personnel lying between those for the artillery and infantry personnel. Figure 3-1 illustrates the average hearing thresholds as a function of frequency for

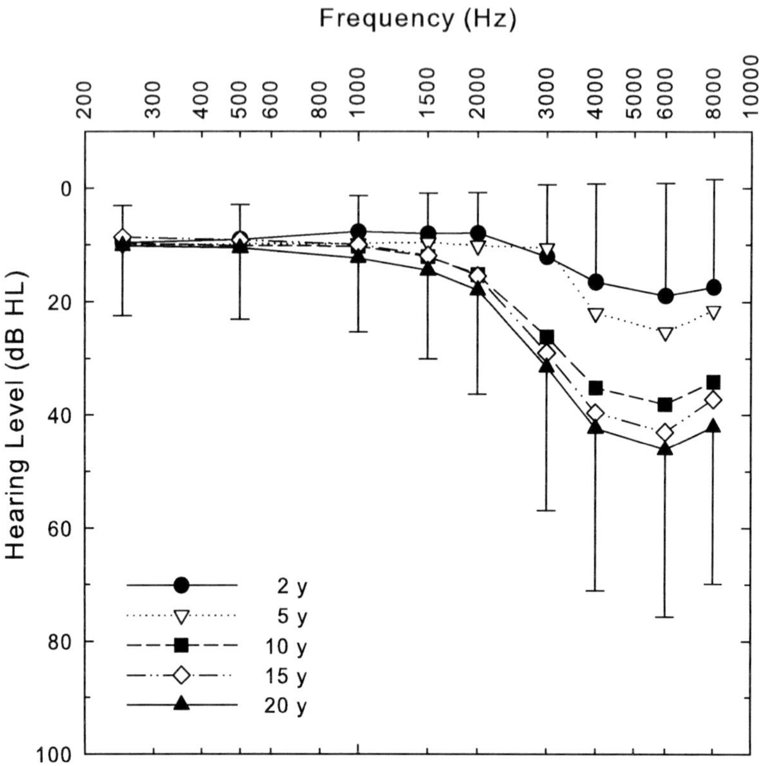

FIGURE 3-1 Mean hearing thresholds (left ear) for Army enlisted men serving in 1974 in the infantry, by frequency and length-of-service group. Error bars for one standard deviation toward better hearing are shown at each frequency for the 2-year length-of-service group. Error bars for one standard deviation toward worse hearing are shown at each frequency for the 20-year length-of-service group. SOURCE: Walden et al. (1975).

infantry personnel in 1974 (Walden et al., 1975). The pilot study (Walden et al., 1971) found similar hearing levels among a mixed population of officers and enlisted personnel from these three branches.

The data plotted in Figure 3-1 are the mean hearing thresholds in the left ear at each frequency tested, by length-of-service (LOS) categories. In this study, the left-ear data represent the "worse ear," a pattern observed in most of the reports reviewed and one that is common with exposure to gunfire. The difference between left and right ears, however, was seldom more than a few decibels.

The audiometric configuration shows the notch at higher frequencies that serves as a hallmark of noise-induced hearing loss. The hearing thresholds are poorest at 6000 Hz, a pattern seen in the other reports reviewed by the committee and reported previously for military noise exposures (e.g., Gravendeel and Plomp, 1959; Ylikoski and Ylikoski, 1994). There is a fairly broad maximum for the hearing loss that spans from about 3000 Hz to 8000 Hz. The pattern can be seen for all lengths of service, with worse thresholds at nearly all frequencies for those with longer service. The pattern of greater differences in hearing thresholds among groups with shorter lengths of service, up to 10–15 years, and smaller differences among groups with longer service, has also been commonly observed in studies of industrial noise-induced hearing loss (see Chapter 2).

Frequency-specific data from the 1970s were also available for certain personnel in the Navy (Robertson et al., 1978b) and Air Force (Sutherland and Gasaway, 1978). The Navy data were for enlisted personnel in 16 occupational specialties (referred to in the Navy as ratings), representing about 20 percent of all rating categories. The Air Force data (median thresholds) were from audiometric testing done for the hearing conservation program in 1975–1976. The Navy data were reported by length of service, whereas the Air Force report used age. In both the Navy and Air Force data, maximum hearing thresholds were at 6000 Hz, with a fairly broad maximum beginning at 3000 Hz. An upturn at 8000 Hz, a characteristic of the noise notch, could be seen in the Navy data, but Air Force personnel were not tested at that frequency. The average hearing thresholds were higher with greater length of service or at older ages. However, the thresholds for Air Force personnel were consistently better than those for Army and Navy personnel in comparable age/LOS groups.

Unfortunately, measures of hearing thresholds obtained from military personnel after the 1970s no longer included 8000 Hz as a test frequency. As a result, it is unclear whether a "notch" exists in the average data, given that the notch typically occurred at 6000 Hz when threshold measurements included 8000 Hz. This situation is illustrated by the data in Figure 3-2, which depicts the average hearing thresholds through 6000 Hz from Bohnker et al. (2002) for 51,643 individuals enrolled in the Navy's hearing conservation program in 1995–1999. A noise notch is not apparent in these average data.

Summary of Average Hearing-Threshold Data for the 1970s, 1980s, and 1990s

For the most part, large-scale studies of hearing loss from multiple branches of the military using reliable and valid measurement procedures and conditions were not available until the 1970s. In nearly all of these studies, data were gathered from military personnel grouped according to age, length of service, or both. In the few reports that used only length of

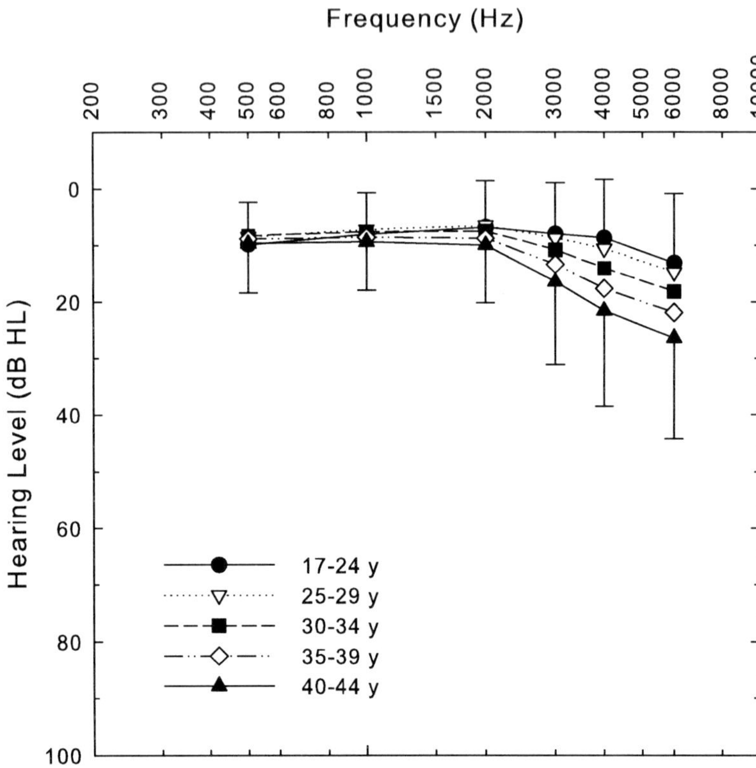

FIGURE 3-2 Mean hearing thresholds (left ear) for enlisted men enrolled in the Navy hearing conservation program, by frequency and age group, 1995–1999. Error bars for one standard deviation toward better hearing are shown at each frequency for the age group 17–24 years. Error bars for one standard deviation toward worse hearing are shown at each frequency for the age group 40–44 years. Data for age groups of 45–49 years and 50 years or older are not shown in this figure due to smaller sample sizes compared to those age groups depicted.
SOURCE: Bohnker et al. (2002).

service, age at entry into military service was assumed to be 18 years and length of service was assumed to be correlated with age. To assess average hearing thresholds of military personnel in various age groups, it is necessary to compare these results to the average hearing thresholds of similar age groups from the general population.

Accordingly, the committee examined age-specific comparisons between the average hearing thresholds for certain groups of military personnel and those of two reference groups of the same average age. As noted in Chapter 2, there are essentially two types of reference datasets available on age-related hearing loss. One is based on an unscreened sample that may include people who have a history of certain types of otologic disease or noise exposure. The other type is a screened sample that is designed to exclude such individuals. For an unscreened reference point, the committee used data drawn from the results of a U.S. Public Health Service (USPHS) survey completed in 1962, whose participants were a nationally representative sample (n = 6,672) of the civilian adult population in the United States at that time (Glorig and Roberts, 1965).[5] The better-ear thresholds from the 1962 USPHS study comprise database B from the ISO-1999 standard, but most of the studies of military personnel did not report better-ear thresholds. As a result, the reference set of data from the 1962 USPHS study was used to derive an average amount of hearing loss for both ears for each age group. The screened sample used was database A from the ISO-1999 standard (ISO, 1990). For both the screened and unscreened reference data, the specific measure used for comparison was an average high-frequency threshold, calculated as the arithmetic average of the mean or median (depending on the source) thresholds at 3000, 4000, and 6000 Hz, for both ears combined. This average was selected to focus on the frequencies most closely associated with noise-induced hearing loss.

When determining the effects of industrial or military noise on hearing, one would prefer that the reference or comparison data be obtained from a non-noise-exposed sample that was otherwise identical to the noise-exposed sample. Military personnel are subjected to a hearing-related screening that is less rigorous than that in studies of "screened" samples, but possibly more rigorous than that of "unscreened" samples (e.g., Robertson et al., 1978b; AMSARA, 2002). Thus, it is unclear which reference group would be the most appropriate to use for comparison to the hearing thresholds measured in military personnel.

One approach to resolving this is to compare the hearing thresholds of recruits prior to military noise exposure to those thresholds observed in screened and unscreened samples of the same age. An initial comparison found that the thresholds for men, ages 18–24 years, in the unscreened data were comparable to those of small groups of Army (n = 246) and Navy (n = 121) recruits of similar age (Walden et al., 1971; Robertson et al., 1978b). Similar thresholds were also reported for a larger group of Army recruits (n = 3,534) at a later date (Ohlin, 1992), suggesting no marked change over this period in the baseline hearing thresholds for Army personnel and, therefore, correspondence with the thresholds from the 1962 Public Health Service study. The average thresholds for both the Army and

[5]The reported thresholds were corrected to reflect the values for audiometric zero that were adopted in 1969 (ANSI, 1969).

Navy recruits, as well as the corresponding age group from the 1962 USPHS study, were approximately 5–10 dB HL in the low and mid frequencies, increasing to about 15–20 dB HL at 6000 Hz. The average thresholds for a screened sample of this age would basically be 0 dB HL at all frequencies. Thus, at entry into military service, these samples of recruits from the Army and Navy appear to have average hearing thresholds equivalent to those of a separate group of unscreened young adults of similar age from the 1962 USPHS study.

Figure 3-3 displays data on age-specific average high-frequency hearing thresholds in various groups of Army personnel in the 1970s and early and late 1980s, along with the thresholds for the two sets of reference data. The data from the two relatively large studies ($n \sim 3{,}000$) in the 1970s (Walden et al., 1971; Walden et al., 1975) show average high-frequency thresholds that are similar for men in the infantry, artillery, and armor branches, all of whom were considered to have had high noise doses. Those thresholds exceed the unscreened reference thresholds by greater amounts at older ages, a pattern consistent with noise-induced hearing loss in excess of expected age-related changes alone. A similar study (Ohlin, 1992) examined data from 1989 for enlisted men from the same three branches who were enrolled in the Army's hearing conservation program. Average thresholds were lower at every age compared with the 1970s data, but exceeded the unscreened reference levels by 3–5 dB. Factors contributing to the change in thresholds between 1974 and 1989 may have included not only differences in noise exposure or use of hearing protection, but also unidentified selection effects in hearing conservation registry data and demographic differences arising from a higher percentage of black soldiers in the population in 1989 (Ohlin, 1992; Henselman et al., 1995). Hearing thresholds for blacks have been observed to be lower than those for whites (Royster et al., 1978).

Figure 3-3 also includes data from studies with smaller sample sizes from the early 1980s. Army engineers ($n = 209$) in various occupational specialties generally had thresholds similar to the 1970s data (Chandler and Fletcher, 1983). Average high-frequency thresholds for two small groups of Army aviators, however, were generally similar to those in the unscreened comparison (Walden et al., 1971; Peters and Ford, 1983). Another study used Army hearing conservation registry data for 1989 to compare, by race, hearing thresholds of enlisted men presumed to be exposed to high noise levels (i.e., in the armor, artillery, and infantry branches; $n = 39{,}006$) or to low noise levels (e.g., personnel in administration, supply and services, public affairs; $n = 18{,}730$) (Henselman et al., 1995). These data represented 25 percent of soldiers in the noise-exposed branches and 18 percent of the soldiers in the non-noise-exposed occupational groups. Because the results were reported in terms of average age-corrected thresholds at 1000, 2000, 3000, and 4000 Hz, they were not comparable to the data from the studies

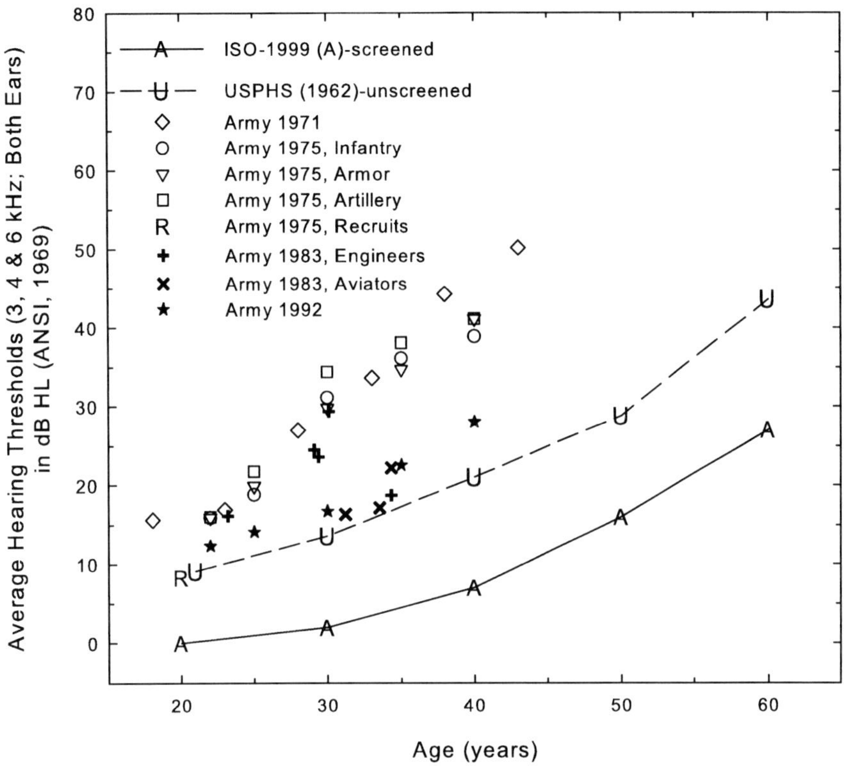

FIGURE 3-3 Average high-frequency thresholds for both ears (arithmetic average of mean values at 3000, 4000, and 6000 Hz in both ears), by age, for selected groups of Army personnel during the 1970s, 1980s, and 1990s. The Army data plotted are for a 1971 pilot study (◇); a sample of recruits (R) and enlisted men in the infantry (○), armor (∇), and artillery (□) branches in 1974; small groups of engineers (+) and aviators (x) in the early 1980s; and enlisted men from the infantry, armor, and artillery enrolled in the hearing conservation program in 1989 (★). Average high-frequency thresholds for men from screened (A; ISO-1999, 1990) and unscreened (U; Glorig and Roberts, 1965) reference groups are also shown.
SOURCES: Glorig and Roberts (1965); Walden et al. (1971); Walden et al. (1975); Chandler and Fletcher (1983); Peters and Ford (1983); ISO (1990); Ohlin (1992).

described above and displayed in Figure 3-3. For all soldiers, the difference in age-corrected average thresholds between those with high and low noise exposure was less than 5 dB, regardless of length of service.

Figure 3-4 shows data for certain Navy personnel from the 1970s (n = 3,050) and 1990s (n = 51,643). The earlier data (Robertson et al., 1978b) for enlisted personnel in 16 occupational specialties show average high-frequency thresholds that exceed those for the unscreened reference. With a

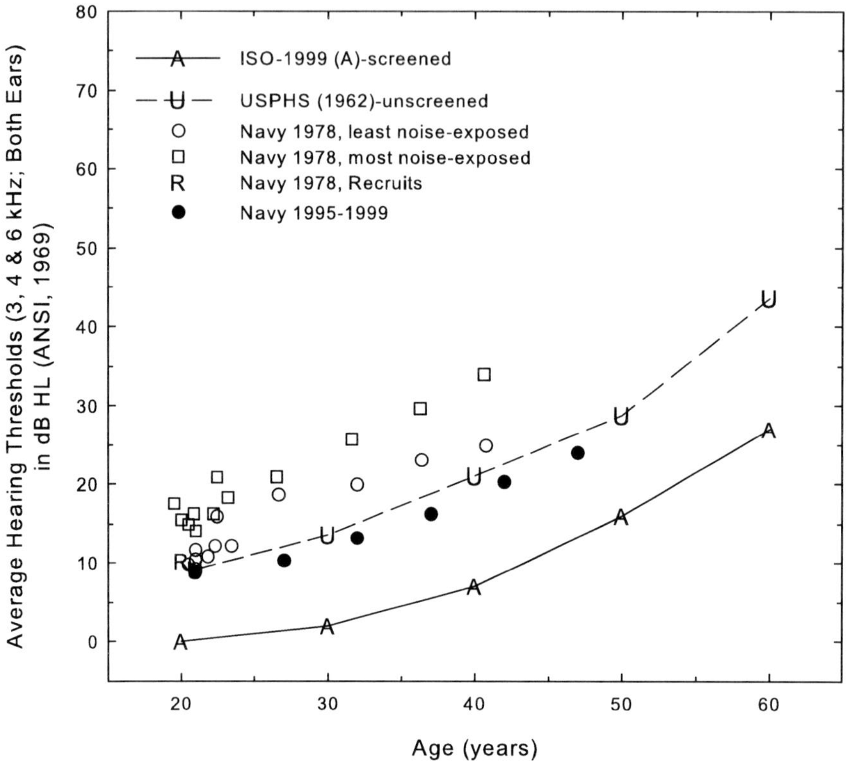

FIGURE 3-4 Average high-frequency thresholds for both ears for selected Navy enlisted personnel in the 1970s and for Navy enlisted men enrolled in the hearing conservation program in 1995–1999 (●). The 1970s data are for eight occupational specialties considered to have high noise exposure (□) and eight considered to have low noise exposure (○), plus recruits (R). Average high-frequency thresholds for men for screened (A; ISO-1999, 1990) and unscreened (U; Glorig and Roberts, 1965) reference groups are also shown.
SOURCES: Glorig and Roberts (1965); Robertson et al. (1978b); ISO (1990); Bohnker et al. (2002).

few exceptions, however, the thresholds for eight occupational specialties thought to have the least noise exposure are within 3 to 5 dB of those for the unscreened reference. The eight specialties thought to have the most noise exposure generally had higher average thresholds than the low-noise groups, but Robertson and colleagues noted more overlap between the high- and low-noise groups than had been anticipated. The thresholds for older Navy personnel, even for the presumed high-noise job specialties, are better than those observed in the 1970s Army data (see Figure 3-3). In more recent data for enlisted men in the Navy's hearing conservation program during the period 1995–1999 (Bohnker et al., 2002), average high-frequency thresholds are slightly better than those in the unscreened sample.

The available data from both the 1970s and the 1990s for Marine Corps personnel (Figure 3-5) (Goldenberg, 1977; Bohnker et al., 2002) and Air Force personnel (Figure 3-6) (Sutherland and Gasaway, 1978; Thomas, 1995) show average high-frequency thresholds that were similar to or somewhat better than those for the unscreened sample. All the Air Force data are for participants in the hearing conservation program. The more recent Marine Corps data are also for participants in the hearing conservation

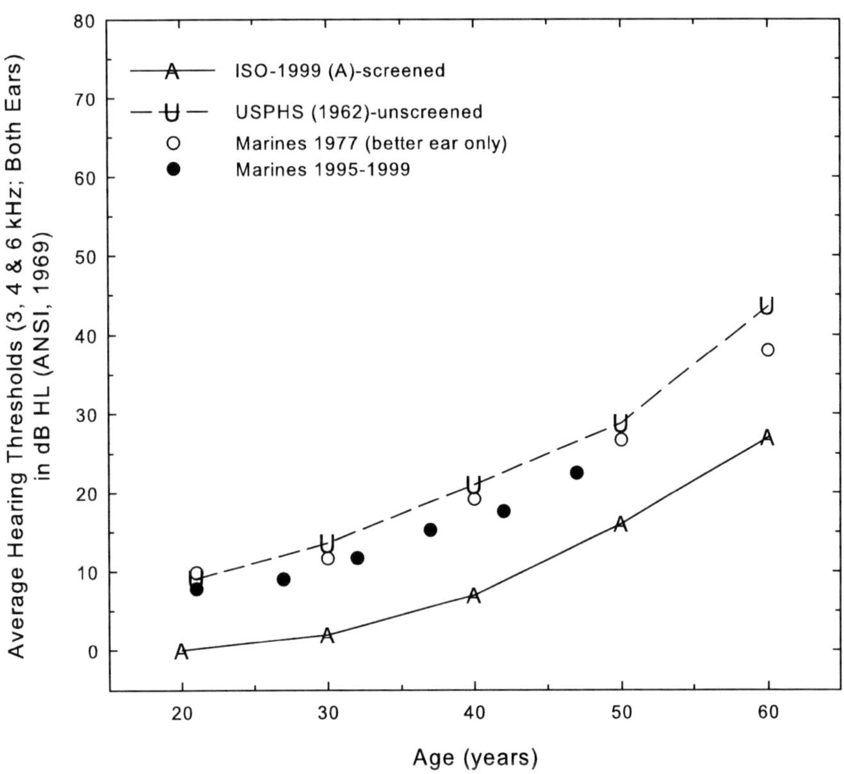

FIGURE 3-5 Average high-frequency thresholds for the better ear for selected Marine Corps personnel (officers and enlisted men) in the 1970s (○) and for both ears for enlisted Marine Corps men in the hearing conservation program in 1995–1999 (●). Average high-frequency thresholds (both ears) for men for screened (A; ISO-1999, 1990) and unscreened (U; Glorig and Roberts, 1965) reference groups are also shown.
SOURCES: Glorig and Roberts (1965); Goldenberg (1977); ISO (1990); Bohnker et al. (2002).

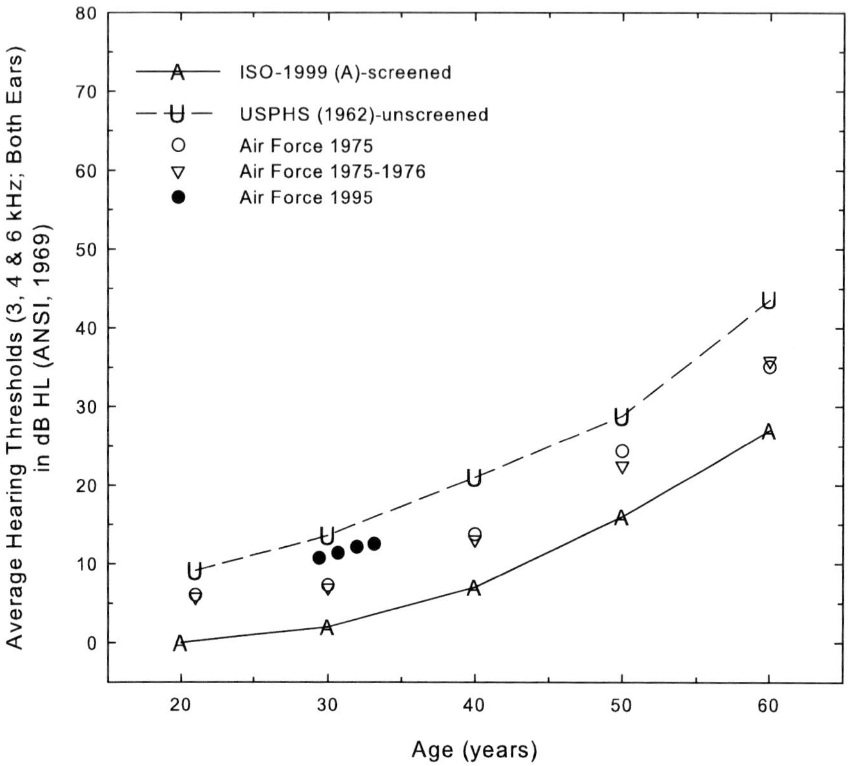

FIGURE 3-6 Average high-frequency thresholds for Air Force personnel enrolled in the hearing conservation program in 1975 (○), 1975–1976 (∇), and 1995 (●). Average high-frequency thresholds for men for screened (A; ISO-1999, 1990) and unscreened (U; Glorig and Roberts, 1965) reference groups are also shown. SOURCES: Glorig and Roberts (1965); Sutherland and Gasaway (1976, 1978); ISO (1990); Thomas (1995).

program. The earlier data (available for the better ear only) are for Marine Corps officers and enlisted personnel tested at a single base. The committee was surprised by the comparability of the Marine Corps thresholds to the unscreened comparison and the contrast with the higher thresholds seen in the 1970s data for Army personnel in infantry, artillery, and armor specialties. The committee had no information on baseline thresholds of Marine Corps personnel, the effect of the inclusion of officers in the Marine Corps study population (versus only enlisted personnel in the Army study), or unidentified selection factors that might have affected the characteristics of the populations tested. Information was also lacking on the distribution of thresholds (see below) and the difference between the better-ear thresholds reported for the Marines and their average thresholds for both ears (the measure reported in Army study). These or other factors might help explain the difference between the data from the Marine Corps and the Army with regard to average high-frequency hearing loss.

EVIDENCE BASED ON VARIATIONS IN HEARING THRESHOLDS WITHIN GROUPS

The average thresholds do not reflect the variability within the group or the nature of the distribution of the data, so the committee also examined data describing the distribution of thresholds and other measures of variability within groups. As the data on hearing thresholds deviate increasingly from a normal distribution, typical measures of central tendency, such as the mean, become less representative of the group. For example, a set of normally distributed thresholds and a set of thresholds distributed with either positive or negative skewness may each have the same mean value, but the two underlying distributions would provide different interpretations with regard to the proportion of each sample who have noise-induced hearing loss. Generalizing from mean thresholds to thresholds for individuals is essentially impossible.

The committee's work was limited by the lack of data available to examine the range of hearing thresholds within a given population. The limited data offering some indication of the distribution of hearing thresholds in military populations are reviewed here. This section begins with a discussion of the distributions of pure-tone thresholds. It then turns to a review of the proportion of individuals in the military identified with various categories of hearing loss. It concludes with a review of the annual incidence of STS among military personnel. In all of these analyses, the goal is to get a clearer picture of the proportion of military personnel who have or are developing noise-induced hearing loss.

Individual Variations of Hearing Thresholds in Military Personnel

Standard-deviation error bars, such as those shown in Figure 3-1 for the mean data for the groups of Army infantry personnel with the shortest and longest lengths of service, provide some indication of the range of the middle two-thirds of the distribution of individual hearing thresholds. That is, assuming a normal distribution of hearing thresholds, 68 percent of the individual data will be within one standard deviation of the mean threshold. Thus, although the average high-frequency hearing threshold for Army infantry personnel from 1974 with 2 years of service is slightly below 20 dB HL (Figure 3-1), the range representing the middle two-thirds of the group is from about 0 dB HL to 40 dB HL. As noted, however, such estimates are based on an assumption of a normal distribution, and the distributions of pure-tone thresholds at each frequency are usually *not* normal, most often being negatively skewed. The negatively skewed distributions of pure-tone hearing thresholds may be attributable in part to the restricted range of hearing thresholds that can be measured below 0 dB HL due to limitations in the equipment or facilities.

The Air Force report (Sutherland and Gasaway, 1978) on hearing thresholds from 99,318 military personnel enrolled in the hearing conservation program, during the period June 1975 through May 1976, was one of the few reports to include the distribution of pure-tone thresholds at each frequency for each age group. In Figure 3-7, the solid line represents the median hearing threshold at each frequency for personnel ages 45–54 years. The dashed lines represent the 10th and 90th percentiles of the distribution

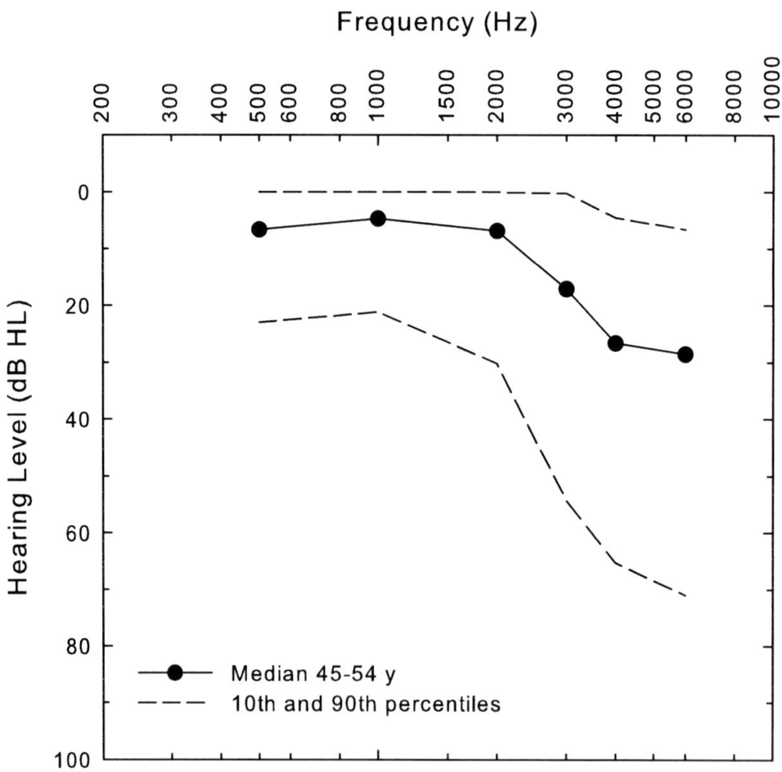

FIGURE 3-7 Hearing thresholds (left ear), by frequency, for personnel ages 45–54 years in the Air Force hearing conservation program, 1975–1976. The solid line and filled circles are the median pure-tone thresholds (50th percentile) at each frequency. The dashed lines represent the 10th and 90th percentiles.
SOURCE: Sutherland and Gasaway (1978).

of hearing thresholds at each frequency. At 4000 Hz and 6000 Hz, the median thresholds were approximately 28 dB HL, corresponding to a mild degree of hearing loss. However, the upper dashed line indicates that 10 percent of the members of this age group had thresholds that were about 5 dB HL or better at 4000 Hz and 6000 Hz, well within normal limits. The lower dashed line indicates that another 10 percent had thresholds of about 70 dB HL or worse, a severe hearing loss at these frequencies.

Is this broad range of hearing thresholds for 45- to 54-year-old military personnel in the Air Force unique to noise-induced hearing loss or typical for this age group? To address this question, Figure 3-8 provides a more detailed look at some of the results from this same Air Force report (Sutherland and Gasaway, 1978). In this figure, the distributions of pure-tone thresholds at 3000, 4000, and 6000 Hz for Air Force military personnel are shown by the solid lines in the left, center, and right columns, respectively. Distributions are shown for four of the five age groups included in the Air Force report. For the oldest group included in that report, 55–64 years of age, the sample size (n = 364) was much smaller than for the other age groups and not considered to be representative. As a result, these data were excluded from Figure 3-8. For comparison, similar distributions have been displayed in Figure 3-8 for the pure-tone thresholds for men from the 1962 USPHS data (dashed lines). At all three frequencies, 5 dB was subtracted from the hearing-threshold-level categories to correct the USPHS data to the ANSI (1969) standard values used in the Air Force report.[6] For both the Air Force and the USPHS data, the distributions are of hearing thresholds obtained from the left ear.

Several general trends are apparent in these data. First, every distribution is negatively skewed with a much greater spread of hearing thresholds above the mean than below it. Second, for both sets of distributions, the amount of the spread in hearing thresholds increases as the age of the group increases. Third, although there is some evidence of an upward shift in the distribution of hearing thresholds for the military personnel in the Air Force relative to the USPHS data in the younger age groups, this has largely disappeared in the older age groups. Finally, at the younger age groups where the two distributions differ, the differences are confined primarily to differences in the distribution of normal-hearing thresholds (thresholds less than 25 dB HL).

To recap, the variability of hearing thresholds across military personnel in the Air Force was shown in Figure 3-7 to be quite large for the age group comprised of 45- to 54-year-olds, the group with the greatest amount of high-frequency hearing loss and a substantial sample size (n = 3,340). The

[6]The actual correction factors from ASA (1951) to ANSI (1969), including allowance for the difference in headphones used, are 5.6, 4.8, and 4 dB at 3000, 4000, and 6000 Hz, respectively (Sutherland and Gasaway, 1978).

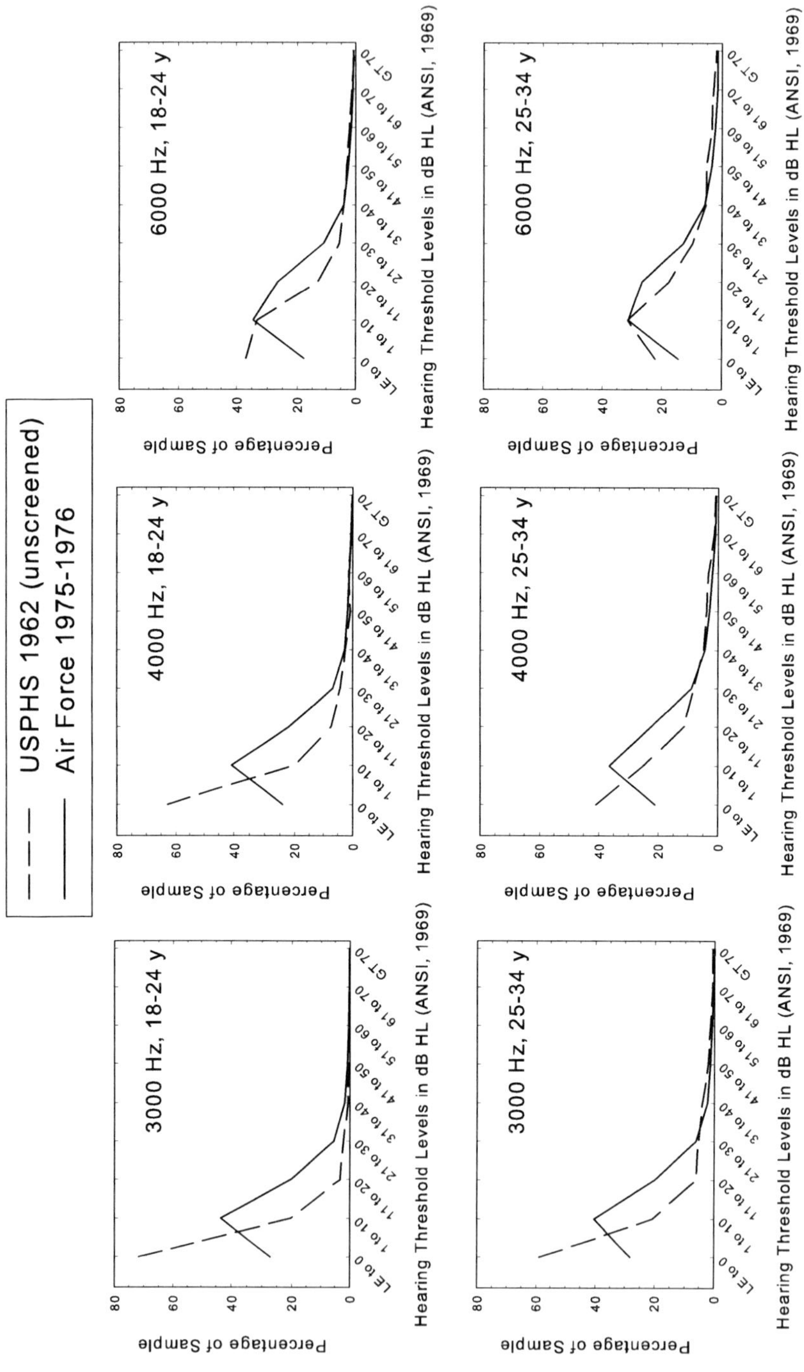

USPHS 1962 (unscreened)
Air Force 1975-1976
3000 Hz, 18-24 y
4000 Hz, 18-24 y
6000 Hz, 18-24 y
3000 Hz, 25-34 y
4000 Hz, 25-34 y
6000 Hz, 25-34 y
Percentage of Sample
0
20
40
60
80
LE to 0
1 to 10
11 to 20
21 to 30
31 to 40
41 to 50
51 to 60
61 to 70
GT 70
Hearing Threshold Levels in dB HL (ANSI, 1969)

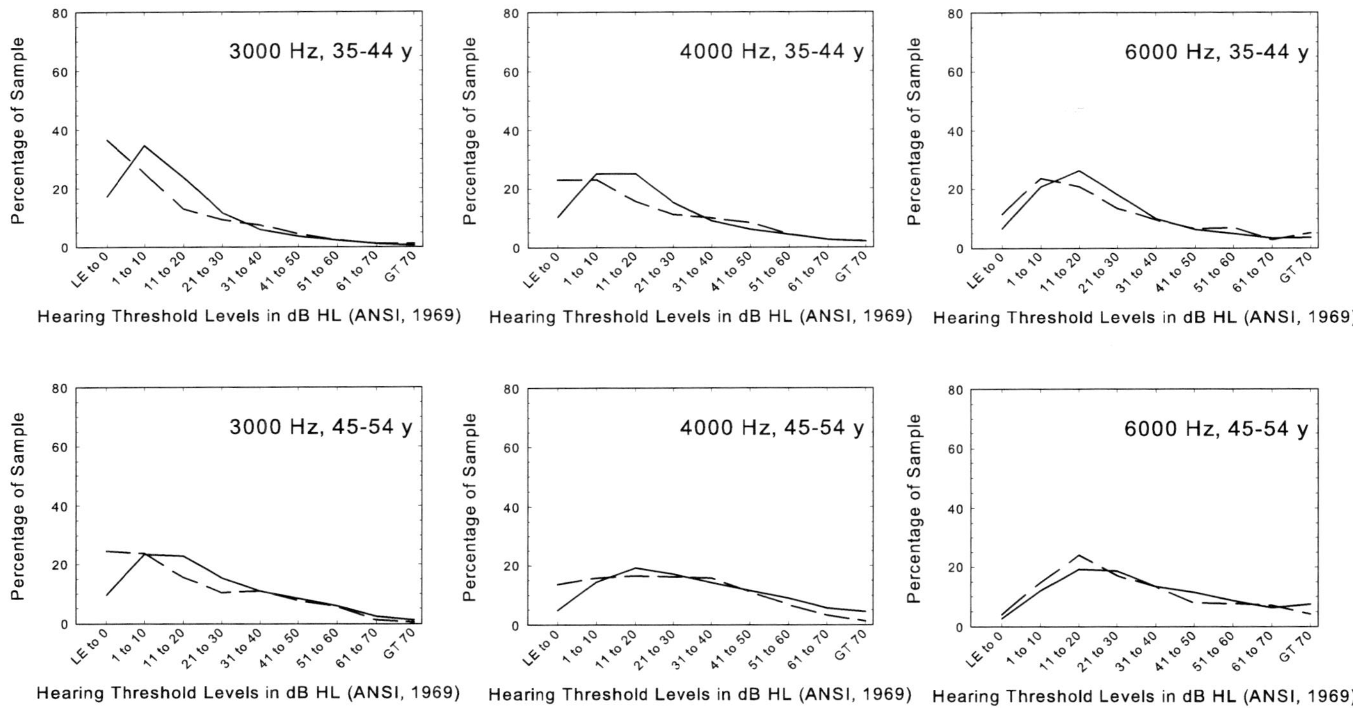

FIGURE 3-8 Distributions of pure-tone thresholds from Air Force personnel enrolled in the hearing conservation program in 1975–1976 (solid lines) and from an unscreened sample of men in the general population (dashed lines) at 3000 (left column), 4000 (center column), and 6000 Hz (right column). Each row depicts the distributions for separate age groups from 18–24 years (top) to 45–54 years (bottom). All data are for hearing thresholds from the left ear.
SOURCES: Glorig and Roberts (1965); Sutherland and Gasaway (1978).

comparison of distributions of hearing thresholds for this age group from the Air Force to comparable data from the 1962 USPHS study in Figure 3-8, however, suggests that the observed variability in hearing thresholds across individuals in the Air Force is similar to that of an unscreened sample of the general population. Again, this raises the issue of the appropriate reference group to which one should compare thresholds from military personnel. Considering the average-threshold data for both ears presented previously for the Air Force (Figure 3-6) and the distribution of individual pure-tone thresholds for the left ear for the Air Force (Figure 3-8), military personnel in the Air Force in the 1970s do not appear to have hearing thresholds that differ substantially from an unscreened sample of the general population. Unfortunately, there are no other data available on the distribution of pure-tone thresholds for military personnel, especially for groups for whom the average hearing thresholds appear to be substantially worse than an unscreened sample of the general population (e.g., Army data from the 1970s, Figure 3-3).

Although other reports of additional data on the distributions of pure-tone thresholds in military personnel were not available, some reports did provide percentages for individuals in each of several age groups who had hearing thresholds greater than some criterion amount of hearing loss. Goldenberg (1977), for example, reported on the percentage of better-ear hearing thresholds that were greater than 25 dB HL among 11,580 Marine Corps personnel whose hearing was tested at one base during a 13-month period. These percentages for Marine Corps personnel are displayed by the dashed lines in Figure 3-9 for pure-tone frequencies of 3000, 4000, and 6000 Hz and each of five age groups. Percentages at lower frequencies were all at or below about 10 percent, even for the oldest age group, and are not of interest here. The solid lines in Figure 3-9 provide comparison percentages from the 1962 USPHS unscreened sample for better-ear pure-tone thresholds greater than 25 dB HL (adjusted to ANSI (1969) standards) at the same three frequencies and for the same five age groups.

Although there is excellent agreement between the two sets of data for the two younger age groups, similar to what had been observed previously in the average data for Army and Navy recruits, for the three older age groups, the percentage of Marines with hearing thresholds greater than 25 dB HL at each frequency exceeds that of the unscreened sample from the general population by 15–25 percent. Thus, although the average-threshold data (better ear) for this same sample of Marines were generally consistent with average data from the USPHS unscreened sample (both ears) (Figure 3-5), the limited distributional data available from this same report (Figure 3-9) suggest there were substantially more Marines over age 35 years who had at least a mild hearing loss (> 25 dB HL) in the high frequencies in their better ear than in corresponding age groups from an unscreened sample of the general population.

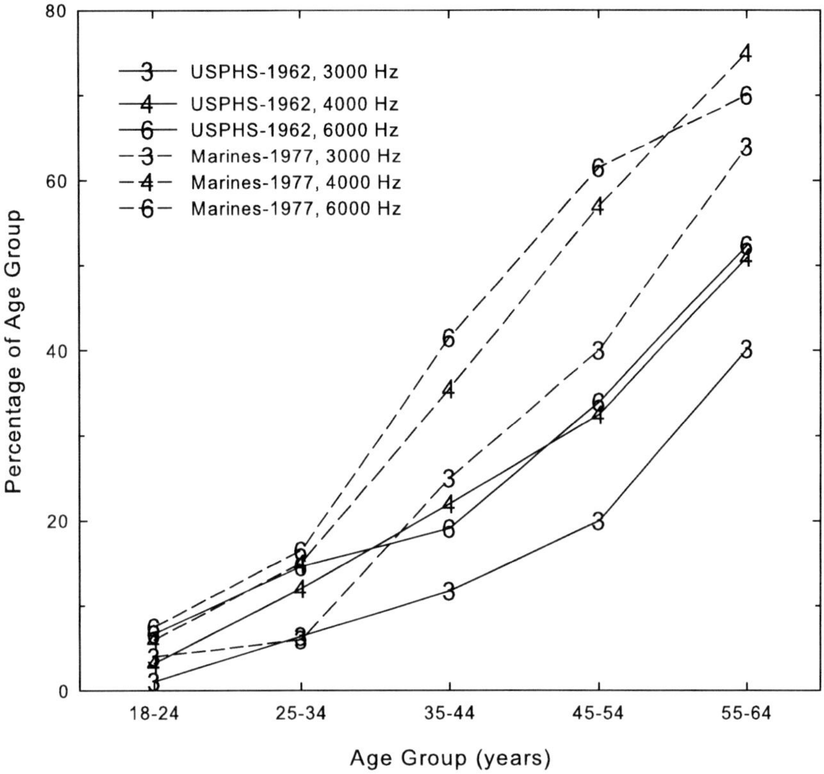

FIGURE 3-9 Percentages of each age group, from 18–24 years (left) to 55–64 years (right) having hearing thresholds greater than 25 dB HL at each of three pure-tone frequencies: 3000 Hz (3), 4000 Hz (4), and 6000 Hz (6). Dashed lines represent data from the better ear of Marines in the 1970s and the solid lines represent data for the better ear of men in an unscreened sample of the general population from the 1962 U.S. Public Health Service survey.
SOURCES: Glorig and Roberts (1965); Goldenberg (1977).

Another study, by Robertson and colleagues (1978b), measured hearing thresholds in 3,050 Navy military personnel and defined a significant high-frequency hearing loss as an average threshold at 3000, 4000, and 6000 Hz ≥ 30 dB HL. Among personnel with more than 5 years of service, 37 percent of those in the high-noise occupational specialties and 23 percent of those in the low-noise occupational specialties had a significant high-frequency hearing loss. Unfortunately, there are no comparable percentages from the 1962 USPHS sample to which these values from the Navy can be compared to determine if such percentages are greater than expected in the general population.

The Army and the Air Force include criteria for hearing in a classification system used to characterize the medical fitness of service members. For hearing, these ratings, known as profiles, vary from H-1 (closest to normal hearing) to H-4 (the most severe hearing loss). The Army's current criteria for an H-1 profile are an average threshold in each ear of ≤ 25 dB HL for the frequencies 500, 1000, and 2000 Hz; no individual threshold > 30 dB HL at these frequencies; and a threshold of ≤ 45 dB HL at 4000 Hz (Department of the Army, 2003). Thus, having an H-1 profile means having normal or near-normal hearing for low and middle frequencies (500–2000 Hz) in both ears, but moderate hearing losses are possible at 4000 Hz and profound hearing losses are possible at 6000 and 8000 Hz. Thus, H-1 classification is not synonymous with "normal" hearing (see Chapter 1).

Data for the Army from the 1970s for personnel in the infantry, armor, and artillery indicate that 20 to 30 percent were classified as H-2 or worse (Walden et al., 1971; Walden et al., 1975). Among a small group of recruits who had not begun basic training (n = 246), about 3 percent had H-2 hearing or worse. In a group of recruits who had completed their basic training (n = 255), 6 percent had H-2 hearing or worse. More recent data from the DOEHRS hearing conservation (DOEHRS-HC) database showed that from 1982 through 2003, only about 8 to 10 percent of Army personnel in the hearing conservation program were classified as H-2 or worse (U.S. Army Center for Health Promotion and Preventive Medicine, 2004b). The data suggest lower rates of hearing loss over the past 20 years compared to the mid-1970s, but the two populations are not necessarily comparable. For Air Force personnel in the hearing conservation program for the period 2000–2003, about 10 to 12 percent of those in the hearing conservation program were classified as H-2 hearing or worse (Air Force Hearing Conservation Registry, 2004a).

Given that the H-1 classification permits moderate-to-profound high-frequency hearing loss, the percentages of personnel who do not meet the H-1 standard most likely underestimate the prevalence of hearing loss at high frequencies. Such losses are consistent with noise exposure, but from the H classification system alone, it is not possible to determine their etiology. In addition, these overall percentages do not account for any differences in the underlying age distributions or noise-exposure histories of the populations under consideration.

Threshold Shifts

The committee also examined reports on cases of STS and permanent threshold shift (PTS) provided by the hearing conservation programs of the Army, Navy, Air Force, and Marine Corps. Although the definition of STS used by the services has varied over time, the purpose of the measure is to identify for follow-up individuals who demonstrate a clinically significant change in hearing thresholds relative to an earlier baseline (reference) measurement. An STS should be followed up with up to two additional measurements of hearing thresholds after a prescribed period of quiet. If the STS remains, or if the individual is not retested, the STS is classified as a PTS and a new reference is established to be used in subsequent assessments of STS.

Reports of cases of STS are available for thousands of military personnel enrolled in the services' hearing conservation programs, in some cases across several years. Data were available from the Army for 1983–2003 (U.S. Army Center for Health Promotion and Preventive Medicine, 2004c), from the Air Force for 1989–2004 (Air Force Hearing Conservation Registry, 2004c), and for the Navy and Marine Corps for 1999–2004 (Navy Environmental Health Center, 2004a). During this period, definitions of STS changed in 1987 and 1999. For example, for the period up to 1987, the Army defined STS as a change of at least 20 dB at 1000, 2000, 3000, or 4000 Hz or a change greater than 10 dB in the average hearing loss at 2000, 3000, and 4000 Hz in either ear. No age corrections were applied for the comparison with the reference thresholds. From 1987 to 1999, however, the Army used the same criteria, except that age corrections were applied. Applying age corrections would be expected to reduce the incidence of STS (and PTS), all else being equal, but a steady increase in reported STS occurred from 1987 through 1999. As of 1999, the Army (and the other branches of the military) eliminated age correction and the 20-dB individual-frequency criterion was reduced to 15 dB.

The data from the Army show that roughly 6 to 8 percent of the audiograms obtained from military personnel in the hearing conservation program demonstrated positive STS each year from 1983 to 1987. The percentage of tested personnel who were found to have an STS has progressed steadily since 1987, and by 1999, the STS percentage had reached about 18 percent. It has remained at about that level since 1999. Since then, the STS percentage for the Navy and Marine Corps hearing conservation programs declined from levels of about 22–25 percent to levels of about 15–18 percent. In general, STS percentages have been lowest for the Air Force hearing conservation program, ranging from about 7 to 13 percent over the 15-year period for which data are available. Reports of PTS have generally followed similar patterns, with the Air Force again having the lowest percentages.

Other STS data have also been reported. For example, the STS percentages for the Air Force hearing conservation program started at about 23 percent in 1975 and declined steadily to about 14 percent in 1979 (Department of the Air Force, 1980). However, a more complex definition of STS was employed in which the criterion amount of threshold change considered to be significant varied with the severity of hearing loss demonstrated in the reference thresholds.

Wolgemuth et al. (1995) reported on STS incidence for a large sample (n = 12,492) of Navy personnel in the hearing conservation program of the Atlantic Fleet from 1987 to 1990. An overall STS incidence of 29 percent was reported using STS definitions equivalent to those used by the Army. This level is about three times greater than that for the Army from the same 1987–1990 period. It is not clear, however, whether age corrections were applied to the Navy data, as was the case for the Army. If not, this could account for some of the difference in STS incidence. In addition, the Army data are provided only for positive STS cases, and it is unclear if this was also true for the data from the Navy.

A recent analysis conducted for the Navy (Shaw and Trost, 2005) used STS as an outcome variable to examine the effect of noise on hearing of Navy enlisted personnel (n ~ 251,000) during the period 1979–2004. STS was established based on the difference between a sailor's earliest and latest hearing tests under the hearing conservation program. It was defined as an average change of 10 dB or more in thresholds at 2000, 3000, and 4000 Hz in either ear. In the absence of direct data on individual noise exposures, the analysis used data on occupational specialties and time spent assigned to various types of ships or other assignments to assess the combined effects on the risk of STS. The statistical analysis controlled for age, race, and gender.

Overall, 11 percent of the study population was found to have an STS, which is about half the annual incidence values reported in DOEHRS-HC for the Navy's hearing conservation program from 1999 to 2004. Shaw and Trost (2005) identified some variables that can influence the amount of STS observed. They reported that the time assigned to surface warships (e.g., aircraft carriers, battleships, destroyers, amphibious ships), for example, had a greater impact on STS than time assigned to other surface ships, certain submarine duties, or shore duties. Assignments to submarine engine rooms and Air Wings produced no greater risk than shore assignments, possibly because of ready access to and use of hearing protection. The probability of an STS reached 0.46 if the individual spent the entire study assigned to a surface war ship, a level significantly higher than the probability of 0.27 for individuals assigned entirely to shore duty.

The foregoing review of STS data suggests that a substantial number of individuals may be experiencing declines of at least 10 dB in their hearing while in the military. It is not possible, however, to discern the causes underlying such changes (especially in the absence of age corrections), the time over which such shifts occurred (e.g., 1 year or 20 years), or the resulting degree of hearing loss associated with the shift (a shift from 0 to 15 dB HL, both levels reflecting normal hearing, or a shift from 20 to 35 dB

HL, from normal to mild hearing loss). Moreover, hearing thresholds at 6000 Hz are not considered in any definition of STS, yet the average data indicate that hearing loss among military personnel is likely to be greatest at this frequency.

These data must be interpreted with considerable caution for other reasons as well. The DOEHRS-HC system is still experiencing difficulty in collecting data from each service. Percentages of individuals in the hearing conservation programs for whom reports of annual audiograms have been submitted to the DOEHRS-HC system have varied but have averaged only about 45 percent for the Army since 1998 and 55 percent for the Air Force since 2000 (Air Force Hearing Conservation Registry, 2004b; U.S. Army Center for Health Promotion and Preventive Medicine, 2004a; also see Chapter 5). Similar problems have been described for the Army data for the years preceding the implementation of the DOEHRS-HC system (Ohlin, 2004b). Moreover, the apparent incidence of PTS reflects not only threshold shifts that persist after retesting, but any cases in which the follow-up testing was not done, likely leading to an overestimate of the true incidence of PTS.

EPIDEMIOLOGICAL STUDIES OF NOISE-INDUCED HEARING LOSS IN INDIVIDUALS WITH PRIOR MILITARY SERVICE

The Vietnam Experience Study, conducted by the Centers for Disease Control (CDC, 1988a,b), included examination of hearing status among a random sample of male veterans who had enlisted in the U.S. Army during the period 1965 through 1971. The hearing levels of the men who had served in Vietnam (n = 2,490) were compared with those of men who had not served in Vietnam (n = 1,972). The mean age of the study participants in each group was 37 years. High-frequency hearing loss was defined as an average threshold at 3000, 4000, and 6000 Hz ≥ 51 dB HL. Individual cases were classified as positive or negative with regard to the presence of high-frequency hearing loss based on this definition. Six covariates were included in the logistic regression analysis that was performed on these data: age at enlistment, race, year of enlistment, enlistment status (volunteer versus draftee), score on a general technical test, primary military occupation, and preservice hearing loss.

Of the veterans who served in Vietnam, 18 percent met the hearing-loss criterion for their left ears, compared with 13 percent of those who were not in Vietnam. Smaller percentages had qualifying hearing losses in their right ears or in both ears. Overall, the Vietnam veterans were 40 percent more likely to have high-frequency hearing loss in either ear alone or both ears than the veterans who had not served in Vietnam. Follow-up analyses indicated that this was driven primarily by those Vietnam veterans with military tactical occupational specialties (e.g., infantry, artillery, armor), who were 2.5 times more likely than non-Vietnam veterans to have high-frequency hearing loss. Those veterans who served in Vietnam in nontactical occupational specialties were not at significantly greater risk for hearing loss than otherwise similar veterans who did not serve in Vietnam.

These results illustrate the importance of exposure conditions, including combat, for identifying an elevated risk for hearing loss among Vietnam veterans. These data reflect hearing thresholds 15–20 years after service in Vietnam and include any effects of subsequent noise exposures or any selection effects that may have resulted in differences between the groups in subsequent noise exposure or survival.

Noe et al. (2002) compared hearing thresholds among veterans and nonveterans using data from a community-based cohort study of older adults in Beaver Dam, Wisconsin. The average age for both groups was approximately 65 years. Hearing loss was defined as a pure-tone average hearing threshold of 25 dB HL or greater for 500, 1000, 2000, and 4000 Hz, and cases were then classified as being either positive or negative with regard to hearing loss. Covariates included in the subsequent logistic-regression analysis included age, longest held occupation, history of head injury, and smoking. Veterans ($n = 999$) were not found to be at greater risk for hearing loss than nonveterans ($n = 588$). In addition, mean hearing thresholds for the two groups were similar at all measured frequencies from 500 through 8000 Hz. This analysis does not attempt to address differences related to noise exposure, only differences associated with prior military service. Nonveterans, too, are subject to occupational noise exposure, and in this study prior military service as such is not associated with an increased risk of hearing loss in a population of older adults.

FINDINGS

In the more than 60 years since the U.S. entrance into World War II, over 25 million people have served in the U.S. armed forces. Their experiences, in five different services and at least five major conflicts, as well as peacetime eras, have exposed many to loud noise. These noise exposures are likely to have varied widely, even within similar occupational specialties and eras. Data and analyses to document and quantify noise exposures of military personnel during this period, as well as to document and quantify their hearing thresholds and permanent changes in those thresholds over the course of military service, are not available.

The committee found only a limited number of studies on which to base its findings, and those studies were primarily for the period since 1970. Among these were cross-sectional studies showing patterns of hearing loss consistent with noise exposure, but no longitudinal studies that could provide reliable data on changes in individuals' hearing thresholds over the course of military service.

The available studies were not designed to be representative of a service as a whole and only rarely of a particular occupational group. Together, these factors made it impossible to generalize findings from these studies to broader populations of military service members or veterans or to personnel serving in other time periods. Furthermore, the variability of individual responses to noise exposure precludes using the average hearing thresholds reported for groups of study participants to estimate the hearing loss of individuals.

FINDING: The evidence is sufficient to conclude that certain military personnel from World War II to the present have exhibited hearing thresholds while in the military that are typical of noise-induced hearing loss.

FINDING: The evidence is not sufficient to reach conclusions regarding the number or proportion of service members, overall or in specific occupational groups or eras since World War II, who have experienced noise-induced hearing loss while in the military.

FINDING: The evidence is not sufficient to determine the probability of acquiring noise-induced hearing loss associated with service in the military, or in specific branches of the military, for a given individual. The probability of acquiring noise-induced hearing loss can only be determined precisely with well-controlled, longitudinal epidemiological studies.

FINDING: The evidence is sufficient to conclude that, in the absence of audiograms obtained at the beginning and end of military service, it is difficult or impossible to determine with certainty how much of a specific individual's hearing loss was acquired during military service.

REFERENCES

Air Force Hearing Conservation Registry. 2004a. DOEHRS Data Repository: Air Force Hearing Profile by Pay Grade, 2000–2004. Data provided to the Institute of Medicine Committee on Noise-Induced Hearing Loss and Tinnitus Associated with Military Service from World War II to the Present, Washington, DC.

Air Force Hearing Conservation Registry. 2004b. DOEHRS Data Repository: Hearing Conservation Compliance Report, 2000–2003. Data provided to the Institute of Medicine Committee on Noise-Induced Hearing Loss and Tinnitus Associated with Military Service from World War II to the Present, Washington, DC.

Air Force Hearing Conservation Registry. 2004c. DOEHRS Data Repository: Positive Significant Threshold Shift, 1999–2004. Data provided to the Institute of Medicine Committee on Noise-Induced Hearing Loss and Tinnitus Associated with Military Service from World War II to the Present, Washington, DC.

AMSA (Army Medical Surveillance Activity). 1998. *Tri-Service Reportable Events: Guidelines and Case Definitions.* [Online]. Available: http://amsa.army.mil/TriServiceRE/Jul98TriServREGuide.pdf [accessed April 24, 2005].

AMSARA (Accession Medical Standards Analysis and Research Activity). 2002. *Annual Report 2002.* Washington, DC: Walter Reed Army Institute of Research.

ANSI (American National Standards Institute). 1969. *Specification for Audiometers (ANSI S3.6-1969).* New York: Acoustical Society of America.

ANSI. 1996. *ANSI S3.44 Determination of Occupational Noise Exposure and Estimation of Noise-Induced Hearing Impairment.* New York: Acoustical Society of America.

ASA (American Standards Association). 1951. *American Standard Specification for Audiometers for General Diagnostic Purposes.* Pub. No. Z24.5-1951. New York: ASA.

Bohnker BK, Page JC, Rovig G, Betts LS, Muller JG, Sack DM. 2002. U.S. Navy and Marine Corps Hearing Conservation Program, 1995–1999: Mean hearing thresholds for enlisted personnel by gender and age groups. *Military Medicine* 167(2):132–135.

Carmichael JC. 1955. Noise health hazards in the Air Force. *Public Health Reports* 70(12):1251–1256.

CDC (Centers for Disease Control). 1988a. Health status of Vietnam veterans. I. Psychosocial characteristics. The Centers for Disease Control Vietnam Experience Study. *Journal of the American Medical Association* 259(18):2701–2707.

CDC. 1988b. Health status of Vietnam veterans. II. Physical health. The Centers for Disease Control Vietnam Experience Study. *Journal of the American Medical Association* 259(18):2708–2714.

CHABA (National Academy of Sciences–National Research Council Committee on Hearing, Bioacoustics, and Biomechanics). 1968. *Proposed Damage-Risk Criterion for Impulse Noise (Gunfire).* Report of Working Group 57 (Ward WD, ed.). Washington, DC: National Academy of Sciences.

Chandler D (Walter Reed Army Medical Center). 2005. RE: Seek your help with technical review. E-mail to L. Joellenbeck, Institute of Medicine, June 20.

Chandler DW, Fletcher JL. 1983. Hearing levels in U.S. Army engineers. *Journal of Audiology Research* 23(1):23–32.

Crowder L (Navy Environmental Health Center). 2005. RE: Request technical review of draft report sections. E-mail to L. Joellenbeck, Institute of Medicine, June 9.

Department of the Air Force. 1980. *Hearing Conservation Program Status Report: January–June 1980.* Washington, DC: Department of the Air Force.

Department of the Air Force. 1982. *AF Regulation 161-35: Hazardous Noise Exposure.* Washington, DC: Department of the Air Force.

Department of the Air Force. 1994. *AFOSH Standard 48-19: Hazardous Noise Program.* Washington, DC: Department of the Air Force.

Department of the Army. 2003. Section 3-30: Hearing conservation program. In: *TRADOC Regulation 350-6: Training: Enlisted Initial Entry Training (IET) Policies and Administration.* Washington, DC: Department of the Army.

Department of the Navy. 1983. Chapter 18: Hearing conservation and noise abatement. In: *OPNAVINST 5100.23B: Navy Occupational Safety and Health Program Manual.* Washington, DC: Department of the Navy.

Department of Veterans Affairs. 2005. *America's Wars.* [Online]. Available: http://www1.va.gov/opa/fact/amwars.html [accessed May 2005].

DoD (Department of Defense). 1978. *Department of Defense Instruction 6055.3: Hearing Conservation.* Washington, DC: Department of Defense.

DoD. 1987. *Department of Defense Instruction 6055.12: Department of Defense Hearing Conservation Program.* Washington, DC: Department of Defense.

DoD. 1996. *Department of Defense Instruction 6055.12: DoD Hearing Conservation Program.* Washington, DC: Department of Defense.

DoD. 1997. *MIL-STD-1474D: Department of Defense Design Criteria Standard—Noise Limits.* Washington, DC: Department of Defense.

DoD. 1998. *Department of Defense Instruction 6055.1: DoD Safety and Occupational Health (SOH) Program.* Washington, DC: Department of Defense.

DoD. 2004. *Department of Defense Instruction 6055.12: DoD Hearing Conservation Program.* Washington, DC: Department of Defense.

Fairman TM, Johnson DL. 1979. Noise dosimeter measurements in the Air Force. *Aviation, Space, and Environmental Medicine* 50(11):1150–1157.

Gasaway DC. 1986. Noise levels in cockpits of aircraft during normal cruise and considerations of auditory risk. *Aviation, Space, and Environmental Medicine* 57(2):103–112.

Gasaway DC. 1988. Occupational hearing conservation in the military. In: Lipscomb DM, ed. [Jacobson JT, advisory ed.]. *Hearing Conservation in Industry, Schools, and the Military.* Boston: Little, Brown. Pp. 243–262.

Gasaway DC. 2002. *Donald C. Gasaway's Aircraft Noise Compendia.* Washington, DC: Department of the Air Force.

Glorig A. 1952. Field investigations on military personnel as regards sound. In: University of Michigan School of Public Health, Institute of Industrial Health. *The Acoustical Spectrum. Sound—Wanted and Unwanted.* Ann Arbor: University of Michigan Press. Pp. 99–106.

Glorig A, Roberts J. 1965. Hearing Levels of Adults by Age and Sex, United States, 1960–1962. *Vital and Health Statistics.* Series 11, No. 11. Washington, DC: Department of Health, Education, and Welfare.

Glorig A, Wheeler D, Quiggle R, Grings W, Summerfield A. 1957. *1954 Wisconsin State Fair Hearing Survey. Statistical Treatment of Clinical and Audiometric Data.* Los Angeles: American Academy of Ophthalmology and Otolaryngology.

Goldenberg RA. 1977. Hearing loss in the Marine Corps. *Transactions—American Academy of Ophthalmology and Otolaryngology* 84(6):991–996.

Gravendeel DW, Plomp R. 1959. The relation between permanent and temporary noise dips. *American Medical Association-The Archives of Otolaryngology* 69(6):714–719.

Henselman LW, Henderson D, Shadoan J, Subramaniam M, Saunders S, Ohlin D. 1995. Effects of noise exposure, race, and years of service on hearing in U.S. Army soldiers. *Ear and Hearing* 16(4):382–391.

ISO (International Organization for Standardization). 1990. *ISO 1999: Acoustics—Determination of Occupational Noise Exposure and Estimation of Noise-Induced Hearing Impairment.* Geneva, Switzerland: ISO.

Johnson DL, Nixon CW. 1974. Simplified methods for estimating hearing protector performance. *Sound and Vibration* 8(6):20–27.

Jordan DT Jr, Jones CM. 1983. *Noise Exposure Study of Reese AFB Personnel Working on the Flightline in Selected Adjacent Shops.* OEHL 83-220EH165HNA. Brooks Air Force Base, TX: U.S. Air Force Occupational and Environmental Health Laboratory.

Kardous CA, Willson RD. 2004. Limitations of using dosimeters in impulse noise environments. *Journal of Occupational and Environmental Hygiene* 1(7):456–462.

Marshall L, Carpenter S. 1988. *Hearing Levels of 416 Sonar Technicians.* NSMRL Report No. 1123. Groton, CT: Naval Submarine Medical Research Laboratory.

McConnell W. 2004. Hearing Conservation in the Coast Guard. Presentation to the Institute of Medicine Committee on Noise-Induced Hearing Loss and Tinnitus Associated with Military Service from World War II to the Present, Meeting I. Washington, DC. May 12.

McConnell W (U.S. Coast Guard). 2005. Technical Review of Draft Section on Coast Guard. Comments provided to the Institute of Medicine Committee on Noise-Induced Hearing Loss and Tinnitus Associated with Military Service from World War II to the Present, Washington, DC.

Navy Environmental Health Center. 2004a. DOEHRS Data Repository: Positive Significant Threshold Shift, 1999–2004. Data provided to the Institute of Medicine Committee on Noise-Induced Hearing Loss and Tinnitus Associated with Military Service from World War II to the Present, Washington, DC.

Navy Environmental Health Center. 2004b. *Navy Medical Department Hearing Conservation Program Procedures*. NECH-TM 6260.51.99-2. Portsmouth, VA: U.S. Navy, Bureau of Medicine and Surgery.

Nixon CW. 1998. *A Glimpse of History: The Origin of Hearing Conservation Was in the Military?* Wright-Patterson Air Force Base, OH: Air Force Research Laboratory.

Noe CM, Cruickshanks KJ, Nondahl D, Wiley TL, Wilson RH. 2002. *Prevalence of Hearing Loss Among the Veteran Population in the Beaver Dam (WI) Cohort (abstract)*. Third Rehabilitation Research and Development Conference, Arlington, VA. February 10–12.

Ohlin D. 1992. *15 Years Revisited: The Prevalence of Hearing Loss Among Selected U.S. Army Branches*. Hearing Conservation Special Study No. 51-01-PM82-93. Aberdeen Proving Ground, MD: U.S. Army Environmental Hygiene Agency.

Ohlin D. 2004a. DoD and Army Data and Databases. Presentation to the Institute of Medicine Committee on Noise-Induced Hearing Loss and Tinnitus Associated with Military Service from World War II to the Present, Meeting II. Washington, DC. July 19.

Ohlin D (U.S. Army Center for Health Promotion and Preventive Medicine). 2004b. Requested DOEHRS-HC Reports for Committee on Noise-Induced Hearing Loss and Tinnitus Associated with Military Service. E-mail to J. Durch, Institute of Medicine, July 23.

Ohlin D (U.S. Army Center for Health Promotion and Preventive Medicine). 2005. RE: Seek Help with STS Definitions. E-mail to L. Joellenbeck, Institute of Medicine, March 25.

Peters LJ, Ford H. 1983. *Extent of Hearing Loss Among Army Aviators at Fort Rucker, Alabama*. USAAMRL 83-12. Fort Rucker, AL: U.S. Army Aeromedical Research Laboratory.

Ritter DC, Perkins JL. 2001. Noise-induced hearing loss among U.S. Air Force cryptolinguists. *Aviation, Space, and Environmental Medicine* 72(6):546–552.

Robertson RM, Maxwell DW, Williams CE. 1978a. *The Landing Signal Officer: Auditory Aspects*. NAMRL-1255. Pensacola, FL: Naval Aerospace Medical Research Laboratory.

Robertson RM, Page JC, Williams CE. 1978b. *The Prevalence of Hearing Loss Among Selected Navy Enlisted Personnel*. NAMRL-1251. Pensacola, FL: Naval Aerospace Medical Research Laboratory.

Robertson RM, Greene JW, Williams CE, Maxwell DW. 1990. *Noise Exposure of Naval Communication Station Radio Operators: A Field Study*. NAMRL Technical Memorandum 90-1. Pensacola, FL: Naval Aerospace Medical Research Laboratory.

Royster LH, Thomas WG, Royster JD, Lilley D. 1978. Potential hearing compensation cost by race and sex. *Journal of Occupational Medicine* 20(12):801–806.

Shaw GB, Trost RP. 2005. *Statistical Analysis of Hearing Loss Among Navy Personnel*. CRM D0011228.A2. Alexandria, VA: CNA.

Solomon LN, Fletcher JL. 1958. Survey of hearing losses among armor personnel. *Journal of Speech and Hearing Disorders* 1(4):350–358.

Sutherland HC, Gasaway DC. 1976. *Hearing Levels of Noise-Exposed U.S. Air Force Personnel Compared to Those in the Total U.S. Population*. Report SAM-TR-76-27. Brooks Air Force Base, TX: US Air Force School of Aerospace Medicine.

Sutherland HC, Gasaway DC. 1978. *Current Hearing Threshold Levels for Noise-Exposed U.S. Air Force Personnel. One Year's Reportings*. Report SAM-TR-78-39. Brooks Air Force Base, TX: U.S. Air Force School of Aerospace Medicine.

Thomas JW. 1995. *The Application of Audiometric Data Base Analysis to Selected Air Force Bases.* Master's thesis, University of North Carolina, Chapel Hill.

U.S. Army Center for Health Promotion and Preventive Medicine. 1999. *Technical Guide 181: Noise Dosimetry and Risk Assessment.* Aberdeen Proving Ground, MD: U.S. Army Center for Health Promotion and Preventive Medicine.

U.S. Army Center for Health Promotion and Preventive Medicine. 2004a. DOEHRS Data Repository: Army Hearing Conservation Program Compliance Reports, 1988–2003. Data provided to the Institute of Medicine Committee on Noise-Induced Hearing Loss and Tinnitus Associated with Military Service from World War II to the Present, Washington, DC.

U.S. Army Center for Health Promotion and Preventive Medicine. 2004b. DOEHRS Data Repository: Army Hearing Profile Reports, 1975–2003. Data provided to the Institute of Medicine Committee on Noise-Induced Hearing Loss and Tinnitus Associated with Military Service from World War II to the Present, Washington, DC.

U.S. Army Center for Health Promotion and Preventive Medicine. 2004c. DOEHRS Data Repository: Hearing Conservation Program Positive STS Report, 1983–2003. Data provided to the Institute of Medicine Committee on Noise-Induced Hearing Loss and Tinnitus Associated with Military Service from World War II to the Present, Washington, DC.

U.S. Army Center for Health Promotion and Preventive Medicine. 2004d. *Noise Levels of Common Army Equipment.* [Online]. Available: http://chppm-www.apgea.army.mil/hcp/NoiseLevelsPrint.htm [accessed April 21, 2004].

Walden BE, Worthington DW, McCurdy HW. 1971. *The Extent of Hearing Loss in the Army: A Survey Report.* Washington, DC: Walter Reed Army Medical Center.

Walden BE, Prosek RA, Worthington DW. 1975. *The Prevalence of Hearing Loss Within Selected U.S. Army Branches.* Washington, DC: Walter Reed Army Medical Center.

Weisman W (U.S. Air Force Surgeon General's Office). 2005. Technical Review of Draft Section on Air Force Noise Measurements and Dosimetry. Comments provided to the Institute of Medicine Committee on Noise-Induced Hearing Loss and Tinnitus Associated with Military Service from World War II to the Present, Washington, DC.

Wolgemuth KS, Luttrell WE, Kamhi AG, Wark DJ. 1995. The effectiveness of the Navy's hearing conservation program. *Military Medicine* 160(5):219–222.

Yankaskas K. 2001. Landing on the Roof: Flight Operation Noise. Presentation at the International Military Noise Conference, Baltimore, MD. April 25.

Yankaskas K. 2004. Shipboard Noise and Hearing Conservation Issues. Presentation to the Institute of Medicine Committee on Noise-Induced Hearing Loss and Tinnitus Associated with Military Service from World War II to the Present, Meeting IV, Washington, DC. December 7.

Yankaskas KD, Shaw MF. 1999. Landing on the roof: CVN noise. *Naval Engineers Journal* 111(4):23–34.

Yarington CT. 1968. Military noise induced hearing loss: Problems in conservation programs. *Laryngoscope* 78(4):685–692.

Ylikoski ME, Ylikoski JS. 1994. Hearing loss and handicap of professional soldiers exposed to gunfire noise. *Scandinavian Journal of Work, Environment & Health* 20(2):93–100.

4

Tinnitus

Tinnitus is a perceived sound that cannot be attributed to an external sound source (Eggermont, 2003). It is a subjective phenomenon, perceivable only by the person who is experiencing it.[1] The committee was asked to review the evidence regarding noise levels that can cause tinnitus and other risk factors for tinnitus.

This chapter first provides a brief overview of the features of tinnitus, its impact on individuals with the condition, and approaches to its clinical assessment and treatment. Some of the issues that arise in studying tinnitus are noted, and basic data on its occurrence in the general population are presented. The major portion of the chapter focuses on a review of epidemiological data on the relationship between tinnitus and noise exposure, hearing loss, and other risk factors. Important features of studies reviewed by the committee are summarized in Table D-6 in Appendix D. The chapter

[1]Tinnitus is distinct from other acoustic events that can be generated in the head or neck regions and reach perception. Some of these events include vascular pulsations, palatal and intratympanic myoclonus, patulous Eustachian tube, jugular outflow syndrome, and cervical crepitus. Although these effects have been referred to as "objective tinnitus" (see Hazell, 1995), the term "somatosounds" (Anonymous, 1981) better describes these phenomena. Use of the term somatosounds provides a clearer distinction between acoustic events generated within the body and the completely subjective perception of tinnitus.

The condition referred to as "pulsatile tinnitus" can be a somatosound or a neural event induced within the brain, usually by a dilated arterial loop affecting the auditory nerve (De Ridder et al., 2005). However, this phenomenon is not related to tinnitus associated with noise exposure, and the specifics of this distinction are beyond the scope of the committee's report.

concludes with a discussion of tinnitus in the context of military service and a proposal for monitoring tinnitus among U.S. military personnel.

BACKGROUND

Tinnitus is often referred to as "ringing in the ears," reflecting a characterization of the sound that individuals commonly report. Other descriptions of the perceived sound include buzzing, hissing, whistling, and humming (e.g., Alberti, 1987; Stouffer and Tyler, 1990; Mrena et al., 2002). Reports of other characteristics of tinnitus also vary. Tinnitus induced by noise exposure, for example, is often described as high-pitched (e.g., Melinek et al., 1976; Man and Naggan, 1981; Cahani et al., 1983; Alberti, 1987; Chermak and Dengerink, 1987; Stouffer and Tyler, 1990), whereas tinnitus associated with Ménière's disease[2] has been described as a low-pitched sound (NRC, 1982). Tinnitus can be transient or persistent. Some studies define persistent or prolonged tinnitus as lasting at least 5 minutes (e.g., Coles, 1984; Parving et al., 1993; Palmer et al., 2002; Sindhusake et al., 2003b). Persistent tinnitus can be perceived continuously (all or most of the time) or occasionally. A given episode of tinnitus may also resolve, with new episodes possible in the future.

Tinnitus is considered a symptom rather than an illness (NRC, 1982). It is associated with many conditions, including noise exposure and noise-induced hearing loss. Among the other conditions that may cause or be accompanied by tinnitus are presbycusis, Ménière's disease, otosclerosis, head injury, cerebellar-pontine angle tumors, otitis media, meningitis, dental disorders, and exposure to certain medications (e.g., salicylates, aminoglycoside antibiotics, and some chemotherapy agents) (Lockwood et al., 2002).

It is not always possible to identify a precipitating cause of tinnitus. A survey of tinnitus patients found that only 54 percent attributed their tinnitus to a particular cause (Stouffer and Tyler, 1990). The onset of tinnitus is described by some as gradual and by others as sudden (Axelsson and Barrenas, 1992). In a population-based study of older adults, 55 percent of participants with tinnitus reported a gradual onset, 24 percent reported a sudden onset, and the remainder did not know (Sindhusake et al., 2003b). Uncertainty about the onset of tinnitus can make the identification of a precipitating cause challenging.

Individuals differ in their susceptibility and reaction to tinnitus. The reasons for these differences are not known but are likely related both to

---

[2]Ménière's disease is a disorder of the inner ear affecting balance and hearing (the audiovestibular system), characterized by abnormal sensations of movement (vertigo), dizziness, loss of hearing in one or both ears, and tinnitus.

the magnitude and quality of the perceived tinnitus and to the psychological makeup, life experiences, and current stress factors of the individual.

The mechanisms underlying tinnitus are not completely understood. Generally, it is reasonable to presume that the involvement of central components of the auditory system "results in" the perception of sound. In addition, if there is an emotional reaction to the tinnitus, other areas of the central nervous system that are involved in emotionally charged events, such as the amygdala, are activated (e.g., Aggleton and Mishkin, 1986; LeDoux, 2000; see Cacace, 2004, for a review). The actual site of the origin of the tinnitus could be anywhere in the auditory system but likely includes the auditory periphery in many if not most cases. Other areas in the brain that relate to vision, touch, and movement can also affect tinnitus in some instances (Baguley, 2002; Cacace, 2003; Eggermont, 2003). As tinnitus is usually accompanied by hearing loss, similar mechanisms are likely involved.

Experimental studies of noise-induced tinnitus present a relatively homogeneous and consistent body of research. (This is in contrast to pharmacological manipulations to induce tinnitus, which result in more varied effects.) In these studies, noise exposure results in cochlear hair cell damage (see Chapter 2). Perhaps surprisingly, spontaneous neural activity arising from the auditory nerve is lost or significantly diminished following noise damage. However, increases in spontaneous neural activity have been found in brainstem and cortical regions. For example, following exposure to intense sound, the dorsal cochlear nucleus has been implicated in consistently producing and/or modulating hyperactive neural activity, which may serve as a trigger or generator site for tinnitus (Kaltenbach et al., 2005). A common hypothesis for tinnitus in this instance is a "release from inhibition," and in some instances, this may contribute to reactive changes (reorganization) at more central locations in the auditory system. Thus, tinnitus might be initiated by a discontinuity in the spontaneous activity across auditory nerve fibers with different characteristic frequencies, which may result in the reduction of lateral inhibition and produce changes in tonotopic maps in the auditory cortex (e.g., Muhlnickel et al., 1998; Salvi et al., 2000; Eggermont, 2003; see Eggermont and Roberts, 2004, for a review). This supposes that the effects of noise exposure and subsequent hearing loss disrupt the delicate balance between excitation and inhibition in the central auditory pathways.

The possibility that the onset of noise-induced tinnitus might be delayed by months has been raised because studies in laboratory animals have shown that degenerative processes initiated by the noise exposure continue in central auditory pathways after termination of the exposure (Kim et al., 1997; Morest et al., 1998). Although degenerative changes in afferent pathways will most likely not affect auditory thresholds, it is possible that they could contribute to other central processes such as tinnitus. The time required for this reorganization might vary across individuals and potentially could be a long-term process. However, as the interval between a noise exposure and the onset of tinnitus lengthens, the possibility that tinnitus will be triggered by other factors increases. A more complete understanding of the mechanisms by which tinnitus is generated will be needed before the existence of delayed onset of noise-induced tinnitus can be confirmed or rejected.

Impact of Tinnitus

Most people with tinnitus report few problems, but for some individuals, tinnitus can be a life-altering experience. The adverse effect of tinnitus can impair psychological well-being and the ability to function in social and professional settings (Stephens and Hallam, 1985; Wilson et al., 1991; Tyler, 1993; Meric et al., 1998). For those affected, problems occur with their emotional health, hearing, sleep, and concentration (Axelsson and Sandh, 1985; Mrena et al., 2002; Tyler et al., 2004). Tinnitus has also been reported to induce fear, frustration, anger, irritability, and anxiety (Erlandsson et al., 1991; Halford and Anderson, 1991; Dineen et al., 1997; Andersson et al., 2001; Mrena et al., 2002; reviewed in Erlandsson, 2000). Depression or depressive symptoms may be contributing factors for some individuals who experience serious distress from tinnitus (Erlandsson, 2000; Dobie, 2003). A potential link to suicide has been suggested for severe cases, but depression or other relevant factors may exist (Jacobson and McCaslin, 2001). For some people, the impact of tinnitus may be worse than the impact of an accompanying hearing loss (Salmivalli, 1967; Axelsson and Barrenas, 1992; Mrena et al., 2002).

Clinical Assessment and Treatment

Currently, the assessment of tinnitus involves psychoacoustical measures and self-reporting by patients. Psychoacoustical assessment of tinnitus attributes, such as pitch and loudness, and validated questionnaires offer a standardized basis for characterizing the perceived magnitude of tinnitus and its impact. In addition, these measures can help in identifying treatment needs, selecting treatment approaches, and monitoring the status of tinnitus over time. In psychoacoustical testing, patients compare the perceived pitch and loudness of their tinnitus with defined sounds presented under controlled conditions (for reviews, see Tyler, 2000; Vernon and Meikle, 2003; Henry, 2004). Self-report questionnaires widely used to assess the impact of tinnitus include the Tinnitus Handicap Questionnaire (Kuk et al., 1990), the Tinnitus Reaction Questionnaire (Wilson et al., 1991), and the Tinnitus

Handicap Inventory (Newman et al., 1996). (For reviews, see Erlandsson, 2000; Tyler, 1993, 2000; Noble, 1998; Wilson and Henry, 2000.) See Box 4-1 for examples of questions from these assessment instruments. The quality, pitch, and loudness of tinnitus influence the subjective annoyance (as they do for all sounds), but other factors can also be important. The degree of annoyance caused by tinnitus is influenced by both the characteristics of the tinnitus and an individual's personality (Axelsson and Sandh, 1985; Dauman and Tyler, 1992).

Some people with tinnitus do not seek treatment, but others do. Axelsson and Ringdahl (1989) found that of the survey respondents who described their tinnitus as always present, 25 percent felt an "urgent" need for treatment. Of those who experienced tinnitus "often," 6 percent reported an urgent need for treatment. In a population-based study of older adults, 37 percent of those who had tinnitus had sought care and 6 percent received some form of treatment (Sindhusake et al., 2003b). Among a group of workers with tinnitus and newly detected hearing loss, 14 percent consulted a physician (Phoon et al., 1993).

For people with tinnitus who do seek treatment, a thorough evaluation includes a comprehensive medical examination of body systems with spe-

BOX 4-1
Sample Questions from Questionnaires Used to Assess the Impact of Tinnitus

Tinnitus Handicap Questionnaire (Kuk et al., 1990)
Subjects rate their agreement with 27 items using a scale from 0 (strongly disagree) to 100 (strongly agree).

- I do not enjoy life because of tinnitus
- Tinnitus causes me to avoid noisy situations
- I think I have a healthy outlook on tinnitus

Tinnitus Reaction Questionnaire (Wilson et al., 1991)
Subjects rate 26 items, using a 5-point scale (0 = not at all; 1 = a little of the time; 2 = some of the time; 3 = a good deal of the time; 4 = almost all of the time).

- My tinnitus has made me unhappy
- My tinnitus has made me feel tense
- My tinnitus has interfered with my ability to work

Tinnitus Handicap Inventory (Newman et al., 1996)
Subjects report their endorsement of 25 items (yes, sometimes, no).

- Because of your tinnitus is it difficult for you to concentrate?
- Does your tinnnitus make you angry?
- Do you feel that you have no control over your tinnitus?

cial attention to the head, neck, cranial nerves, medications, drug use, and psychiatric history (e.g., depression) (Tyler and Babin, 1986; Shulman, 1997; Dobie, 2003). Tinnitus associated with certain medical conditions (e.g., tumors and some infections) may resolve when those conditions are treated. No current form of treatment can eliminate tinnitus arising from many other causes, but some treatments may lessen the adverse impact of tinnitus. Treatment approaches being used include counseling, counseling combined with sound therapies, antidepressants and other medications, electrical stimulation at the ear, and transcranial magnetic stimulation over specific areas of the brain (e.g., Hazell and Wood, 1981; Hazell, 1995; Dobie, 1999; Dauman, 2000; Sweetow, 2000; Vernon and Meikle, 2000; Wilson and Henry, 2000; Young, 2000; Jastreboff and Jastreboff, 2003; Plewnia et al., 2003; Rubinstein et al., 2003; Tyler and Cacace, 2004; Tyler et al., 2004; Kleinjung et al., 2005).

ISSUES IN STUDYING TINNITUS

A significant advance in exploring mechanisms of tinnitus has been research on the development of behavioral animal models of tinnitus (e.g., Jastreboff et al., 1988a,b; Bauer et al., 1999; Bauer and Broznoski, 2001; Heffner and Harrington, 2002; Bauer, 2003; Lobarinas et al., 2004; Moody, 2004). Experimental work has been instrumental in gaining insight into the anatomical loci and neural codes for noise-induced tinnitus (e.g., increases in spontaneous neural activity, bursting, synchronous discharges among different neurons) by combining behavioral verification with detailed physiologic recordings (e.g., Kaltenbach et al., 2004a; for a review, see Kaltenbach et al., 2004b).

Experimental studies have also been conducted with humans, but such studies must be designed to protect subjects from permanent injury. Temporary tinnitus has been elicited in response to well-defined noise exposures (e.g., Loeb and Smith, 1967; Chermak and Dengerink, 1987; George and Kemp, 1989). It is not clear, however, that this transient tinnitus is comparable to the more persistent form that is the source of concern in most epidemiological and clinical studies. Furthermore, the experimental noise exposures may not be representative of "real-world" noise exposures.

Epidemiological studies are the most common means of assessing the prevalence of tinnitus in the population and determining the factors that are associated with higher rates of tinnitus. However, these observational studies do not permit random assignment of subjects to exposures of interest. Most studies of tinnitus are cross-sectional rather than longitudinal. Documentation of past exposures to noise and other risk factors is rarely available. With retrospective reports, the timing and magnitude of those exposures or their comparability across study participants cannot be determined.

Furthermore, with no objective basis for detecting or characterizing tinnitus, studies have used various definitions of tinnitus and different methods to elicit reports on whether and how tinnitus was experienced.

OCCURRENCE OF TINNITUS

Little is known about the incidence of tinnitus, that is, the number of new cases that develop in a population during a given period. Tinnitus can arise without a distinctive precursor (Alberti, 1987; Stouffer and Tyler, 1990; Axelsson and Prasher, 2000), and retrospective assessments of the timing of its onset are vulnerable to recall errors. A rare prospective, population-based assessment of the incidence of tinnitus among older adults (ages 48–92 years at baseline) in Beaver Dam, Wisconsin, found that 6 percent developed tinnitus over a 5-year period (Nondahl et al., 2002). The incidence of tinnitus in this older population was not associated with age or gender.

Several reports offer some perspective on the prevalence of tinnitus in the United States (see Table 4-1). Data from national surveys during the 1990s (Adams et al., 1999; Hoffman and Reed, 2004) showed prevalence rates for adults (ages 18 or 20 years and older) ranging from 4 to 8 percent. An earlier prevalence estimate from a nationally representative health examination survey conducted in the United States in 1960–1962 found that 6 percent of adults (ages 18–79 years) reported what was classified as "severe" tinnitus and 27 percent had "mild" tinnitus (Roberts, 1968).

The study of older adults in Beaver Dam (Nondahl et al., 2002) found that 9 percent of men and 8 percent of women reported "significant tinnitus," which was defined in the study as tinnitus experienced during the past year that the respondent rated as at least moderately severe or as causing problems getting to sleep. Overall, 2 percent of the study participants reported that their tinnitus was severe (Nondahl et al., 2002). With an alternative definition of tinnitus ("buzzing, ringing, or noise in your ears in the past year that usually lasts longer than 5 minutes"), the prevalence estimate was 18 percent at a 5-year follow-up of this population (Nondahl et al., 2004). Studies in Europe and Australia, using various definitions of tinnitus and study populations of various ages, have produced prevalence estimates ranging from about 10 to 15 percent in people 20 years of age and older and from 14 to 30 percent in people 50 years of age and older (see Table 4-1).

Whether age is an independent risk factor for tinnitus is unclear. The prevalence of tinnitus is generally higher at older ages, but the rates at the oldest ages are not always the highest (see Table 4-1). A multivariate analysis of a subset of the data collected in a 1994–1995 study of U.S. adults found significant age-related differences in the prevalence of tinnitus. Compared with those ages 20–24 years, the likelihood of having tinnitus was greatest for persons ages 65–74 years (odds ratio [OR] 4.08, 95% confidence interval [CI] 3.25–5.12) (Hoffman and Reed, 2004). In contrast, an analysis of Norwegian data on the prevalence of "bothersome tinnitus" showed no significant age-related differences for women and a significant difference for men only for the age group 45–54 years (OR 1.30, 95% CI 1.06–1.60) (Hoffman and Reed, 2004). In these cross-sectional studies, the possibility of underlying differences among age cohorts must be considered along with any effects of aging or tinnitus risk factors, such as hearing loss, that may themselves be related to age. Multivariate analyses are also available from two community-based studies limited to older adults. One study (study population ages 48–92 years) found a significant reduction in the likelihood of having tinnitus with increases in age (OR 0.84, 95% CI 0.78–0.90) (Nondahl et al., 2002). The other study (study population ages 55 years and older) found no significant difference in tinnitus prevalence by age (Sindhusake et al., 2003b).

TINNITUS AND NOISE EXPOSURE

Tinnitus may occur following a single exposure to high-intensity impulse noise, long-term exposure to repetitive impulses, long-term exposure to continuous noise, or exposure to a combination of impulses and continuous noise (Loeb and Smith, 1967; Chermak and Dengerink, 1987; Metternich and Brusis, 1999; Temmel et al., 1999; Stankiewicz et al., 2000; Mrena et al., 2002). To assess the association between tinnitus and noise exposure, the committee reviewed reports from studies of the general population, workers in high-noise environments, military personnel, and persons who experienced acoustic trauma. These studies are described in the next several sections. The discussion concludes with summary observations drawn from across these study populations and the committee's findings regarding tinnitus and exposure to noise in general, as well as impulse noise in particular.

Tinnitus and Noise Exposure in the General Population

Population-based studies provide a mixed picture of the association between noise exposure and tinnitus. In such studies, quantitative data on noise doses or noise levels are not available. Noise exposure is commonly represented by indicators such as occupation or qualitative characterizations of the level of workplace noise.

A multivariate analysis of U.S. data from a special supplement to the 1994 National Health Interview Survey found that veterans of military service had an increased likelihood of having chronic tinnitus (has lasted for at least 3 months) compared with those who had not served in the military (OR 1.29, 95% CI 1.17–1.43) (Hoffman and Reed, 2004). In this study

TABLE 4-1 Prevalence of Tinnitus in Adults by Age Group, from Selected Studies

Study	N
U.S. Health Examination Survey, 1960–1962	6,672
National Study of Hearing, United Kingdom, 1980–1986	34,050
Gothenberg, Sweden, 1989	2,556
U.S. National Health Interview Survey: Hearing Supplement, 1990	59,343
U.S. National Health Interview Survey: Disability Supplement, 1994–1995	99,435
Epidemiology of Hearing Loss Study, Beaver Dam, Wisconsin, 1993–1995	3,737
Nord Trøndelag Hearing Loss Study, Norway, 1996–1998	47,410
Blue Mountains Hearing Study, Australia, 1997–1999	2,015

NOTES: Definitions of tinnitus used in each study:

U.S. Health Examination Survey: Noticed ringing in the ears or bothered by other funny noises at any time over the past few years and bothered "quite a bit" by the noise (severe tinnitus). Age groups are 18–34, 35–44, 45–54, 55–64, 65–74, 75–79.

National Study of Hearing: "Prolonged spontaneous tinnitus" that lasts for more than 5 minutes and occurs not only after loud sounds.

Gothenberg, Sweden: An ear noise that occurs often or always and sounds like a peep, chirping, roaring, wind blowing in the trees, etc.

U.S. National Health Interview Survey, Hearing Supplement: Having been bothered by ringing in the ears or other funny noises in the head in the past 12 months.

population, 29 percent of veterans and 15 percent of nonveterans reported having chronic tinnitus. Similar rates were seen in data from a 1990 supplement to the National Health Interview Survey: 21 percent of veterans and 14 percent of nonveterans reported "bothersome" tinnitus (Hoffman and Reed, 2004). (Measures of the statistical significance of the difference were not presented.) In age- and gender-adjusted comparisons between occupational groups, the 1994 data showed a marginally significant elevation in the prevalence of tinnitus among skilled and unskilled workers compared with professionals (OR 1.18, 95% CI 1.00–1.39), but no significant differences between professionals and other occupational groups. These analyses of the National Health Interview Survey data used responses obtained only from direct self-reporting on tinnitus and excluded proxy responses. In the 1990 data, one-third of the reports had come from proxy respondents. The effect on the results of any selection biases associated with reporting (the respondents themselves or proxy respondents) is not known.

Prevalence (%)								
Age (years)								
20–29	30–39	40–49	50–59	60–69	70–79	≥80	≥50	Total Adult
3	4	6	9	12	11	—	10	6
6	7	10	13	16	14	14	14	10
8	6	9	19	20	21	—	20	14
5	6	7	10	13	13	14	12	8
1	2	4	6	8	9	8	8	4
—	—	—	7	10	9	6	8	—
10	10	12	17	20	24	23	20	15
—	—	—	28	33	31	25	30	—

U.S. National Health Interview Survey, Disability Supplement: Now have a ringing, roaring, or buzzing in the ears that has lasted for at least 3 months (chronic tinnitus).

Epidemiology of Hearing Loss Study: Buzzing, ringing, or noise in the ears in the past year of at least moderate severity and/or tinnitus that caused difficulty in falling asleep (significant tinnitus).

Nord Trøndelag Hearing Loss Study: Bothered by ringing in the ears.

Blue Mountains Hearing Study: Experienced any prolonged (lasting 5 minutes or longer) ringing, buzzing, or other sounds in the ears or head within the past year.

SOURCES: Roberts (1968); Sindhusake et al. (2003b); Hoffman and Reed (2004).

The data from Norway on bothersome tinnitus showed a significantly higher likelihood of having the condition among both men and women who had been exposed to loud noise at work ("difficult to have a conversation") for more than 5 hours per week (Tambs et al., 2003; Hoffman and Reed, 2004). The odds ratio for men exposed for more than 15 hours per week was 1.70 (95% CI 1.53–1.87) compared with men who had not been exposed to loud noise at work. Frequent exposure to impulse noise significantly (and similarly) increased the likelihood of having tinnitus for both men and women. For men, the odds ratio for a history of exposure to impulse noise was 1.78 (95% CI 1.61–1.96).

Tinnitus was significantly related to occupational noise exposure among working-age British men (Palmer et al., 2002). The prevalence of tinnitus was 13 percent for those who had worked in a noisy job for more than 10 years and 5 percent for men with no noise exposure: an age-adjusted rate ratio of 2.6 (95% CI 2.0–3.4). Earlier British data also showed rates of tinnitus that were about twice as high among persons reporting having worked in a noisy environment for more than 6 months (24 percent) as among those who had not worked in noise (14 percent) (Medical Research Council's Institute of Hearing Research, 1981). The pattern was similar across three broad age groups (< 40 years, 40–60 years, and > 60 years), with tinnitus rates of 22–33 percent for those who had worked in noise and 11–18 percent for those who had not.

Among the older adults in the Beaver Dam study, major occupation, history of occupational noise exposure, and hunting history were not associated with a higher likelihood of having tinnitus (prevalence) or developing tinnitus (incidence) (Nondahl et al., 2002). It is important to note that many people were retired at the time of the initial examination and therefore were no longer exposed to occupational noise. In the similarly designed Australian study, however, the prevalence of tinnitus was significantly related to the severity of work-related noise exposure (Sindhusake et al., 2003a). Compared with unexposed participants, the relative risk of having tinnitus was 1.39 (95% CI 1.13–1.72) for participants exposed to "tolerable noise" and 1.53 (95% CI 1.13–2.06) for those exposed to high levels of occupational noise ("unable to hear speech"). Similarly, a Swedish study of two cohorts of older men (Rosenhall and Karlsson, 1991) found a significant association at age 70 years between tinnitus and more than 10 years of occupational noise exposure.

Tinnitus and Noise Exposure in Industrial Populations

Studies of industrial populations offer an opportunity to quantify noise levels in the workplace and in some cases to estimate workers' noise doses. Three studies drew on large datasets from audiometric surveillance programs. Only one small study reported a comparison with workers who had not been exposed to noise. Some studies excluded workers with other risk factors for tinnitus, including evidence of ear disease.

Among 110,647 noise-exposed Austrian industrial workers, 7 percent reported having tinnitus (Neuberger et al., 1992). Workers with a history of ear disease and other risk factors for tinnitus were included in the analysis. Minimum noise exposure was 4-hour daily exposure to levels greater than 85 dBA for at least 6 months. Median noise levels were 90 dBA, with 6 percent of workers exposed to levels higher than 100 dBA. No association was found between the prevalence of tinnitus and the type of noise exposure (i.e., continuous versus impact/impulse noise). Although 7 percent of the workers were described as exposed to impact noise, it was noted that impact noise may not have been distinguished from continuous noise if unweighted peak levels were less than 145 dB (Neuberger et al., 1992).

Two other analyses of audiometric surveillance records excluded workers with a history of ear disease or head injury. A group of 33,168 workers in British Columbia were given annual evaluations because their noise exposure was an 8-hour time-weighted average of 85 dBA or more (Chung et al., 1984). The prevalence of tinnitus was 7 percent. Among 38,725 noise-exposed workers in Australia, 18 percent reported experiencing tinnitus and 10 percent responded that they might have tinnitus (Gabriels et al., 1996). The workers were exposed to 8-hour-equivalent noise doses of 90 dBA or greater or peak noise levels of 140 dB (lin) or higher.

A retrospective review of records for a group of 138 men monitored in a steel foundry's hearing conservation program showed that 28 percent reported experiencing tinnitus at least once at an annual audiometric assessment over a 20-year period (Griest and Bishop, 1996). At the time the study was conducted, 17 percent reported experiencing tinnitus at least several times a week and 39 percent reported never experiencing tinnitus. The noise exposures for these workers were 8-hour time-weighted averages of 85–101 dBA for at least 10 years.

High rates of tinnitus were seen in at least two studies of workers in industries with high levels of impact noise. Few workers in these two study populations used hearing protection devices. In a group of 88 Egyptian forge hammering workers, whose duration of noise exposure ranged from 9 to 25 years, 77 (88 percent) reported having tinnitus (Kamal et al., 1989). The background noise level was 90 to 94 dBA. The median peak hammer noise ranged from 112 to 139 dBA, with an irregular pattern of 20 to 50 impacts per minute. In a group of 261 Polish drop-forge operators who had been employed an average of 10 years, 70 percent reported having tinnitus (Sulkowski et al., 1999). The peak sound pressure level (SPL) was 135 dB SPL, with an average of 20 impulses per minute. This study included a comparison with an unexposed age-matched control group from the same facility. Among the controls, only 4 percent had tinnitus.

Tinnitus and Noise Exposure in Military Populations

Only a few studies have reported on the prevalence of tinnitus in samples of military populations that might be considered representative, and information on noise exposures is limited. In one such study of a random sample of 2,200 Israeli soldiers, 14 percent reported having tinnitus (Attias et al., 2002), but no information was available on individuals' levels of noise exposure. Among the 204 officers in a Swedish infantry regiment, 17 percent reported experiencing tinnitus (Christiansson and Wintzell, 1993). The rate ranged from 11 percent among the officers 30 years of age and younger to 24 percent among those over age 50. The prevalence of tinnitus was 26 percent among those who had been exposed to heavy-weapons fire (e.g., mortars, recoilless rifles) and significantly higher than the rate of 5 percent among those exposed to gunfire from only smaller arms (Christiansson and Wintzell, 1993). The peak sound pressure level for a typical smaller weapon was about 155–160 dB SPL and up to 185 dB SPL for heavier weapons. The amount of exposure to weapons fire was not described. The authors also cautioned that the association between tinnitus and heavy weapons fire might be affected by recall bias.

In an age-stratified random sample of 699 Finnish Army officers, 34 percent reported experiencing "occasional" tinnitus and 9 percent continuous tinnitus (Ylikoski and Ylikoski, 1994). Tinnitus was significantly correlated with exposure to a greater number of weapons impulses (from weapons fired either personally or by others) and, in contrast to the Swedish study, more strongly correlated with exposure to impulses from small-caliber than large-caliber weapons (Ylikoski and Ylikoski, 1994). The scale of the exposure was not directly specified, but it was estimated to be more than 200,000 impulses for at least some officers.

The committee found no published studies on tinnitus and noise exposure among U.S. military personnel. A study of hearing loss among Navy enlisted personnel (n = 3,530), however, did find that 7–10 percent of the subjects reported having or having had "head noises" (Robertson et al., 1978).

Unpublished data on noise exposure and tinnitus reported by U.S. military personnel on post-deployment health assessment questionnaires (Department of Defense [DD] Form 2796) were provided to the committee by the Army Medical Surveillance Activity (AMSA, 2004).[3] The data were for responses to questions on exposure to loud noise during deployment (response options: no, sometimes, often) and whether the respondent developed "ringing in the ears" during the deployment, is currently experiencing it, or both (AMSA, 2004; Rubertone, 2004, 2005). The questionnaires are completed by officers and enlisted personnel from each of the services within a period from 30 days before to 30 days after leaving a deployment location (Rubertone, 2004). The data provided to the committee were for questionnaires processed from June 2003 through October 2004.

Overall, 11 percent of those submitting forms (n = 440,451) reported developing tinnitus during deployment, experiencing it at the time they completed the form, or both (see Table 4-2). Among those who reported they were often exposed to loud noise, nearly 22 percent reported experi-

[3]For the post-deployment health assessment questionnaire, deployment refers to postings outside the United States in support of military operations.

TABLE 4-2 Percentage of U.S. Military Personnel Completing Post-Deployment Health Assessment Questionnaires Who Reported Tinnitus, by Reported Exposure to Loud Noise During Deployment, 2003–2004

		Exposure to Loud Noise During Deployment		
Reports on Ringing in Ears	Total (*n* = 440,451)	No (*n* = 159,725)	Sometimes (*n* = 120,928)	Often (*n* = 159,798)
No ringing	89	97	91	78
Ringing in ears				
Developed during deployment	8	2	6	15
Now*	2	1	2	4
During deployment and now	1	0	1	3
Total	100%	100%	100%	100%

* "Now" refers to the time at which the questionnaire was completed (within 30 days before or after the end of deployment).
SOURCE: AMSA (2004).

encing tinnitus. By comparison, only 3 percent of those who reported not being exposed to loud noise during deployment reported experiencing any tinnitus. Most of the tinnitus reports were for cases that developed during deployment but did not persist. It may be noteworthy that 4 percent of those who were often exposed to loud noise reported tinnitus "now" (within 30 days before or after the end of deployment), but not during deployment. Among those who had tinnitus at the time they completed the questionnaire, rates remained highest in the group that reported often being exposed to loud noise.

With these data, it was also possible to examine differences among the services in reports of tinnitus (see Table 4-3). The percentages reporting having developed tinnitus during deployment or having it at the time the questionnaire was completed were highest for Army personnel and lowest for Air Force personnel, regardless of reported noise exposure. The percentages for Marine Corps personnel were higher than those for Navy personnel, except for the groups reporting no exposure to loud noise. The patterns across the services and categories of noise exposure persisted when only reports of tinnitus present at the time the questionnaire was completed were considered (see Table 4-4). No statistical analyses of these data were available nor is it known if the proportion of service members completing the survey differed across service branches.

TABLE 4-3 Percentage of U.S. Military Personnel Completing Post-Deployment Health Assessment Questionnaires Who Reported Any Tinnitus, by Military Service and Reported Exposure to Loud Noise During Deployment, 2003–2004

Exposure to Loud Noise During Deployment	Army (%)	Navy (%)	Air Force (%)	Marine Corps (%)
No	4	2	1	1
Sometimes	12	5	3	8
Often	26	12	8	22
Total reporting any tinnitus	14	6	4	12

SOURCE: AMSA (2004).

TABLE 4-4 Percentage of U.S. Military Personnel Completing Post-Deployment Health Assessment Questionnaires Who Reported Current Tinnitus, by Military Service and Reported Exposure to Loud Noise During Deployment, 2003–2004

Exposure to Loud Noise During Deployment	Army (%)	Navy (%)	Air Force (%)	Marine Corps (%)
No	1	1	< 1	< 1
Sometimes	3	1	1	2
Often	9	4	3	6
Total reporting current tinnitus	5	2	1	3

NOTE: "Current tinnitus" refers to reports of having tinnitus at the time the questionnaire was completed (within 30 days before or after the end of deployment).
SOURCE: AMSA (2004).

Tinnitus and Acoustic Trauma

Exposure to gunfire and explosive detonations may cause acoustic trauma accompanied by tinnitus. A study of 83 survivors of the 1995 Oklahoma City bombing found that 67 percent developed tinnitus or perceived a change in existing tinnitus within days of the blast and that 59 percent had tinnitus at an initial evaluation completed within 5 months of the blast (Van Campen et al., 1999). Other cross-sectional studies have reported on tinnitus among soldiers treated for acoustic trauma. In a group of 81 Austrian soldiers, 84 percent had tinnitus (Temmel et al., 1999). For a group of Israeli soldiers, each ear was analyzed independently, and tinnitus was reported for 62 percent of the patients' ears (Melinek et al., 1976). Their ears were evaluated separately because of differences between ears in the degree of acoustic trauma. In another group of Israeli soldiers who were treated for acoustic trauma, 81 of the 102 patients (79 percent) reported having tinnitus (Man and Naggan, 1981). Patients with a history of ear disease were excluded in both Israeli studies. The Oklahoma City study recruited participants from the community without regard to their audiological status. The studies of military personnel were based on patient populations, which may mean that the study subjects had more severe problems than others who were not treated.

Summary of Evidence on Tinnitus and Noise Exposure

Summarized here is the evidence from the reports and data described above on tinnitus and noise exposure in general. This is followed by a closer look at impulse noise in particular.

Noise Exposure and Noise Levels

Several cross-sectional studies of community, industrial, and military populations show that tinnitus rates are significantly higher for subjects with longer exposure to occupational noise or exposure to higher levels of occupational noise. Two cross-sectional, community-based studies produced estimates that occupational noise exposure was associated with a 40–70 percent increase in the likelihood of having tinnitus (Sindhusake et al., 2003a; Hoffman and Reed, 2004). However, in one study of older adults, occupational noise exposure was not associated on a cross-sectional or longitudinal basis with an greater likelihood of having or developing tinnitus (Nondahl et al., 2002). Several cross-sectional studies (Medical Research Council's Institute of Hearing Research, 1981; Rosenhall and Karlsson, 1991; Sulkowski et al., 1999; Palmer et al., 2002), as well as the data provided to the committee on post-deployment health assessments (AMSA, 2004; Rubertone, 2004, 2005), showed higher rates of tinnitus among those reporting longer exposure to occupational noise or exposure to higher levels of occupational noise, but provided no statistical estimates of the effect on the likelihood of having tinnitus. The studies of larger samples of industrial workers provided no comparisons between exposed and unexposed workers, making it impossible to evaluate the effect of the noise exposure on tinnitus (Chung et al., 1984; Neuberger et al., 1992; Gabriels et al., 1996).

Conclusions about tinnitus prevalence rates or the likelihood of having tinnitus in noise-exposed populations are not possible because of differences in the characteristics of the study populations and the definitions of tinnitus.

Detailed measurements of noise exposures are available for some industrial workers, but in other studies, data on noise exposures are indirect and often retrospective and qualitative. Retrospective estimates of noise exposure are subject to recall bias and other measurement problems that allow for possible misclassification of participants' exposures. The studies reviewed also vary in their exclusion criteria, with some including and others excluding subjects who had a history of ear disease or other conditions that have an association with tinnitus that is independent of noise exposure. Prospective studies that can monitor noise exposures, the presence of other tinnitus risk factors, and the onset of tinnitus are needed to develop estimates of the risk of tinnitus associated with different levels of noise exposure.

FINDING: The evidence is sufficient to conclude that noise doses associated with hearing loss are likely to be associated with tinnitus.

FINDING: The evidence was not sufficient to reach conclusions regarding the specific number or proportion of service members, overall or in specific branches or occupational groups, who report that tinnitus began or was exacerbated by noise exposure during military service.

Impulse Noise

It has been suggested that exposure to impulse noise increases the risk of developing tinnitus. The data reviewed by the committee are mixed. One large cross-sectional, community-based study found a significant increase in the likelihood of having tinnitus among those who reported exposure to impulse noise (Hoffman and Reed, 2004). Among a population of older adults, however, a history of hunting was not associated with the differences in prevalence or incidence of tinnitus (Nondahl et al., 2002). Small studies of workers with high exposure to impact noise showed a high prevalence of tinnitus (70 percent and 88 percent) (Kamal et al., 1989; Sulkowski et al., 1999). One study included a comparison showing low tinnitus prevalence (4 percent) in an age-matched comparison group not exposed to noise (Sulkowski et al., 1999). However, a large study using data from a national hearing conservation database found no significant difference in tinnitus associated with exposure to impulse or continuous noise (Neuberger et al., 1992). In small cross-sectional studies of military personnel exposed to impulse noise from weapons (Christiansson and Wintzell, 1993; Ylikoski and Ylikoski, 1994), the prevalence of tinnitus was lower than the rates seen in the two small studies of workers. Furthermore, the associations between tinnitus and exposure to weapons fire from smaller and larger weapons were inconsistent, and comparisons were not made with groups not exposed to weapons fire.

High rates of tinnitus among acoustic trauma patients suggest that impulse/impact noise is likely to precipitate tinnitus, but these studies may not be an appropriate basis for judging the magnitude of the tinnitus hazard associated with noise exposures that do not produce acoustic trauma. Individuals who are being treated for acoustic trauma may not be comparable to individuals who experienced acoustic trauma without receiving treatment or individuals exposed to the same or similar noise who did not experience acoustic trauma. Finally, assessment of the data is also hampered by limited information about differences among the study populations in their use of hearing protection, as it is in some studies of noise-induced hearing loss.

FINDING: There is limited or suggestive evidence that exposure to impulse noise is associated with a greater likelihood of having tinnitus compared with exposure to steady-state noise.

TINNITUS AND HEARING LOSS

The committee reviewed studies of national or community populations and military populations that assessed the relation between tinnitus and hearing loss. Some analyses determined hearing loss on the basis of hearing thresholds measured in audiometric testing, but others relied on qualitative assessments of hearing derived from self-reports of hearing difficulties and use of hearing aids. The qualitative assessments of hearing problems may not be sensitive to hearing loss that is limited to high frequencies, which may be the case for noise-induced hearing loss.

Several studies in community, industrial, and military populations have found varying rates of tinnitus among individuals with better and worse hearing, but the prevalence of tinnitus was consistently higher with some degree of hearing loss than with normal hearing (e.g., Salmivalli, 1967; Roberts, 1968; Parving et al., 1993; Ylikoski and Ylikoski, 1994; Coles, 1996; Gabriels et al., 1996; Griest and Bishop, 1996; Melinek et al., 1976; Attias et al., 2002; Palmer et al., 2002; Hoffman and Reed, 2004), except in a small study of acoustic trauma patients, which included six individuals who had tinnitus and normal hearing (Temmel et al., 1999).

Estimates of the association between the prevalence of tinnitus and measured hearing loss are also available. Analysis of data from the community-based Norwegian study from the mid-1990s showed that adults with hearing loss had a substantially greater likelihood of having tinnitus compared with those with normal hearing (Hoffman and Reed, 2004). With pure-tone averages > 25 dB HL and ≤ 40 dB HL (average threshold in the worse ear for frequencies 500, 1000, 2000, and 4000 Hz), the odds ratio for the presence of tinnitus was 2.84 (95% CI 2.55–3.16) for men and

2.78 (2.45–3.15) for women. With pure-tone averages > 40 dB HL, the odds ratio for the presence of tinnitus increased to 4.18 (95% CI 3.66–4.77) for men. For women with this degree of hearing loss, the odds ratio reached 5.40 (95% CI 4.67–6.24). In the National Study of Hearing in the United Kingdom in the 1980s, the likelihood of having tinnitus among those with high-frequency hearing thresholds > 80 dB HL was 27 times higher than for those with thresholds < 10 dB HL (Coles, 1996).

The Beaver Dam study (Nondahl et al., 2002) found that 12 percent of older adults with a pure-tone hearing loss (defined as average threshold in the worse ear > 25 dB HL for the same frequencies as the Norwegian study) had tinnitus, compared with 5 percent of those with normal hearing. Pure-tone hearing loss emerged among the factors tested as having the strongest association with having tinnitus (OR = 3.90; 95% CI 2.89–5.27). Hearing loss also significantly increased the risk of developing tinnitus during a 5-year follow-up period (OR = 1.83; 95% CI 1.21–2.75). A similar study of older adults in Australia found that tinnitus was reported by 35 percent of persons with a hearing loss (defined as in the U.S. study) and 27 percent of those with normal hearing (Sindhusake et al., 2003b, 2004). In contrast to the U.S. study, however, incrementally worse hearing had only a modest association with the prevalence of tinnitus (relative risk 1.11; 95% CI 1.04–1.17).

Qualitative reports of hearing difficulties are less precise but have still shown a consistent association with higher rates of tinnitus (e.g., Parving et al., 1993; Palmer et al., 2002). In data from the 1994 National Health Interview Survey, the likelihood of having tinnitus was about six times higher for those reporting either that they could not hear normal conversation or that they could not hear loud noises than for those with no or only slight trouble hearing (Hoffman and Reed, 2004).

In sum, at least four cross-sectional studies in the general population show that measured or perceived hearing loss is associated with a higher prevalence of tinnitus (Coles, 1996; Nondahl et al., 2002; Sindhusake et al., 2003b, 2004; Hoffman and Reed, 2004). Many other epidemiological studies show a higher prevalence of tinnitus for those with measured or perceived hearing loss than for those without hearing loss (e.g., Salmivalli, 1967; Roberts, 1968; Melinek et al., 1976; Parving et al., 1993; Ylikoski and Ylikoski, 1994; Coles, 1996; Gabriels et al., 1996; Griest and Bishop, 1996; Attias et al., 2002; Palmer et al., 2002; Hoffman and Reed, 2004). One population-based longitudinal study of older adults also showed that an existing hearing loss increased the risk for developing tinnitus over a 5-year period (Nondahl et al., 2002). Conclusions about the strength of the association between hearing loss and tinnitus prevalence rates are not possible because of differences in the characteristics of the study populations and the definitions of hearing loss and tinnitus. Additional prospective studies, including studies of younger adults, are needed to learn more about the relation between hearing loss and the incidence of tinnitus.

FINDING: The evidence is sufficient to conclude that hearing loss (hearing thresholds greater than 25 dB HL at one or more audiometric frequencies between 250 and 8000 Hz) is associated with a higher prevalence of tinnitus.

FINDING: The evidence is not sufficient to determine precisely the magnitude of the risk of tinnitus associated with hearing loss.

OTHER RISK FACTORS

Clinical and epidemiological studies have shown that tinnitus is associated with exposure to a variety of factors in addition to noise and hearing loss. The committee identified one study that examined the combined effect of noise exposure and smoking on the prevalence of tinnitus (Palmer et al., 2004). In this British postal survey (n = 10,418), smoking did not have a significant effect on the prevalence of persistent tinnitus (present "most or all of the time in the past 12 months") after age and duration of occupational noise exposure were taken into account. Other studies have examined tinnitus and smoking without specific consideration of possible interactions with noise exposure. Some found no association between smoking and tinnitus (Chung et al., 1984; Nondahl et al., 2002; Sindhusake et al., 2003a), but a multivariate analysis of data for Norwegian adults found a significantly higher prevalence of tinnitus for both men and women who had smoked for at least 5–15 years, compared with those who had never smoked (Hoffman and Reed, 2004).

Studies of the effect of noise in combination with other factors on the prevalence of tinnitus were not identified. Reviewed briefly here are findings on the association between tinnitus and factors other than noise and hearing loss. The statistically significant associations are summarized in Table 4-5.

A prospective study of the incidence of tinnitus in older adults found a substantial increase in risk associated with otosclerosis and more modest increases in risk associated with a history of head injury (Nondahl et al., 2002). In the same study, higher concentrations of serum cholesterol were associated with a marginally significant increase in the risk of tinnitus. Cross-sectional, population-based studies have found statistically significant associations with a higher prevalence of tinnitus for a history of head injury (Nondahl et al., 2002; Hoffman and Reed, 2004), severe neck injury (Sindhusake et al., 2003a), and cardiovascular disease (Nondahl et al.,

TABLE 4-5 Health and Socioeconomic Factors Associated with a Significant Increase or Decrease in Incidence or Prevalence of Tinnitus

Significant Risk Factors	Incidence of Tinnitus: Epidemiology of Hearing Loss Study, Beaver Dam, Wisconsin OR (95% CI)
Otosclerosis	8.85 (1.42–55.14)
Total serum cholesterol	1.042 (1.004–1.080)
Cardiovascular disease	
History of head injury	1.55 (1.08–2.22)
Hospitalized for head injury	
History of middle ear infection	
History of sinus infection	
History of severe neck injury	
History of migraine	
Self-reported health status:	
Excellent	
Very good	
Good	
Fair	
Poor	
Smoking	
Never	
0 to < 5 yrs	
≥ 5 and < 15 yrs	
≥ 15 yrs	
Annual income	
$0–9,999	
$10,000–19,999	
$20,000–34,999	
$35,000–49,999	
$50,000+	
Residence	
Northeast	
Midwest	
South	
West	

NOTES: OR = odds ratio, RR = relative risk, CI = confidence interval.
SOURCES: Nondahl et al. (2002); Sindhusake et al. (2003a); Hoffman and Reed (2004).

Prevalence of Tinnitus			
U.S. National Health Interview Survey: Disability Supplement OR (95% CI)	Nord Trøndelag Hearing Loss Study, Norway (men only) OR (95% CI)	Epidemiology of Hearing Loss Study, Beaver Dam, Wisconsin OR (95% CI)	Blue Mountains Hearing Study, Australia RR (95% CI)
		1.45 (1.05–2.00)	
		1.56 (1.21–2.02)	
	1.43 (1.26–1.62)		
			1.35 (1.05–1.73)
			1.30 (1.08–1.57)
			1.34 (1.04–1.73)
			1.28 (1.06–1.56)
0.58 (0.52–0.65)			
0.79 (0.72–0.87)	0.78 (0.68–0.88)		
1.00	1.00		
1.63 (1.46–1.80)	1.28 (1.14–1.43)		
2.36 (2.10–2.65)	1.50 (1.04–2.16)		
	1.00		
	1.26 (1.00–1.58)		
	1.22 (1.07–1.38)		
	1.19 (1.09–1.30)		
1.56 (1.36–1.78)			
1.22 (1.08–1.39)			
1.20 (1.06–1.36)			
1.05 (0.92–1.20)			
1.00			
1.00			
1.20 (1.03–1.39)			
1.16 (1.00–1.35)			
1.48 (1.26–1.72)			

2002). Middle ear infection, sinus infection, and migraine have also been associated with higher rates of tinnitus in older adults (Sindhusake et al., 2003a). In cross-sectional studies of noise-exposed workers, higher rates of tinnitus have been found among subjects with a history of head injury or ear diseases compared to those without a history of those conditions (e.g., Neuberger et al., 1992; Phoon et al., 1993).

Fair or poor self-reported health status was associated with a significantly higher prevalence of tinnitus compared with reported good health (Hoffman and Reed, 2004). The same analyses found a significant reduction in tinnitus risk for persons reporting very good or excellent health. Clinical evidence shows that certain medications can induce tinnitus, but at the population level, studies of older adults did not show a significantly higher prevalence of tinnitus associated with past exposure to ototoxic medications (Sindhusake et al., 2003a) or regular use of aspirin (325 mg at least twice a week for more than 3 months) (Nondahl et al., 2002; Cruickshanks, 2005).

In an analysis of the sample of National Health Interview Survey data, modest increases in the likelihood of having tinnitus were associated with lower income and residence in other regions compared to the Northeast. Education level was not associated with significant differences in tinnitus prevalence (Hoffman and Reed, 2004).

TINNITUS AND U.S. MILITARY PERSONNEL

Tinnitus that is determined to have been incurred during or aggravated by military service ("service connected") is compensable as a disability by the Department of Veterans Affairs. At the end of fiscal year 2003, tinnitus was the third most common disability, with 242,610 cases among veterans receiving compensation for any form of service-connected disability (Veterans Benefits Administration, 2004a). Until May 1999, it was necessary to establish that persistent tinnitus had resulted from service-connected head injury, concussion, or acoustic trauma (Henry et al., 2004; Veterans Benefits Administration, 2004b). Since then, recurrent tinnitus associated with any condition incurred in or aggravated by military service has been compensable (Henry et al., 2004; Veterans Benefits Administration, 2004b).

Despite the fact that tinnitus is compensable, the committee found little indication that the services monitor the presence or absence of tinnitus among military personnel during active duty. The forms that summarize service members' medical histories (DD Form 2807-1, DD Form 2807-2, Standard Form [SF] 93) and the results of their periodic physical examinations (DD Form 2808, SF 88) capture information on hearing, including pure-tone thresholds and use of hearing aids, but do not include questions about tinnitus. Similarly, the forms used by the services' hearing conservation programs (see Chapter 5 and Appendix G) to record the results of audiometric testing and reports on the use and distribution of hearing protection devices include no question about the presence or absence of tinnitus. However, the opportunity does exist for reports of tinnitus volunteered by service members or solicited by medical or audiology personnel to be recorded on these forms.

Perhaps the only current source of limited but explicit documentation of tinnitus is the post-deployment health assessment questionnaire (DD Form 2796). A requirement for post-deployment health assessments was established in 1997 (DoD, 1997), and the current version of the assessment form with the questions on noise exposure and tinnitus dates from April 2003. Only personnel who are deployed would have the opportunity to complete the form.

Without systematic documentation of the tinnitus status of military personnel, it is not possible to determine whether service members have tinnitus at the time they enter or leave military service or when during military service tinnitus might have developed. In the absence of systematic information about tinnitus, it is not possible for the services to identify tinnitus hazards that might be different from those for noise-induced hearing loss.

The committee strongly suggests that the Department of Defense add monitoring of tinnitus to both its routine health surveillance and the audiologic surveillance that is part of the services' hearing conservation programs. One approach might be the addition of questions to the forms used to report medical history (DD Form 2807-1) or the results of medical examinations (DD Form 2808) and to the forms used by the hearing conservation programs to record reference audiograms (DD Form 2215) and periodic and termination audiograms (DD Form 2216). Tinnitus surveillance might begin with audiologic assessments at entry into and separation from military service. Basic questions, such as "Do you have any ringing or buzzing in your ears or head?" and "When did you first notice the ringing or buzzing?" could be considered as a way to begin to obtain information about tinnitus during military service.

At this time, however, there is no generally accepted and validated question, or set of questions, that can be recommended for use in tinnitus surveillance. Factors that may deserve consideration in defining tinnitus for a monitoring program include persistence of episodes (e.g., longer than 5 minutes), regularity of occurrence (e.g., perceived all the time, perceived several times per week, perceived only after exposure to loud noise), and level of annoyance. The Department of Defense and the military services should explore the merits of various approaches to tinnitus surveillance.

Tinnitus and hearing loss are both associated with exposure to high levels of noise. Although the factors that result in tinnitus are not as easy to predict as those that result in hearing loss, doses of noise (and risk factors) that are considered hazardous for hearing may also be tinnitus hazards. Consequently, the precautions regarding noise-exposure limits and the use of hearing protection may help prevent tinnitus as well as hearing loss. The scope of the hearing conservation programs of the military services could appropriately be broadened to encompass the prevention of tinnitus as well as hearing loss—thus "hearing loss and tinnitus prevention programs."

REFERENCES

Adams PF, Hendershot GE, Marano MA. 1999. Current estimates from the National Health Interview Survey, 1996. *Vital and Health Statistics*, Series 10, No. 200. Hyattsville, MD: National Center for Health Statistics.

Aggleton JP, Mishkin M. 1986. The amygdala: Sensory gateway to the emotions. In: Pultchik R, Kellerman H, eds. *Emotion, Theory, Research, and Experience*. Vol. 3. New York: Academic Press. Pp. 281–289.

Alberti PW. 1987. Tinnitus in occupational hearing loss: Nosological aspects. *Journal of Otolaryngology* 16(1):34–35.

AMSA (Army Medical Surveillance Activity). 2004. Responses to Selected Questions on DD 2796 Post Deployment Health Assessments, Since 01-Jun-2003. Data provided to the Institute of Medicine Committee on Noise-Induced Hearing Loss and Tinnitus Associated with Military Service from World War II to the Present, Washington, DC.

Andersson G, Vretblad P, Larsen HC, Lyttkens L. 2001. Longitudinal follow-up of tinnitus complaints. *Archives of Otolaryngology–Head and Neck Surgery* 127(2):175–179.

Anonymous. 1981. Definition and classification of tinnitus. In: Evered D, Lawrenson G, eds. *Tinnitus*. Ciba Foundation Symposium 85. London, England: Pitman Books Ltd. Pp. 300–302.

Attias J, Reshef I, Shemesh Z, Salomon G. 2002. Support for the central theory of tinnitus generation: A military epidemiological study. *International Journal of Audiology* 41(5):301–307.

Axelsson A, Barrenas M-L. 1992. Tinnitus in noise-induced hearing loss. In: Dancer AL, Henderson D, Salvi RJ, Hamernik RP, eds. *Noise-Induced Hearing Loss*. St. Louis, MO: Mosby Year Book. Pp. 269–276.

Axelsson A, Prasher D. 2000. Tinnitus induced by occupational and leisure noise. *Noise Health* 2(8):47–54.

Axelsson A, Ringdahl A. 1989. Tinnitus—A study of its prevalence and characteristics. *British Journal of Audiology* 23(1):53–62.

Axelsson A, Sandh A. 1985. Tinnitus in noise-induced hearing loss. *British Journal of Audiology* 19(4):271–276.

Baguley DM. 2002. Mechanisms of tinnitus. *British Medical Bulletin* 63:195–212.

Bauer CA. 2003. Animal models of tinnitus. *Otolaryngology Clinics of North America* 36(2):267–285.

Bauer CA, Brozoski TJ. 2001. Assessing tinnitus and prospective tinnitus therapeutics using a psychophysical animal model. *Journal of the Association for Research in Otolaryngology* 2:54–64.

Bauer CA, Brozoski TJ, Rojas R, Boley J, Wyder M. 1999. Behavioral model of chronic tinnitus in rats. *Otolaryngology–Head and Neck Surgery* 121:457–462.

Cacace A. 2003. Expanding the biologic basis of tinnitus: Crossmodal origins and the role of neuroplasticity. *Hearing Research* 175:112–132.

Cacace AT. 2004. The limbic system and tinnitus. In: Snow JB, Jr., ed. *Tinnitus: Theory and Management*. Hamilton, Ontario, Canada: BC Decker. Pp. 162–170.

Cahani M, Paul G, Shahar A. 1983. Tinnitus pitch and acoustic trauma. *Audiology* 22(4):357–363.

Chermak GD, Dengerink JE. 1987. Characteristics of temporary noise-induced tinnitus in male and female subjects. *Scandinavian Audiology* 16(2):67–73.

Christiansson BA, Wintzell KA. 1993. An audiological survey of officers at an infantry regiment. *Scandinavian Audiology* 22(3):147–152.

Chung DY, Gannon RP, Mason K. 1984. Factors affecting the prevalence of tinnitus. *Audiology* 23(5):441–452.

Coles RR. 1984. Epidemiology of tinnitus: (1) Prevalence. *Journal of Laryngology and Otology* 9(Suppl.):7–15.

Coles RRA. 1996. Epidemiology, aetiology and classification. In: Reich GE, Vernon JA, eds. *Proceedings of the Fifth International Tinnitus Seminar 1995*. Portland, OR: American Tinnitus Association.

Cruickshanks K. 2005. Comments. E-mail to J. Durch, Institute of Medicine, June 13.

Dauman R. 2000. Electrical stimulation for tinnitus suppression. In: Tyler RS, ed., *Tinnitus Handbook*. San Diego, CA: Singular Publishing Group. Pp. 377–398.

Dauman R, Tyler RS. 1992. Some considerations on the classification of tinnitus. In: Aran J-M, Dauman R, eds. *Proceedings of the Fourth International Tinnitus Seminar, Bordeaux, France*. Amsterdam, The Netherlands: Kugler & Ghedini Publications. Pp. 225–229.

De Ridder D, Ryu H, De Mulder G, Van de Heyning P, Verlooy J, Moller A. 2005. Frequency specific hearing improvement in microvascular decompression of the cochlear nerve. *Acta Neurochirurgica* 147(5):495–501.

Dineen R, Doyle J, Bench J. 1997. Audiological and psychological characteristics of a group of tinnitus sufferers, prior to tinnitus management training. *British Journal of Audiology* 31(1):27–38.

Dobie RA. 1999. A review of randomized clinical trials in tinnitus. *Laryngoscope* 109(8): 1202–1211.

Dobie RA. 2003. Depression and tinnitus. *Otolaryngologic Clinics of North America* 36(2): 383–388.

DoD (Department of Defense). 1997. *Department of Defense Instruction 6490.3: Implementation and Application of Joint Medical Surveillance for Deployments*. Washington, DC: Department of Defense.

Eggermont JJ. 2003. Central tinnitus. *Auris Nasus Larynx* 30 (Suppl 1):S7–S12.

Eggermont JJ, Roberts LE. 2004. The neuroscience of tinnitus. *Trends in Neuroscience* 27:676–682.

Erlandsson SI. 2000. Psychological profiles of tinnitus patients. In: Tyler RS, ed. *Tinnitus Handbook*. San Diego, CA: Singular Thompson Learning. Pp. 25–57.

Erlandsson SI, Rubinstein B, Axelsson A, Carlsson SG. 1991. Psychological dimensions in patients with disabling tinnitus and craniomandibular disorders. *British Journal of Audiology* 25(1):15–24.

Gabriels P, Monley P, Guzeleva D. 1996. Noise-exposed workers: Is tinnitus being ignored? In: Reich GE, Vernon JA, eds. *Proceedings of the Fifth International Tinnitus Seminar 1995*. Portland, OR: American Tinnitus Association. Pp. 373–380.

George RN, Kemp S. 1989. Investigation of tinnitus induced by sound and its relationship to ongoing tinnitus. *Journal of Speech and Hearing Research* 32(2):366–372.

Griest SE, Bishop PM. 1996. Evaluation of tinnitus and occupational hearing loss based on 20-year longitudinal data. In: Reich GE, Vernon JA, eds. *Proceedings of the Fifth International Tinnitus Seminar 1995*. Portland, OR: American Tinnitus Association. Pp. 381–394.

Halford JB, Anderson SD. 1991. Anxiety and depression in tinnitus sufferers. *Journal of Psychosomatic Research* 35(4-5):383–390.

Hazell JWP. 1995. Models of tinnitus: Generation, perception, clinical implications. In: Vernon JA, Møller AR, eds. *Mechanisms of Tinnitus*. Boston: Allyn and Bacon. Pp. 57–72.

Hazell JWP, Wood SM. 1981. Tinnitus masking—a significant contribution to tinnitus management. *British Journal of Audiology* 15(4):223–230.

Heffner HE, Harrington IA. 2002. Tinnitus in hamsters following exposure to intense sound. *Hearing Research* 170:83–95.

Henry JA. 2004. Audiological assessment. In: Snow JB, Jr., ed. *Tinnitus: Theory and Management*. Hamilton, Ontario, Canada: B.C. Decker. Pp. 220–236.

Henry JA, Schechter MA, Regelein RT, Dennis KC. 2004. Veterans and tinnitus. In: Snow JB, Jr., ed. *Tinnitus: Theory and Management*. Hamilton, Ontario, Canada: B.C. Decker. Pp. 337–355.

Hoffman HJ, Reed GW. 2004. Epidemiology of tinnitus. In: Snow JB, Jr., ed. *Tinnitus: Theory and Management*. Hamilton, Ontario, Canada: B.C. Decker. Pp. 16–41.

Jacobson GP, McCaslin DL. 2001. A search for evidence of a direct relationship between tinnitus and suicide. *Journal of the American Academy of Audiology* 12(10):493–496.

Jastreboff PJ, Jastreboff MM. 2003. Tinnitus retraining therapy for patients with tinnitus and decreased sound tolerance. *Otolaryngology Clinics of North America* 36(2):321–336.

Jastreboff PJ, Brennan JF, Coleman JK, Sasaki CT. 1988a. Phantom auditory sensation in rats: An animal model for tinnitus. *Behavioral Neuroscience* 102:811–822.

Jastreboff PJ, Brennan JF, Sasaki CT. 1988b. An animal model for tinnitus. *Laryngoscope* 98: 280–286.

Kaltenbach JA, Zacharek MA, Zhang, Frederick S. 2004a. Activity in the dorsal cochlear nucleus of hamsters previously tested for tinnitus following intense tone exposure. *Neuroscience Letters* 355:121–125.

Kaltenbach JA, Zhang J, Zacharek MA. 2004b. Neural correlates of tinnitus. In: Snow JB, Jr., ed. *Tinnitus: Theory and Management*. Hamilton, Ontario, Canada: B.C. Decker. Pp. 141–160.

Kaltenbach JA, Zhang J, Finlayson P. 2005. Tinnitus as a plastic phenomenon and its possible neural underpinnings in the dorsal cochlear nucleus. *Hearing Research* 206:200–226.

Kamal AA, Mikael RA, Faris R. 1989. Follow-up of hearing thresholds among forge hammering workers. *American Journal of Industrial Medicine* 16(6):645–658.

Kim J, Morest DK, Bohne BA. 1997. Degeneration of axons in the brainstem of the chinchilla after auditory stimulation. *Hearing Research* 103:169–191.

Kleinjung T, Eichhammer P, Langguth B, Jacob P, Marienhagen J, Hajak G, Wolf SR, Strutz J. 2005. Long-term effects of repetitive transcranial magnetic stimulation (rTMS) in patients with chronic tinnitus. *Otolaryngology–Head and Neck Surgery* 132:566–569.

Kuk FK, Tyler RS, Russell D, Jordan H. 1990. The psychometric properties of a tinnitus handicap questionnaire. *Ear and Hearing* 11(6):434–445.

LeDoux JE. 2000. Emotion circuits in the brain. *Annual Review of Neuroscience* 23:155–184.

Lobarinas E, Sun W, Cushing R, Salvi R. 2004. A novel behavioral paradigm for assessing tinnitus using schedule-induced polydipsia avoidance conditioning (SIP-AC). *Hearing Research* 190(1-2):109–114.

Lockwood AH, Salvi RJ, Burkhard RF. 2002. Tinnitus. *New England Journal of Medicine* 347(12):904–910.

Loeb M, Smith RP. 1967. Relation of induced tinnitus to physical characteristics of the inducing stimuli. *Journal of the Acoustical Society of America* 42(2):453–455.

Man A, Naggan L. 1981. Characteristics of tinnitus in acoustic trauma. *Audiology* 20(1): 72–78.

Medical Research Council's Institute of Hearing Research. 1981. Epidemiology of tinnitus. In: Evered D, Lawrenson G, eds. *Tinnitus*. Ciba Foundation Symposium 85. London, England: Pitman Books Ltd. Pp. 16–34.

Melinek M, Naggan L, Altman M. 1976. Acute acoustic trauma—a clinical investigation and prognosis in 433 symptomatic soldiers. *Israel Journal of Medical Sciences* 12(6):560–569.

Meric C, Gartner M, Collet L, Chery-Croze S. 1998. Psychopathological profile of tinnitus sufferers: Evidence concerning the relationship between tinnitus features and impact on life. *Audiology and Neuro-Otology* 3(4):240–252.

Metternich FU, Brusis T. 1999. [Acute hearing loss and tinnitus caused by amplified recreational music.] *Laryngorhinootologie* 78(11):614–619.

Moody DB. 2004. Animal models of tinnitus. In: Snow JB, Jr., ed. *Tinnitus: Theory and Management*. Hamilton, Ontario, Canada: B.C. Decker. Pp. 80–95.

Morest DK, Kim J, Potashner SJ, Bohne BA. 1998. Long-term degeneration in the cochlear nerve and cochlear nucleus of the adult chinchilla following acoustic overstimulation. *Microscopy Research and Technique* 41:205–216.

Mrena R, Savolainen S, Kuokkanen JT, Ylikoski J. 2002. Characteristics of tinnitus induced by acute acoustic trauma: A long-term follow-up. *Audiology and Neuro-Otology* 7(2): 122–130.

Muhlnickel W, Elbert T, Taub E, Flor H. 1998. Reorganization of auditory cortex in tinnitus. *Proceedings of the National Academy of Sciences* 95(17):10340–10343.

Neuberger M, Korpert K, Raber A, Schwetz F, Bauer P. 1992. Hearing loss from industrial noise, head injury and ear disease. A multivariate analysis on audiometric examinations of 110,647 workers. *Audiology* 31(1):45–57.

Newman CW, Jacobson GP, Spitzer JB. 1996. Development of the tinnitus handicap inventory. *Archives of Otolaryngology–Head and Neck Surgery* 122(2):143–148.

Noble W. 1998. *Self-Assessment of Hearing and Related Functions*. London, England: Whurr.

Nondahl DM, Cruickshanks KJ, Wiley TL, Klein R, Klein BE, Tweed TS. 2002. Prevalence and 5-year incidence of tinnitus among older adults: The Epidemiology of Hearing Loss Study. *Journal of the American Academy of Audiology* 13(6):323–331.

Nondahl DM, Cruickshanks KJ, Wiley TL, Klein R, Klein BE, Tweed TS. 2004. Comment on tinnitus in older adults from the Blue Mountains Study by Sindhusake et al. and comparison with tinnitus data from the Epidemiology of Hearing Loss Study. *International Journal of Audiology* 43(1):60.

NRC (National Research Council). 1982. *Tinnitus—Facts, Theories, and Treatments*. Committee on Hearing, Bioacoustics, and Biomechanics. Working Group 89 (McFadden D, chair). Washington, DC: National Academy Press.

Palmer KT, Griffin MJ, Syddall HE, Davis A, Pannett B, Coggon D. 2002. Occupational exposure to noise and the attributable burden of hearing difficulties in Great Britain. *Occupational and Environmental Medicine* 59(9):634–639.

Palmer KT, Griffin MJ, Syddall HE, Coggon D. 2004. Cigarette smoking, occupational exposure to noise, and self reported hearing difficulties. *Occupational and Environmental Medicine* 61:340–344.

Parving A, Hein HO, Suadicani P, Ostri B, Gyntelberg F. 1993. Epidemiology of hearing disorders. Some factors affecting hearing. The Copenhagen Male Study. *Scandinavian Audiology* 22(2):101–107.

Phoon WH, Lee HS, Chia SE. 1993. Tinnitus in noise-exposed workers. *Occupational Medicine (London)* 43(1):35–38.

Plewnia C, Bartels M, Gerloff C. 2003. Transient suppression of tinnitus by transcranial magnetic stimulation. *Annals of Neurology* 53(2):263–266.

Roberts J. 1968. Hearing status and ear examination: Findings among adults, United States, 1960–1962. *Vital and Health Statistics*. Series 11, No. 32. Washington, DC: Department of Health, Education, and Welfare.

Robertson RM, Page JC, Williams CE. 1978. *The Prevalence of Hearing Loss Among Selected Navy Enlisted Personnel.* NAMRL-1251. Pensacola, FL: Naval Aerospace Medical Research Laboratory.

Rosenhall U, Karlsson AK. 1991. Tinnitus in old age. *Scandinavian Audiology* 20(3):165–171.

Rubertone M (U.S. Army). 2004. RE: Cross Tabulation Help. E-mail to R. Erdtmann, Institute of Medicine, November 4.

Rubertone M (U.S. Army). 2005. RE: Clarification on Tables Provided. E-mail to R. Erdtmann, Institute of Medicine, March 2.

Rubinstein JT, Tyler RS, Johnson A, Brown CJ. 2003. Electrical suppression of tinnitus with high-rate pulse trains. *Otology and Neurotology* 24(3):478–485.

Salmivalli A. 1967. Acoustic trauma in regular army personnel. Clinical audiologic study. *Acta Oto-Laryngologica* (Suppl 222):1–85.

Salvi RJ, Lockwood AH, Burkard R. 2000. Neural plasticity and tinnitus. In: Tyler RS, ed. *Tinnitus Handbook*. San Diego, CA: Singular Thompson Learning. Pp. 123–148.

Shulman A. 1997. Medical evaluation. In: Shulman A, ed. *Tinnitus: Diagnosis/Treatment.* San Diego: Singular Publishing Group. Pp. 253–292.

Sindhusake D, Golding M, Newall P, Rubin G, Jakobsen K, Mitchell P. 2003a. Risk factors for tinnitus in a population of older adults: The Blue Mountains Hearing Study. *Ear and Hearing* 24(6):501–507.

Sindhusake D, Mitchell P, Newall P, Golding M, Rochtchina E, Rubin G. 2003b. Prevalence and characteristics of tinnitus in older adults: The Blue Mountains Hearing Study. *International Journal of Audiology* 42(5):289–294.

Sindhusake D, Golding M, Wigney D, Newall P, Jakobsen K, Mitchell P. 2004. Factors predicting severity of tinnitus: A population-based assessment. *Journal of the American Academy of Audiology* 15(4):269–280.

Stankiewicz C, Przewozny T, Kozlowski J. 2000. [Noise from a car airbag as a cause of acute acoustic trauma.] *Otolaryngologia Polska* 54(6):775–781.

Stephens SD, Hallam RS. 1985. The Crown-Crisp Experiential Index in patients complaining of tinnitus. *British Journal of Audiology* 19:151–158.

Stouffer JL, Tyler RS. 1990. Characterization of tinnitus by tinnitus patients. *Journal of Speech and Hearing Disorders* 55(3):439–453.

Sulkowski W, Kowalska S, Lipowczan A, Prasher D, Raglan E. 1999. Tinnitus and impulse noise-induced hearing loss in drop-forge operators. *International Journal of Occupational Medicine and Environmental Health* 12(2):177–182.

Sweetow R. 2000. Cognitive behavior modification. In: Tyler RS, ed. *Tinnitus Handbook.* San Diego, CA: Singular Publishing Group. Pp. 297–312.

Tambs K, Hoffman HJ, Borchgrevink HM, Holmen J, Samuelsen SO. 2003. Hearing loss induced by noise, ear infections, and head injuries: Results from the Nord-Trøndelag Hearing Loss Study. *International Journal of Audiology* 42(2):89–105.

Temmel AF, Kierner AC, Steurer M, Riedl S, Innitzer J. 1999. Hearing loss and tinnitus in acute acoustic trauma. *Wiener Klinische Wochenschrift* 111(21):891–893.

Tyler RS. 1993. Tinnitus disability and handicap questionnaires. *Seminars in Hearing* 14:377–384.

Tyler RS. 2000. Psychoacoustical measurement. In: Tyler RS, ed. *Tinnitus Handbook.* San Diego, CA: Singular Publishing Group. Pp. 149–180.

Tyler RS, Babin RW. 1986. Tinnitus. In: Cummings CW, Fredrickson JM, Harker L, Krause CJ, Schuller DE, eds. *Otolaryngology—Head and Neck Surgery.* Vol. 4. St. Louis, MO: C.V. Mosby Co. Pp. 3201–3217.

Tyler RS, Cacace AT. 2004. Advances in understanding and treating tinnitus. *Tinnitus Today* 29(1):8–13.

Tyler RS, Noble WS, Preece JP, Dunn C, Witt S. 2004. Psychological treatments for tinnitus. In: Snow JB, Jr., ed. *Tinnitus: Theory and Management.* Hamilton, Ontario, Canada: B.C. Decker. Pp. 314–325.

Van Campen LE, Dennis JM, Hanlin RC, King SB, Velderman AM. 1999. One-year audiologic monitoring of individuals exposed to the 1995 Oklahoma City bombing. *Journal of American Academy of Audiology* 10(5):231–247.

Vernon JA, Meikle MB. 2000. Tinnitus masking. In: Tyler RS, ed. *Tinnitus Handbook.* San Diego, CA: Singular Publishing Group. Pp. 313–356.

Vernon JA, Meikle MB. 2003. Tinnitus: Clinical measurement. *Otolaryngologic Clinics of North America* 36(2):293–305.

Veterans Benefits Administration. 2004a. *Veterans Benefits Administration Annual Benefits Report, FY 2003.* Washington, DC: Department of Veterans Affairs.

Veterans Benefits Administration. 2004b. Written responses to questions from the Institute of Medicine Committee on Noise-Induced Hearing Loss and Tinnitus Associated with Military Service from World War II to the Present, Meeting II, Washington, DC.

Wilson PH, Henry JL. 2000. Psychological management of tinnitus. In: Tyler RS, ed. *Tinnitus Handbook.* San Diego, CA: Singular Publishing Group. Pp. 263–280.

Wilson PH, Henry J, Bowen M, Haralambous G. 1991. Tinnitus reaction questionnaire: Psychometric properties of a measure of distress associated with tinnitus. *Journal of Speech and Hearing Research* 34(1):197–201.

Ylikoski ME, Ylikoski JS. 1994. Hearing loss and handicap of professional soldiers exposed to gunfire noise. *Scandinavian Journal of Work, Environment & Health* 20(2):93–100.

Young DW. 2000. Biofeedback training in the treatment of tinnitus. In: Tyler RS, ed. *Tinnitus Handbook.* San Diego, CA: Singular Publishing Group. Pp. 281–296.

5

Responding to Noise Risks: Hearing Conservation Programs in the Military

The committee's charge to identify when hearing conservation measures were adequate to protect the hearing of service members derives from legislative language. The legislation requested that the committee identify when audiometric measures used by the military became adequate to evaluate individual hearing loss (threshold shift) and when hearing conservation measures to prevent hearing loss were available to service members. The evaluation of hearing conservation programs is not a simple task of either assessing a checklist of necessary components or performing a straightforward analysis of an audiometric database. This chapter describes key aspects of hearing conservation programs and reviews the development and adequacy of programs in the military. Current hearing conservation programs do not include monitoring or prevention of tinnitus. As described in Chapter 4, the relationship between noise exposure and tinnitus is not yet well understood. However, the committee makes the presumption that measures taken to protect against noise-induced hearing loss are likely to help in the prevention of tinnitus. Thus, many of the elements of a hearing conservation program could be applied to prevention of tinnitus as well as hearing loss.

HEARING CONSERVATION PROGRAMS

Overview and Emergence of Key Components

In large part, serious and sustained interest in hearing conservation developed as a result of World War II, when substantial numbers of service members returned home with hearing loss (Gasaway, 1985).[1] In fact, one of the earliest regulations dealing with hearing conservation was issued in 1948 by the Air Force (Department of the Air Force, 1948). Industrial hearing conservation programs began to appear in the late 1940s and early 1950s, with some of the first reported programs established in the aviation and metals industries (Bolger, 1956; Hatton, 1956; Wilkins, 1956; Haluska, 1964). Government noise regulations followed in the late 1960s (U.S. Department of Labor, 1969) and became more prominent and widely enforced with the enactment of the Occupational Safety and Health Act of 1970 and dissemination of the associated regulation on noise in 1971 (OSHA, 1971). However, the Occupational Safety and Heath Administration (OSHA) regulation of 1971 only hinted at details of an occupational hearing conservation program. In 1972, the National Institute for Occupational Safety and Health (NIOSH) published *Criteria for a Recommended Standard: Occupational Exposure to Noise* (NIOSH, 1972). This document described the components of an effective hearing conservation program: sound surveys, noise control, hearing protection, education and training, audiometric monitoring, and recordkeeping. These components were subsequently adopted by OSHA when the Hearing Conservation Amendment (29 C.F.R. 1910.95) was developed (OSHA, 1983). In recent years, program evaluation has been included as an additional component as evaluation tools have been developed (Suter, 1986; NIOSH, 1996; Berger et al., 2000). See Gasaway (1985) for a more complete historical overview of occupational hearing conservation programs and Suter (1988, 2000) for a discussion of the development of the related regulations.

HISTORY OF MILITARY HEARING CONSERVATION PROGRAMS

Overview

The military services' early attention to hearing health focused on rehabilitation. In the 1940s, clinicians in military hospitals and rehabilitation centers noted many patients with severe hearing loss. In anticipation of a large number of hearing-loss casualties from World War II, the Army and Navy established their first aural rehabilitation centers in 1943 and 1944, respectively (Nixon, 1998). The Veterans Administration established similar facilities soon after (Gasaway, 1988). These centers led to the emergence of a cadre of hearing health professionals, including audiologists, who later assumed important roles in prevention-oriented programs in the military

[1]Reports suggest that the service members found to have hearing losses at the conclusion of World War II included many individuals with hearing losses that predated their military service (Nixon, 1998; Bergman, 2002).

(Gasaway, 1985; Donahue and Ohlin, 1993). Figure 5-1 is a time line of major developments in military hearing conservation programs.

The introduction of the jet engine to the military in the late 1940s and early 1950s raised new concerns about noise hazards and helped motivate the development of military hearing conservation programs (Nixon, 1998). The Air Force, the Navy (which also has responsibility for the Marine Corps hearing conservation program), and the Army issued regulations or guidelines on hearing conservation in 1948, 1955, and 1956, respectively (Department of the Air Force, 1948; Department of the Navy, 1955; Department of the Army, 1956).

Subsequent revisions of the military services' noise and hearing conservation documents expanded program elements within each of the services. The Air Force was a leader in these efforts, both in timing and in establishing required program elements, while the Army and Navy programs continued to develop but did not institute mandatory measures. The disparities that existed across the services were noted in a 1977 General Accounting Office report that recommended that the Secretary of Defense adopt consistent policies across the defense agencies (GAO, 1977).

In 1978, a Department of Defense (DoD) Instruction was issued to establish a uniform hearing conservation program, with the goals of eliminating all occupational noise-related hearing loss among DoD personnel (military and civilian) and reducing the costs of compensation (DoD, 1978). The instruction described requirements for sound surveys, the posting of hazardous areas with warning signs, noise abatement, personal hearing protection, education, audiometric testing (preplacement, periodic, and termination audiograms for all personnel exposed routinely to hazardous noise), and recordkeeping (each service was to maintain a hearing conservation data registry) (DoD, 1978). In 1979, DoD introduced standard DoD forms for noise surveys (DD Form 2214), reference audiograms (DD Form 2215), hearing conservation data (DD Form 2216), and biological audiometer calibration checks (DD Form 2217). In the following few years, each of the services published major revisions of their hearing conservation program guidelines to comply with changes mandated by the 1978 DoD Instruction (Department of the Navy, 1979; Department of the Army, 1980; Department of the Air Force, 1982).

Soon thereafter, changes at the federal level were reflected in additional developments in DoD programs. In response to the publication of the OSHA final noise standard, DoD updated its hearing conservation program and policy under a new designation, DoD Instruction 6055.12, in 1987 (DoD, 1987). DoD mandated that the uniformed services comply with OSHA regulations, but the standards and criteria of military programs have been more stringent than those of OSHA (Nixon, 1998). The military services have used lower exchange rates than required by OSHA (the services have

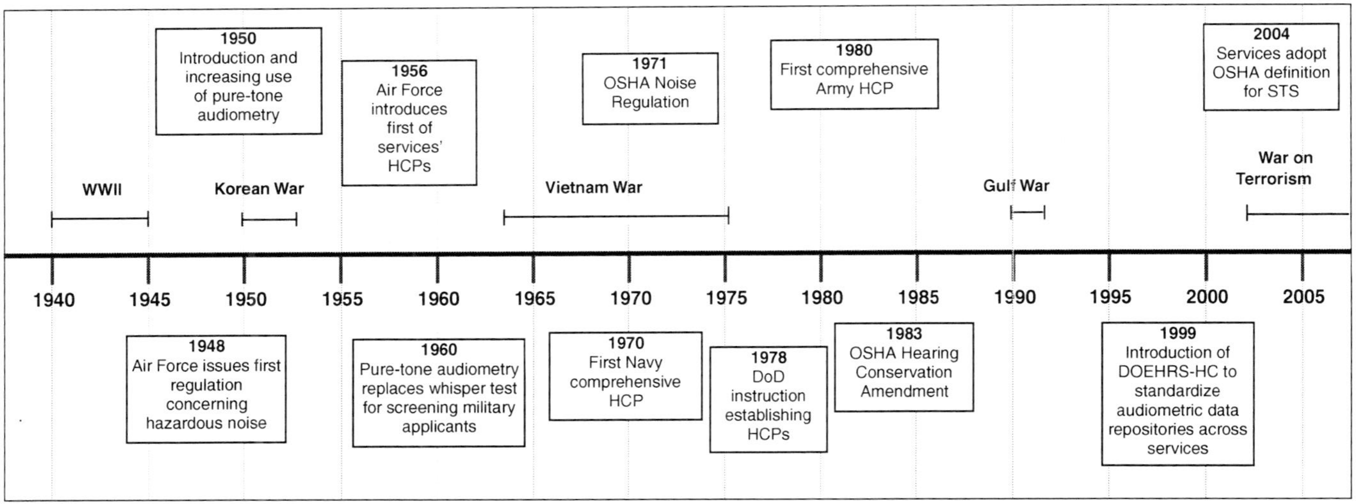

FIGURE 5-1 Time line of major conflicts and milestones in hearing conservation programs.
NOTES: DoD, Department of Defense; DOEHRS-HC, Defense Occupational and Environmental Health Readiness System–Hearing Conservation; HCP, hearing conservation program; OSHA, Occupational Safety and Health Administration; STS, significant threshold shift.
SOURCES: Gasaway (1988); Nixon (1998); Department of Veterans Affairs (2005); Ohlin (2005b).

used 3 or 4 dB, instead of the 5-dB exchange rate specified by OSHA) and currently do not use age corrections in calculations of significant threshold shifts (STSs). Until 2004, the military services' definitions of STS were more sensitive than those of OSHA, identifying more people with possible need for intervention than would have been identified under the OSHA definition. Definitions of STS are discussed in more detail later in this chapter.

In the late 1990s, automated hearing conservation data registries developed by the individual services (the Army and Air Force's Hearing Evaluation Automated Registry Systems and the Navy's Hearing Examination and Audiometric Reporting System, HEARS) were replaced by the hearing conservation component of a new DoD-wide system called the Defense Occupational and Environmental Health Readiness System–Hearing Conservation (DOEHRS-HC). The system is designed to collect, maintain, compare, and report hearing conservation data in DoD (U.S. Army Center for Health Promotion and Preventive Medicine, 2004a). Aspects of this system are discussed in more detail below.

History of Hearing Conservation Programs in the Military Services

This section briefly reviews the history of the Air Force, Navy, Army, and Coast Guard hearing conservation programs. Table 5-1 provides additional information about each of the services' current programs and guiding regulations. The committee's evaluation of these programs is presented separately at the end of the chapter.

Air Force Hearing Conservation Program History

Established as a military service separate from the Army Air Corps in 1947, the Air Force was responsible for the first regulation to protect and conserve hearing in 1948. The brief document required that hearing protection be worn by personnel working in high-level noise, noise measurements be performed to determine degrees of risk, exposure periods be minimized, and audiometric monitoring be performed on people engaged in testing and operating turbojet and rocket engines (Gasaway, 1988). A 1949 update specified noise limits for work areas (85 dB or below in regularly occupied areas) and the availability of protective devices. It required a weekly hearing test for those with high-intensity noise exposures and temporary reassignment if the audiogram indicated a hearing loss of more than 20 dB[2] (Department of the Air Force, 1949).

---

[2]The regulation is not clear regarding the frequencies at which the 20 dB shift is applied.

The 1956 version of Air Force Regulation (AFR) 160-3 established the first hearing conservation program in the services. It included provisions for the six key components of a hearing conservation program described by OSHA more than 20 years later: sound surveys, education, noise control, hearing protection, audiometric monitoring, and recordkeeping. The regulation introduced a standard form (AF Form 1490) to record hearing conservation data and established a central repository for these data. It further established the policy that all military and civilian personnel who enter into or terminate service with the Air Force would receive an audiometric examination as part of their routine physical examination. Personnel assigned to duty or training involving exposure to hazardous noise would be given a follow-up audiogram 90 days after beginning that duty and annually thereafter. Audiograms were to record hearing sensitivity at each specified frequency: 500, 1000, 2000, 3000, 4000, and 6000 Hz. The regulation also established a classification system for hearing based on the degree of hearing loss on the audiogram; the definition of an STS differed depending on an individual's classification. Codes were introduced on audiograms to describe the proximity of work areas to hazardous noise as well as the overall noise level of the area (Department of the Air Force, 1956). The 1956 regulation also established procedures for 15- and 40-hour noise-free follow-up audiograms in response to STSs.

The Air Force revised AFR 160-3 in 1973 with adoption of a criterion for inclusion in the hearing conservation program of an equivalent 8-hour daily exposure of 84 dBA with a 4-dB exchange rate, as well as establishment of detailed exposure criteria for various types of impulse noise (Nixon, 1998). The regulation introduced noise exposure limits specified in terms of the A-weighted level of the noise (dBA) (Nixon, 1998). In late 1974, the Air Force established routine automated handling and storage of audiometric monitoring data to facilitate the use and study of these records (Gasaway, 1988). The next major revision of the Air Force regulation on hazardous noise exposure took place in 1982, updating the regulations to comply with the 1978 DoD Instruction on hearing conservation (DoD, 1978).

The current Air Force hearing conservation program is a decentralized effort, managed at each of the more than 80 Air Force installations around the world, with support provided at the Air Force Institute for Occupational Health at Brooks City-Base, Texas. The Air Force uses the hearing conservation program model of the National Institute for Occupational Safety and Health, with program components such as those listed earlier (Narrigan, 2004). In 2004, more than 156,000 Air Force service members, or 42 percent of the active duty Air Force, and over 14,000 civilian employees, all of whom were considered exposed to hazardous noise, were enrolled in the hearing conservation program (Pluta, 2004; DoD, 2005).

TABLE 5-1 Criteria for Hearing Conservation Programs

Service	Criteria for Hearing Conservation Program Enrollment[a]	Exchange Rate (dB)	STS Definition	STS Follow-up
Air Force	≥ 85 dBA TWA, or exposure to > 140 dBP	3	≥ 10 dB average shift at 2, 3, 4 kHz No age correction	Positive and negative STS require follow-up; for positive STS, f/u 1 and 2 must take place within 30 days of annual audiogram, and f/u 1 must be 14 hours noise free
Navy/ Marine Corps	Routinely exposed to > 84 dBA or > 140 dBP ("routinely" defined as TWA > 84 dBA for more than 2 days/month)	4	≥ 10 dB average shift at 2, 3, 4 kHz Change of ≥ 15 dB in either ear at any test frequency from 1 to 4 kHz considered early warning, requiring verbal counseling and assurance of access to appropriate hearing protection No age correction	Positive and negative STS require follow-up; for positive STS, f/u 1 and 2 must take place within 30 days of annual audiogram, and f/u 1 must be 14 hours noise free
Army	≥ 85 dBA TWA, or exposure to ≥ 140 dBP Exposure to known or suspected ototoxins	3	≥ 10 dB average shift at 2, 3, 4 kHz No age correction	Positive and negative STS require follow-up; for positive STS, f/u 1 and 2 must be 14 hours noise free
Coast Guard	≥ 85 dB TWA for ≥ 30 days per calendar year, or exposure to > 140 dBP; also those with > 35 dB shift in 0.5–3 kHz range	4	≥ 10 dB average shift at 2, 3, 4 kHz	Positive and negative STS require follow-up; for positive STS, f/u with up to two consecutive 14-hour noise-free audiograms

Reference Audiogram	Requirements for Use of Hearing Protection Devices	Guiding Documents
Shall be received prior to duties in hazardous noise Within 30 days of entering a hazardous noise-exposed job; must be 15 hours noise free	≥ 85 dBA TWA, or exposure to > 140 dBP	AFI 48-20, Interim Guidance (2000) AFOSH 161-20 (1991) AFOSH 48-19 (1994)
Required on entry into naval service Hearing tests performed at Military Entrance Processing Stations shall not be used as a baseline hearing test Must be noise free (no noise above 80 dBA) for at least 14 hours	> 84 dBA or > 140 dBP Double protection at > 104dB	OPNAVINST 5100.19D CH-1 Occupational Safety and Health Program Manual for Forces Afloat (2001) OPNAVINST 5100.23F Occupational Safety and Health Program Manual for Forces Ashore (2002) NEHC Technical Manual 6260.51.99-2 (September 2004) MCO 6260.1 (2000) MCO P5100.8 F (1998)
Required at basic training prior to noise exposure	≥ 85 dBA TWA, or exposure to > 140 dBP Double protection at ≥ 104 dB	DA PAM 40-501 (10 December 1998)
Required prior to Coast Guard occupational noise exposure	≥ 85 dB TWA for ≥ 30 days per calendar year, or exposure to > 140 dBP Double protection at > 104 dB	Chapter 4, Coast Guard Safety and Environmental Health Manual (COMDTINST M5100.47) Chapter 12, Coast Guard Medical Manual (COMDTINST M6000.1B)

continued

TABLE 5-1 continued

Service	Criteria for Hearing Conservation Program Enrollment[a]	Exchange Rate (dB)	STS Definition	STS Follow-up
OSHA requirements	≥ 85 dBA (action level)	5	≥ 10 dB average shift at 2, 3, 4 kHz, either ear	Retest to rule out spurious STS is optional
NIOSH recommendations	≥ 85 dBL_{Aeg}	3	≥ 15 dB shift at 0.5, 1, 2, 3, 4, or 6 kHz, either ear, and the same shift at the same test frequency in the same ear on an immediate retest	Immediate retest; if the same, schedule for 30-day confirmation audiogram
DoD requirements	≥ 85 dBA TWA, or impulse noise > 140 dBP	4 (3 strongly encouraged)	≥ 10 dB average shift at 2, 3, 4 kHz, either ear No age correction 15 dB shift at 1, 2, 3, or 4 kHz retained as early warning only	Positive and negative STS require follow-up; for positive STS f/u 1 and 2 must be 14 hours noise free

[a]Criteria concerning airborne high-frequency or ultrasonic noise are not noted here.

NOTES: DoD, Department of Defense; NIOSH, National Institute for Occupational Safety and Health; OSHA, Occupational Safety and Health Administration; STS, significant threshold shift; TWA, time-weighted average.

SOURCES: OSHA (1971, 1983); U.S. Coast Guard (1990, 2003); Department of the Air Force (1991, 1994, 2000); Department of the Army (1998); Department of the Navy (1998, 2000, 2001, 2002); NIOSH (1998); Hall (2001); DoD (2004b); Navy Environmental Health Center (2004b).

Reference Audiogram	Requirements for Use of Hearing Protection Devices	Guiding Documents
Called baseline audiogram and must be established within 12 months of employee's exposure at or above the action level	Optional for ≥ 85 dBA TWA; mandatory for > 90 dBA TWA, and for ≥ 85 dBA TWA for workers with STS	29 C.F.R., Chapter XVII, Part 1910, Subpart G, 36 F.R. 10466, May 29, 1971; Amended 48 F.R. 9776–9785, March 8, 1983
Baseline audiogram to be established within 30 days of enrollment in hearing loss prevention program	Mandatory for > 85 dBA TWA with a 3-dB exchange rate	Criteria for a Recommended Standard: Occupational Noise Exposure (NIOSH, 1998)
Required for all military personnel at basic training prior to noise exposure	Mandatory in "hazardous noise areas" when noise sources are operating, and with exposure to gunfire or artillery fire in test or training situations	DoDI 6055.12 (2004)

Navy and Marine Corps Hearing Conservation History

The Navy is responsible not only for its own hearing conservation program, but also for that of the U.S. Marine Corps. As with the Air Force, concerns about noise from jet aircraft spurred steps toward a hearing conservation program in the Navy. Aircraft carrier crew members must work close to jet aircraft during flight operations and maintenance. A 1952 study of the effects of jet aircraft engine noise on aircraft carrier personnel indicated a likely negative impact on personnel and operations and suggested a larger scope to the jet-engine noise problem than had been understood before (Rosenblith et al., 1952; Nixon, 1998). The study authors recommended that the Navy emulate the hearing protection programs they had observed at several Air Force installations and also make the wearing of hearing protection compulsory in high-noise settings (Rosenblith et al., 1952). They also recommended additional study of the interaction between noise and humans on aircraft carriers. Exposure to high-intensity noise thereafter became a priority concern for the Navy and motivated its role in the formation of the NAS-NRC Committee on Hearing and Bioacoustics (CHABA, as described in Chapter 2).

The first Navy regulation regarding hearing conservation, issued in 1955, formally established the Navy hearing conservation program, but with no requirements for actions (Department of the Navy, 1955). In 1960, the Navy increased civilian staffing for hearing conservation (Nixon, 1998), and in 1970, the Navy issued its first standards, making hearing conservation programs mandatory when noise levels exceeded 90 dBA. It adopted a noise standard of an equivalent 8-hour daily exposure of 90 dBA with a 5-dB exchange rate (Department of the Navy, 1970; Nixon, 1998). In 1976, the Navy Bureau of Medicine and Surgery directed naval activities to discontinue purchasing self-recording audiometers and to limit group testing to four subjects because of problems with the reliability of the hearing tests as well as with program administration and management (Robertson and Williams, 1984).

An important change in the Navy program took place when program responsibility was transferred from the Bureau of Medicine to the Chief of Naval Operations in 1979 (Nixon, 1998), affording the program more visibility. This change coincided with revision of its hearing conservation program requirements to establish exposure limits of 85 dBA for continuous or intermittent noise, and 140 dB peak sound pressure level for impact or impulse noise (Department of the Navy, 1979). For the first time, the Navy acquired 10 military audiologists (Nixon, 1998), and in 1982, an additional 10 positions were added (Page, 2004a, 2005a).

In the early 1980s, the Navy continued to try to improve its ability to collect and store audiometric information. Automated systems were being developed in conjunction with the use of microprocessor-controlled group audiometers. However, as reported in 1984, and continuing until the introduction of DOEHRS-HC in 1999, there remained no efficient way to obtain audiometric data for large numbers of naval service members in order to assess the adequacy of the Navy's hearing conservation programs (Robertson and Williams, 1984; Page et al., 2002). For example, as of 1990, Navy budgetary constraints permitted the storage of no more than 1–2 years of audiometric data at any time (Ridgley and Wilkins, 1991).

The Navy and Marine Corps hearing conservation programs are currently managed by the Navy Occupational and Environmental Health Center in Portsmouth, Virginia, with program oversight the responsibility of the Chief of Naval Operations. The Navy estimates that 285,000 sailors and civilians, and 67,000 members of the U.S. Marine Corps, are currently enrolled in the hearing conservation program (Page, 2005b). Since the mid-1980s the number of Navy audiologists has grown from 37 to 49, with roughly 21 of them military and the remainder civilian (Page, 2005a). The current documents guiding the Navy's program are listed in Table 5-1.

Army Hearing Conservation History

Army efforts in the area of noise research and protection began in the early 1940s, and for decades thereafter, hearing conservation efforts in the Army primarily emphasized identification of noise hazards (Donahue and Ohlin, 1993). The Army's first document describing a hearing conservation program was a technical bulletin issued in 1956 and subsequently revised in 1965 and 1972 (Department of the Army, 1956, 1965, 1972). In 1972, the Army adopted a criterion for hazardous noise exposure of an equivalent 8-hour daily exposure of 85 dBA, with a 5-dB exchange rate (Nixon, 1998). Although all the basic elements of a hearing conservation program were required in a preventive medicine regulation, specific hearing conservation program activities were only recommended (Ohlin, 2005a) and were therefore applied inconsistently at different Army installations. In 1980, the Army issued its first requirements to implement an Army-wide hearing conservation program (Department of the Army, 1980), as directed by the 1978 DoD standards for military hearing conservation programs (DoD, 1978). With this new technical bulletin, hearing protection became mandatory in steady noise when levels were at or above 85 dBA, or when impulse noise levels exceeded 140 dBP. If steady noise was above 108 dBA or impulse noise was over 165 dBP, double protection (both earplugs and ear muffs) was required (Department of the Army, 1980). The exchange rate between noise level and allowable daily exposure was lowered from 5 dB to 4 dB.

In the 1960s, the Army commissioned military audiologists, who came to play an important role in advancing the profession within and outside the military (Nixon, 1998). By 1970, 58 audiology positions had been established to provide a servicewide effort in education and training (Gasaway, 1988). Studies indicating high-frequency hearing loss in combat arms personnel (Walden et al., 1971, 1975) convinced Army leaders to improve the education of personnel regarding hearing protection. In 1971, the Army carried out a pilot study for an effort to collect hearing conservation data, which eventually led to the establishment of the Army's hearing conservation data repository (Nixon, 1998). Another important change in the Army's program was its transition from manual recording of audiometric test results to the Hearing Evaluation Automated Registry System (HEARS) in 1987 (AMSA, 1997).

Technical and administrative guidance for the Army hearing conservation program is provided centrally from the U.S. Army Center for Health Promotion and Preventive Medicine in Aberdeen, Maryland. In 2003, there were 375,186 soldiers, as well as 53,986 civilians employed by the Army, enrolled in the Army's hearing conservation program (U.S. Army Center for Health Promotion and Preventive Medicine, 2004b), representing 77 percent of the active duty Army (DoD, 2004a). Since 1990, the number of military audiologists in the Army has dropped by more than half, from 71 to 31 in 2005 (Gates, 2005). Some have been replaced by civilian contractors, who fill a different role in the hearing conservation program than military audiologists. While civilian audiologists' time is dedicated to clinical roles, uniformed audiologists have additional responsibilities for fieldwork that afford additional opportunities for the education and motivation of service members regarding hearing conservation program goals.

U.S. Coast Guard Hearing Conservation Program History

The U.S. Coast Guard differs from the other four armed services in that it is not ordinarily part of DoD. In times of peace it has operated as part of the Departments of the Treasury (until 1967), Transportation (1967–2003), and Homeland Security (since March 2003), but in times of war or at the direction of the President, it serves under the Navy (U.S. Coast Guard, 1998). By the late 1960s and early 1970s, the Coast Guard hearing conservation program included noise surveys as well as periodic audiometric testing (McConnell, 2004). Regulatory guidance published in 1982 provided hearing conservation recommendations for commercial vessels inspected by the Coast Guard (U.S. Coast Guard, 1982). It described a program similar to that of the other services, but noted the highly mobile nature of the maritime industry and the resultant difficulties in maintaining a program of audiometric testing and recordkeeping.

In the early 1990s, the Coast Guard instituted a program called the Occupational Medical Monitoring Program, intended to facilitate documentation and prevention of occupational hazards. Centrally managed from Coast Guard headquarters, it was determined to be unsuccessful (McConnell, 2004). In 1998, the Coast Guard established a new program called the Occupational Medical Surveillance and Evaluation Program, a physical examination program intended to monitor the health of personnel in jobs designated as having high health risk potential as a result of exposure to chemical or physical agents. The Coast Guard hearing conservation program is one of 14 medical examination protocols in the Occupational Medical Surveillance and Evaluation Program, which helps track audiogram appointments for enrollees to facilitate follow-up.

As with the military services described above, the Coast Guard hearing conservation program requires identification of hazardous noise sources, determination of personnel exposed, application of engineering methods to abate noise, hearing protection and education for those exposed to hazardous noise, and employee monitoring through annual audiometric testing. U.S. Public Health Service officers and Coast Guard line officers serve as Safety and Environmental Health Officers at Coast Guard headquarters and in each of the nine Coast Guard districts. They are responsible for carrying out the sound-level surveys and noise dosimetry necessary to determine which vessels and other work sites necessitate enrollment in the hearing conservation program. Audiometry and other medical aspects of the hearing conservation program are carried out by a different group within the directorate (personal communication, W. McConnell, U.S. Coast Guard, February 10, 2005, and February 23, 2005; McConnell, 2004, 2005). Criteria and guiding documents for the Coast Guard hearing conservation program are shown in Table 5-1.

The Coast Guard does not yet formally participate in DOEHRS-HC. DoD forms 2215 and 2216 or similar forms for recording baseline and monitoring audiogram data are filed in an individual's medical folder.

ASSESSING THE ADEQUACY OF HEARING CONSERVATION PROGRAMS

Criteria for Evaluating Hearing Conservation Program Effectiveness

Although there is general consensus concerning the necessary components of a hearing conservation program (several authors, cited by Royster and Royster, 1990), there is less agreement regarding how to assess the effectiveness of a program. Several approaches have been proposed, among them the use of checklists to assess the presence of important program components and the use of audiometric databases for population and criteria comparisons.

Checklists

Hearing conservation programs can be audited with checklists to determine whether all the necessary components are present. A variety of checklists have been developed (Royster and Royster, 1990; Suter and Franks, 1990; NIOSH, 1996, 2005), including one distributed for use by Army hearing conservation program managers (U.S. Army Center for Health Promotion and Preventive Medicine, 2005). The major limitation of this approach is that a checklist alone cannot evaluate the quality of the implementation of each component (Royster and Royster, 2000). Measures such as the rate of compliance with requirements for annual audiograms are indicators of program activities that are necessary, but not sufficient, for effective programs. The Army found several hearing conservation programs that rated highly on their checklist but nonetheless had high incidences of STS (Byrne and Monk, 1993), and Navy researchers reported a similar finding (Wolgemuth et al., 1995). For a program to succeed, it is essential that the use of hearing protection devices be strictly enforced. Also considered necessary is a "key individual" who has overall responsibility for the program, takes a sincere personal interest in its success, has the full support of management, and has the dedication to motivate employees to be active program participants (Royster and Royster, 1990). These features are difficult to evaluate from outside a program.

Audiometric Database Evaluations

The availability of databases with audiometric monitoring information on the members of a hearing conservation program facilitates, but does not simplify, the evaluation of the program. Dobie (1995) noted several methodological challenges in studying the effectiveness of hearing conservation programs with audiometric data, including the need to take into account the effects of age, the typically slowing course of noise-induced hearing loss (as noted in Chapter 2, most hearing loss associated with noise exposure is observed in the first 10–15 years), and learning as individuals gain familiarity with audiometric testing procedures (discussed in Royster and Royster, 2000).

An American National Standards Institute (ANSI) working group labored for more than a decade to produce a standard on program evaluation and examined more than a dozen proposals for using audiometric data to evaluate hearing conservation program effectiveness. They published a draft standard, but it obtained support sufficient for publication only as a Technical Report (S12.13 TR-2002) rather than as a full standard (Acoustical Society of America, 2002). The procedures proposed by the ANSI Technical Report use audiometric database analysis (ADBA) to evaluate hearing conservation program effectiveness. With this approach, if the year-to-year and year-to-baseline variability of audiometric data exceed certain criteria, the data are deemed too variable to provide useful information regarding changes in hearing thresholds and, therefore, are indicative of an ineffective hearing conservation program. The approach relies on the analysis of audiometric data for persons who have remained in the hearing conservation program long enough to receive at least two annual audiograms.

A weakness seen in the ADBA approach is that the subsamples used for the analyses are not randomly selected. The resulting selection bias may lead to inaccurate assessment of a hearing conservation program because those at high risk of hearing loss may be systematically excluded (Adera et al., 1993, 1995). However, an alternative perspective is that nonrandom selection of samples may help to target and evaluate suspected "worst-case" exposures. In addition, poor agreement has been observed when different ADBA procedures are applied to the same data (Adera et al., 1995). Concerns have also been expressed over the derivation of the numerical ranges for the criteria (Acoustical Society of America, 2002), the potential for bias against audiometric data collected in 5-dB steps (Simpson et al., 1993), and the inability to take into account preexisting hearing loss in the populations evaluated (Simpson et al., 1998).

In recent years, several alternative methods for using audiometric data to evaluate hearing conservation programs have been proposed. One uses comparison of the rate of hearing loss (e.g., as indicated by incidence of STS) in a hearing conservation program to that in an appropriate reference population (Adera et al., 2000a), although this requires availability of the reference population. Another method for hearing conservation program evaluation is time trend analysis, which examines patterns of hearing loss over time in multiple discrete cohorts within a larger database (Adera et al., 2000b).

Use of the percentage of workers showing STS in a given time period to evaluate the effectiveness of a hearing conservation program *without* reference to any comparison population has important limitations. Annual STS rates of 3–6 percent (Morrill and Sterrett, 1981) or 5 percent have been proposed (Franks et al., 1989; Simpson et al., 1994) as achievable by effective programs. However, the effects of variables such as age, sex, race, and previous noise exposure history, as well as merely poor audiometry, may play roles in STS rates, and these would not be taken into account (Melnick, 1984; NIOSH, 1996). Another important concern is that the variability inherent in audiometry is itself sufficiently large to make detection of STS in a noise-exposed population very unlikely (Hetu et al., 1990). Because of the shortcomings inherent in each of the approaches proposed, no standard procedure for evaluating hearing conservation program effectiveness has yet been recognized.

ASSESSING THE ADEQUACY OF HEARING CONSERVATION PROGRAMS IN THE MILITARY

The committee interpreted its charge to "identify when hearing conservation measures were adequate to protect the hearing of service members" as evaluating the effectiveness of these programs over time. This task poses considerable challenges in the absence of a simple or universally accepted means for evaluating hearing conservation programs, as described above. In addition, it requires evaluation of not just a single program at one point in time, but evaluation of several different programs of the military services as they have changed since World War II.

It is important to note that effectiveness or adequacy of the program is not simply a matter of process: implementing a program as required. Instead, it is contingent on outcomes—whether the program is producing the desired results. Royster and Royster (1990, p. 341) provided one definition of effectiveness:

> An effective hearing conservation program provides the noise-exposed population protection from on-the-job noise exposures such that changes in its hearing threshold levels over time are not significantly different from those found in a properly matched control population (a nonindustrial noise-exposed population, or NINEP, exhibiting all of the relevant population characteristics and hearing hazards except for on-the job noise exposure).

Chapter 3 presented some of the data available to the committee to examine changes in hearing thresholds in a few samples of different military service populations over decades from the 1970s to the 1990s, albeit with limitations in the availability of appropriate reference populations. It is difficult to determine the extent to which service members have experienced hearing loss, even for the period since audiometric testing has been required.

In this section the committee reviews additional evidence pertaining to aspects of the process, rather than the outcomes, of military hearing conservation programs since World War II. The committee was unable to carry out an audit of the countless sites at which military hearing conservation programs are implemented; instead it relied on information regarding hearing protection and audiometric monitoring from the published literature and from information, such as that from DOEHRS-HC, provided to the committee by the services. The information available to the committee regarding these features is incomplete but, taken together, may provide some indication of the adequacy of the services' hearing conservation programs over time.

Noise Control

Hazardous noise exposures may occur in various military settings, including industrial-type workplaces (aircraft refurbishing, shipbuilding), as well as military-unique environments, such as combat. Although complete control of hazardous noise in combat is clearly impossible, there are many circumstances in which limiting or reducing hazardous noise at its source is possible. Indeed, the Department of Defense Design Criteria Standard (MIL-STD-1474D), first issued in 1973, emphasizes the importance of incorporating noise control considerations into the design and purchase requirements for military materiel (DoD, 1997). However, as noted in Chapter 3, DoD regulations give priority to maintaining combat readiness and permit tradeoffs between noise reduction and weight, speed, cost, or other factors crucial to the effectiveness of the equipment (DoD, 2004b).

Noise control efforts can range from elaborate engineering measures, to routine maintenance, to isolation of noise sources with insulation. In current operations in Iraq, simple barriers such as sandbags are used where possible to muffle noise from generators or other loud equipment (Chandler, 2004).

In the Navy, silencing technology has been applied in the successive redesign of surface combatant vessels and submarines over the years to provide up to 30 dB reductions in noise levels (Yankaskas and Shaw, 1999; Yankaskas, 2004). In contrast, similar improvements possible on aircraft carriers have not yet been carried out. The 30-year-old carrier design permits intermittent high levels of noise in libraries, passageways, and sleeping berths as well as the high levels measured on flight decks (Yankaskas and Shaw, 1999; Yankaskas, 2004).

Hearing Protection

Control of hazardous noise at its source in military settings is frequently not possible, so that hearing protection, either through administrative controls (e.g., limiting the time a person can spend in hazardous noise) or with protective devices, has been of vital importance. However, administrative controls are difficult to implement in the military, especially in training and on the battlefield, where operational time frames do not adhere to typical workdays. As a result, protection from hearing loss in the military has essentially equated to issuing hearing protection devices.

Recognition of the need for hearing protection has evolved considerably over the past 60 years, as has the quality and effectiveness of the available equipment. Although hearing protection devices were available prior to 1940 (Acton, 1987; Moritz and Bruce, 1994), utilization was essentially nonexistent throughout World War II. A substantial research effort to devise a suitable hearing protection device concluded in 1945 with the development of the Ear Warden V-51R (Shaw and Veneklasen, 1945), a design used by the military through the 1990s. Shaw and Veneklasen (1945) observed that a common form of hearing protection in use by the military during the 1940s was cotton or cotton waste, as also documented by Walpole (1943). Unfortunately, such air-permeable material was inadequate as a noise attenuator. In fact, according to an early Air Force regulation, even the devices available by 1949 (three-sized V-51R, cotton plugs moistened in petroleum jelly or paraffin, and dental acrylic custom earmolds) were "effective only against minimal exposure" (Department of the Air Force, 1949).

The years since World War II have seen some marked improvements in hearing protection. Resilient materials needed as interfaces with the flesh around the ear or in the ear canal were improved in comfort, durability, and dynamic characteristics. An important advancement in earplug technology was introduced in the 1970s—the roll-down slow-recovery foam earplug, which has become a predominant form of hearing protection for both military and industrial users worldwide. In a one-sized product, it offered increased levels of protection, as well as comfort, for most users (Camp et al., 1972; Bailey and Walker, 1979; Shaw, 1979; Schleifer et al., 1984). Another technological development was the introduction of active noise reduction systems into tanker helmets in the 1980s and into Air Force flight helmets in the 1990s. These systems served to attenuate low-frequency noise and thereby enhanced communications (McKinley and Nixon, 1993; Anderson and Garinther, 1997; Mozo and Murphy, 1997). Other electronic products for communication also appeared, though in some high-noise environments, such as the flight decks of aircraft carriers, the same hearing protection ("cranial earmuffs") and communication technology that was used in the 1950s is still in use today (personal communication, J. Page, Naval Environmental Health Center, March, 2005). Developments in hearing protection from World War II to the present are broadly summarized in Table 5-2.

Although gains have been made in the potential noise reduction of hearing protection devices in the past 60 years, the achievable attenuation values have not changed substantially since the 1970s. Table 5-3 summarizes the potential noise reduction provided by the hearing protection devices in current use. Ranges of attenuation are provided because, even in a laboratory setting, performance is highly dependent on the use and fit of the devices. The ranges also allow for the variation in performance between different brands of the same type of device. The use of earplugs together with earmuffs, called dual protection, provides maximum protection. For well-fitted devices, the average attenuation for dual-protection systems is as much as 40–50 dB at frequencies up to 1000 Hz and can be even greater at frequencies at and above 2000 Hz. The amounts of protection that can be provided are adequate in all but the most severe military exposures, such as carrier flight decks.

Regardless of the data measured in the laboratory, it has become clear that the "real-world" performance of devices is quite different as a result of fit and other factors, such as motivation, training, supervision, and enforcement (Berger et al., 1996; Berger, 2000a). A summary chart of noise reduction ratings versus real-world attenuation is presented in Figure 5-2. The data were drawn from 22 field studies, including 1 conducted in a military setting (Smoorenburg et al., 1986).

Even more important than the difference between potential attenuation and the real-world performance of hearing protection devices in the field, however, is the impact of wearing the devices at all. The percentage of time a hearing protection device is used in a noisy environment has a much greater effect on hearing protection than even changes of 5 or 10 dB in the amount of noise reduction the devices provide when assessed in a laboratory. A recent study highlights this issue. Neitzel and Seixas (2005) measured the attenuation of the hearing protection devices in use and also developed verified estimates of actual wearing time. For devices with approximately 20 dB of real-world attenuation, the effective protection, taking into account wearing time, was less than 3 dB. Although the environment they studied was construction, it is likely that many of the same factors apply to the military setting.

In intensive military operations, such as training and combat, the motivation to wear hearing protection may be further limited by concerns that hearing protection devices may jeopardize the wearer's safety. Safety could be compromised when using hearing protection devices by impairing communication or causing service members to miss vital auditory warning signals (sounds of enemy troops, ordnance, and the environment).

In a 1975 survey of 3,000 enlisted men from U.S. Army infantry, armor, and artillery branches, 64 percent reported that they routinely used hearing protection, while 90 percent reported that hearing protective devices were readily available to them (Walden et al., 1975). Nearly half of the soldiers reported that they disliked wearing hearing protection. A smaller study observed only 14 of 34 (41 percent) Army drill instructors using hearing protection on a given day (Loeb et al., 1973). A study of submariners, submarine force workers, and support personnel in the early 1980s found that more than 50 percent of personnel surveyed who worked in noisy environments reported never using hearing protection, with officers less likely to report use of hearing protection (Gwin and Lacroix, 1985). More recently, a study on the use of hearing protection devices in one of the most hazardous noise environments in any industrial or military setting, the aircraft carrier flight deck (where noise levels routinely exceed 140 dBA), found that 47 percent of those surveyed reported *never* wearing double hearing protection even though they were working in

TABLE 5-2 Available Hearing Protection from World War II to the Present

Time Frame	Typical Devices	Comments
World War II	Cotton	Minor amounts of noise reduction
	Fingers	Effective but inconvenient; used by artillery crews to some extent
	Nothing	The standard of the day
1945–late 1950s	Vaseline-impregnated cotton	Messy, modestly effective, better for water protection than noise protection
	V-51R earplug	Initially produced in three sizes; developed just at the end of World War II
	Hard custom earmolds	Easily lost seal, not widely used
	Early circumaural earmuff designs	Initial designs had inadequate cushions and modest attenuation (around 20 dB or less) up to 1000 Hz
	Navy "cranial earmuffs" introduced in mid-1950s and still in use today (circa 2005)	Plastic earmuff cups held in place by fabric head cap with a plastic shell covering the fabric but not enclosing the earmuff cups; inadequate fitting and modest protection
1960s	V-51R earplug	An extra-small and extra-large size added to fit a wider range of ear canals
	Triple-flange earplug	Alternative easier-to-fit design introduced as a two-sized version
	Canal caps (pods on light-weight band)	Modest protection for intermittent environments
	Malleable putty earplug	Not widely used, and ergonomic problems due to required kneading and messiness
	Improved earmuffs	Higher attenuating designs introduced with better cushions and headbands
1970s	Conventional plugs and muffs same as 1960s	Technology essentially mature by this time, but some material improvements such as newer three-sized silicone version of triple-flange plug. Also, color-coded sizing introduced.
	Roll-down slow-recovery foam earplugs	New-concept earplug that provided better protection and comfort, but limited use in military initially
	Tanker helmets and aircraft flight helmets with internal earcups for noise attenuation	Helmets began to provide not only impact protection, but acoustical protection too. Low-frequency attenuation not as good as conventional earmuffs.

1980s	Conventional plugs and muffs same as 1960s and 1970s	No technology advances
	Tanker helmets began to appear with ANR included	ANR in this environment improved communication and protection
1990s	Same as prior decades	Minor technology advancements especially in cosmetics, but performance essentially unchanged
	Widespread use of roll-down slow-recovery foam ear plugs	Most commonly used hearing protection device
	Communication earplugs	Use of earphone in foam earplugs for use in tanker and helicopter applications for enhanced communication under helmet and increased protection
	Widespread use of ANR for tanker helmets and limited application of ANR for aircraft flight helmets	The advantages of ANR began to appear in aircraft applications too
2000–present	Same as prior decades	As before, except that V-51R plug dropped from inventory
	Level-dependent "combat arms" earplugs	New technology provides the ability to protect against weapons and blast noise, but still allow communication and signal detection of lower-level sounds when the impacts are not present

NOTE: ANR = active noise reduction.

SOURCES: Shaw and Veneklasen (1945); Department of the Air Force (1949); Blackstock and Von Gierke (1956); Guild (1966); Gardner and Berger (1994); Mozo and Murphy (1998); Ohlin (2005c); Schulz (2005a); Personal communication, D. Gauger, Bose Corporation, April 2005; personal communication, D. Ohlin, USACHPPM, April 2005.

TABLE 5-3 Representative Minimum and Maximum Mean Attenuation Values of Well-Fitted Hearing Protectors Under Laboratory Conditions, in dB

	Octave-Band Center Frequency (Hz)						
Type of Hearing Protector	125	250	500	1000	2000	4000	8000
Inserted Hearing Protectors							
Foam earplugs (attenuation varies with depth of insertion)	20–40	20–40	25–45	25–45	30–40	40–45	35–45
Premolded earplugs	20–30	20–30	20–30	20–35	25–35	30–45	30–45
Formable (fiberglass/mineral wool)	20–30	20–30	20–30	25–30	25–30	35–40	35–40
Formable (wax-impregnated cotton or silicone)	20–25	20–25	20–25	25–30	30–35	40–45	40–45
Custom-molded earplugs	15–35	15–35	15–35	20–35	30–40	35–45	30–45
Semi-insert earplugs	15–30	15–30	10–30	15–30	25–35	25–45	30–45
Circumaural, Helmet, and Combined							
Earmuffs (with or without communications components)	5–20	10–25	15–40	25–45	30–40	30–40	25–40
Military helmets	0–15	5–15	15–25	15–30	25–40	30–50	20–50
Dual protection (earplugs + earmuffs)	20–40	25–45	25–50	30–50	35–45	40–50	40–50
Active noise reduction (closed-cup systems; identical to conventional muffs above 1 kHz)	15–25	15–30	20–45	25–40	30–40	30–40	25–40
Other Types							
Cotton balls	0–5	0–10	5–10	5–10	10–15	10–20	10–20
Motorcycle helmets	0–5	0–5	0–10	0–15	5–20	10–30	15–35
Air-fed shotblasting helmets	0–5	0–5	0–5	0–15	15–25	15–30	15–25
Finger tips in ear canals	25–30	25–30	25–30	25–30	25–30	30–35	30–35

NOTE: Data are intended to account for brand and testing variability; however, not all manufacturers' reported data or values referenced in the literature will necessarily fall within the ranges cited. All data are from E·A·RCAL Laboratory as reported by Berger (2000a), except for the shotblasting helmets (Price and Whitaker, 1986) and fingers (Holland, 1967).

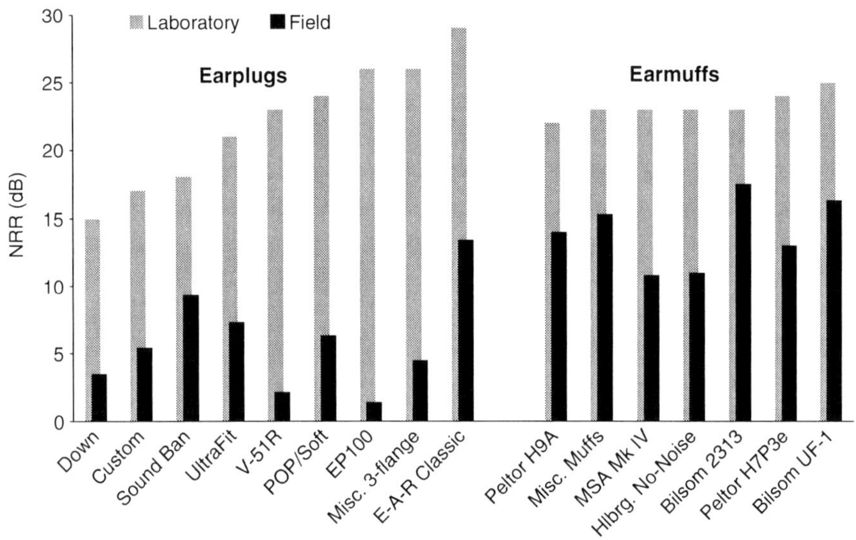

FIGURE 5-2 Comparison of Noise Reduction Ratings published in North America (labeled values based on laboratory tests) to real-world attenuation results derived from 22 studies.
SOURCE: Berger (2000b). Reprinted, with permission, from Berger (1993). Copyright 2000 by E•A•R Company.

mandatory dual-protection environments (Bjorn et al., 2004a). A separate study reported that while all flight deck personnel wear headgear ("cranials") with earmuffs rated at 23 dB, only 1 of the 22 individuals in the study wore dual hearing protection (Rovig et al., 2004).

The evidence, described above, of limited use of hearing protection among personnel in U.S. military units is consistent with findings from studies in other military and industrial work situations. A review of 67 studies published between 1981 and 1999 and providing data on usage of hearing protection devices frequently found that fewer than 50 percent of those who should have been wearing protection reported doing so (Berger, 2000b). Five studies reported on usage of hearing protection in military units from Canada, the United Kingdom, and Israel. From 0 to 42 percent of personnel in these units reported usually wearing hearing protection all the time in combat situations, and from 63 to 89 percent wore it all the time in other settings.

Despite continuing challenges in motivating service members to use hearing protection devices, there is some indication of an emerging willingness among service members to use hearing protectors (Ohlin, 2005d). This is at least in part due to two new and unrelated military initiatives: an extensive effort to solve the extreme noise-exposure problem on aircraft carrier flight decks (Bjorn et al., 2004b), and the introduction of a new type of level-dependent earplug, called the Combat-Arms™ earplug. The earplug provides the ability to hear low-level sounds with less distortion or interference from attenuation than with traditional passive hearing protection devices, while affording protection from blasts and weapons fire (Dancer et al., 1999; U.S. Army Center for Health Promotion and Preventive Medicine, 2004d). Despite these developments, in the coming years it remains likely that the effective use of hearing protectors by service members in combat and other intensive operational settings will be less than that in noncombat and support operations.

FINDING: Compliance with requirements for use of hearing protection devices is crucial for an effective hearing conservation program. There is limited or suggestive evidence to conclude that use of hearing protection devices and the level of real-world hearing protection these devices provide have been and remain not adequate in military hearing conservation programs. However, the studies conducted in U.S. military personnel are generally consistent with studies from other settings that provide additional evidence that the use and real-world protection of hearing protection devices are not adequate.

Audiometric Monitoring

Audiometric monitoring provides some of the most useful information about the effectiveness of hearing conservation programs and for making changes as needed to improve hearing protection. This section reviews the chronology of the availability and use of audiometry, the requirements for entrance and termination audiograms among the military services, and the information available from recent audiometric monitoring as reported through HEARS and DOEHRS-HC.

Whispered Voice Test

Some patients at the military aural rehabilitation facilities established in the late 1940s were not combat casualties, but members of the military accepted for military service with undetected hearing loss (Bergman, 2002). Although audiometers were available at the time, this measurement equipment was not used at the induction centers responsible for the initial processing of personnel. Instead, a conversational speech test or the "whispered voice test" was used to evaluate the hearing ability of recruits. These tests measured the distance at which an individual understood speech at levels that could be understood from 15 or 20 feet away by people with normal hearing (Nixon, 1998). Normal hearing was thus represented by notations such as "15/15" or "20/20." Military applicants were required to have hearing, as measured by the whispered voice test, of 8/15 or better in each ear (Department of Veterans Affairs, 2004). Results of whispered voice tests are extremely variable between testers (Lee, 1998), and early testing environments within the military settings were not likely to be controlled (Bergman, 2002). Without accurate measures of hearing thresholds, some people inducted into the military with hearing loss were classified as normal-hearing individuals. Estimates from two of the military aural rehabilitation centers established in the 1940s suggest 40–65 percent of patients seen had hearing losses predating their military service (Bergman, 2002).

Requirements for Audiograms

During the 1950s and 1960s, measurement of pure-tone thresholds using audiometers became more widespread. In 1956, the Air Force mandated audiometric testing as part of its hearing conservation program, as well as to establish hearing thresholds for all individuals entering Air Force service as part of their routine physical examination (Department of the Air Force, 1956). However, the Air Force's regulations of 1973 and 1982 did not reiterate the requirement for an audiogram immediately upon entrance into service (Department of the Air Force, 1973, 1982). Instead, the 1982 regulation states that the standard procedure in the Air Force is to obtain reference audiograms within 30 days after assignment at the first permanent duty station (Department of the Air Force, 1982). In 1960, the whispered voice test was replaced by pure-tone audiometry at recruit screening centers (Department of Veterans Affairs, 2004). In 1980, the Army published a policy requiring that a reference audiogram be made a part of the preplacement or entrance physical examination (Department of the Army, 1980), and in 1979, the Navy required that all military personnel receive a reference audiogram upon entry into naval service (Department of the Navy, 1979). DoD did not issue a requirement for reference audiograms at basic training prior to noise exposure until 1996 (DoD, 1996).

Despite these requirements, to date, not all service members are administered a reference audiogram upon entrance. In the Army, Fort Sill is the only basic training site conducting universal audiometric examinations during inprocessing (AMSARA, 2002). In the Air Force, baseline audiometric examinations may frequently be administered after basic training, now more than 6 weeks in duration (Pluta, 2004, 2005a). All of the services stipulate that the audiograms used as a screening tool to establish fitness for military service (most collected at the Military Entrance Processing Stations) are not acceptable as reference audiograms (Department of the Air Force, 1982; Department of the Navy, 1979, 1984). This stems from the varying periods of time (up to 18 months) between the military entrance processing examinations and accession to the military as well as concerns about the reliability of these evaluations (Chandler et al., 1986; Niebuhr et al., 2000; Department of Veterans Affairs, 2004). Screening audiograms at military entrance processing stations require manual transcription, allow for variable intervals between calibration, and do not require technician certification or a defined quiet period before the examination (Niebuhr, 2003).

The Air Force was the first service with a requirement that all military and civilian personnel receive an audiometric examination as part of their routine physical examination when they end their Air Force service (Department of the Air Force, 1956). The 1973 and 1982 Air Force regulation updates (Department of the Air Force, 1973, 1982) stated that all personnel whose duties caused them to be routinely exposed to hazardous noise must receive a final audiometric examination within 90 days before the last day of active duty, but they did not specify that all personnel were to receive audiograms at separation from service. In 1980, the Army (Department of the Army, 1980) and in 1979 the Navy and Marine Corps (Department of the Navy, 1979) required hearing tests upon termination of service. In the Coast Guard, termination physical exams included audiograms starting in the early 1990s (McConnell, 2005). However, physical exams are not generally required by the services upon separation unless the last exam or medical assessment is no longer considered current (except for retiring Army service members) (GAO, 2004). Service members may waive the hearing test, however, and many do so rather than face delays in returning home.

Compliance with Requirements for Annual Audiograms

The committee reviewed data regarding compliance with the requirement for annual audiograms for those enrolled in hearing conservation programs. Data were available for the Army from the HEARS and DOEHRS-HC databases for 1989–2003. Air Force data were available only from DOEHRS-HC for 2000–2003. As noted earlier, the DOEHRS system is designed to collect, maintain, compare, and report hearing conservation data within all branches of the DoD. Since its introduction in 1999, the system has experienced repeated changes in contractors, as well as considerable turnover in the audiometric technicians who administer audiograms (Ohlin, 2004a) and several other infrastructure and training problems (Frost, 2004).

Data furnished to the committee on compliance for the Army and the Air Force are plotted in Figure 5-3. Clear peaks and valleys in the data from the Army may, in part, reflect administrative changes in the database. In 1991, for example, a change from reporting the numbers of service members enrolled in the hearing conservation program by medical region to

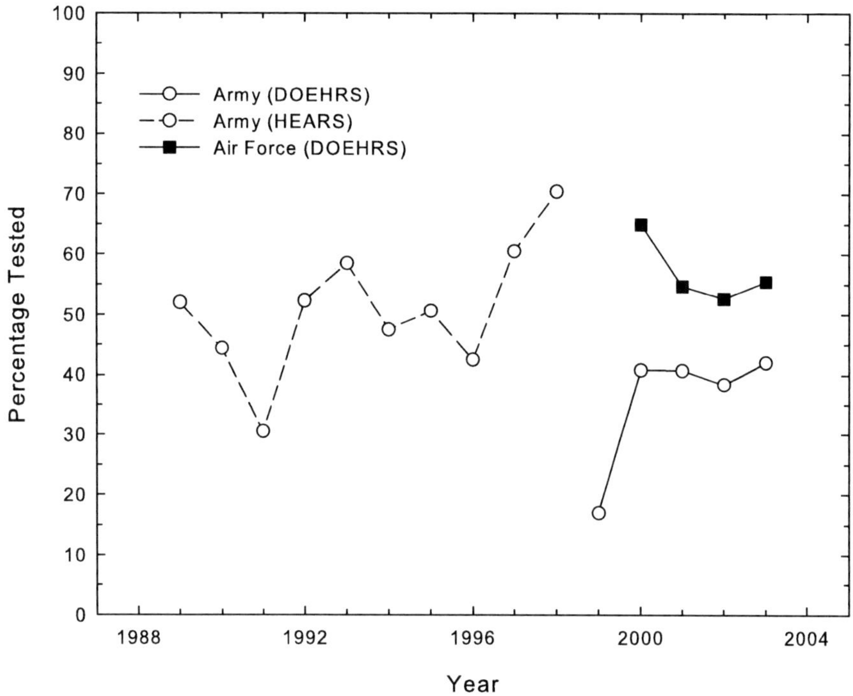

FIGURE 5-3 Percentage of Army and Air Force service members enrolled in hearing conservation programs who received annual audiograms, according to the Army Hearing Evaluation Automated Registry System (HEARS) (1989–1998) and the Defense Occupational and Environmental Health Readiness System–Hearing Conservation (DOEHRS-HC) (1999–2003) data repositories.
SOURCES: Air Force Hearing Conservation Registry (2004a); U.S. Army Center for Health Promotion and Preventive Medicine (2004b).

reporting by individual sites led to a large increase in the number of records reported. In 1998, the Army transitioned its HEARS system to DOEHRS-HC and some of the HEARS data were not available to the new system (Ohlin, 2004b). The denominators (numbers of personnel reported as noise exposed) are often estimates based on unit strengths at the larger installations. Numerators are based on the hearing tests received at the data repository. Compliance can be higher than the reported figures if data from all hearing tests are not received or can be artificially inflated if personnel are tested who are not reported as noise exposed (Ohlin, 2005a). Although there are fluctuations in the Army data from 1988 through 2003, average compliance is estimated to be 45 percent over this period. Since 2000, compliance for the Air Force is approximately 55 percent. Thus, half of the service members who should have annual audiograms are either not being tested or the test results are not being entered into the database. In either case, missing data are serious obstacles to implementing effective hearing conservation programs.

While the Navy participates in DOEHRS-HC, it has not attempted to produce data on compliance with audiometric testing requirements (Page, 2004b). However, Wolgemuth and colleagues (1995) provided some data on compliance for 154 vessels of the U.S. Atlantic Fleet for 1987–1990. A sample of 12,492 audiometric records from the Navy's hearing conservation program, representing sampling rates from 20 to 100 percent per vessel, indicated an average compliance rate for annual audiograms of 81 percent reported across vessels (range: 40–100 percent). Compliance figures were not reported by ship type, so the percentage of personnel represented was not clear. Compliance data were not available for the Coast Guard.

FINDING: Results of annual audiograms are available for approximately half of military service members in hearing conservation programs reporting compliance with testing requirements during the period 1988–2003. Incomplete reporting, lack of compliance with requirements for annual audiograms, or both, severely limit the usefulness of the centralized database and the conclusions that can be drawn from it regarding hearing conservation program effectiveness.

Variability in Military Audiometric Databases

As noted earlier in the chapter, data from reference, periodic, and termination audiograms constitute audiometric databases that can be used to evaluate hearing conservation programs. Thomas (1995) analyzed the Air Force audiometric database, using the ANSI draft standard ADBA protocol (S12.13 1991) to evaluate hearing conservation program effectiveness on the basis of audiometric variability. Under the proposed ANSI criteria, the Air Force hearing conservation program qualified as "unacceptable to marginal." The undesirable levels of variability in the Air Force data may result from normal fluctuations in the responsiveness of individuals, inconsistencies in equipment calibration or testing methods, or actual threshold changes resulting from temporary or permanent hearing loss (Thomas, 1995).

Because methods of audiometric database analysis are best suited for analysis of consecutive audiograms in a stable population with a consistent set of audiometers (Royster and Royster, 2000), a military hearing conservation program may rate poorly using this metric. Service members are typically very mobile—moving from assignment to assignment within the military and then leaving the military within about 4 years (GAO, 1998).

Thus, consecutive annual audiograms are frequently administered in different locations using different audiometers.

Measures of Significant Threshold Shift and Permanent Threshold Shift

Definitions of STS Definitions of STS have changed in the military services over the years, as summarized in Table 5-4. In 2004 (or as the services implemented DOEHRS-HC software distributed in 2004), all services used a definition of STS consistent with that of OSHA (Ohlin, 2005b). OSHA (1983) defines an STS as a change of 10 dB or more in the average pure-tone thresholds at 2000, 3000, and 4000 Hz in either ear compared with the baseline audiogram.

Royster (1992, 1996), Schulz (1994), and Dobie (2005) have analyzed the impact of different definitions of STS on the sensitivity and specificity of the measure. With increasing sensitivity, more hearing conservation program participants are identified as experiencing an STS, and with increasing specificity, fewer participants are unnecessarily designated as warranting follow-up action. Tradeoffs are necessary between the two. Definitions of STS are policy decisions requiring considerations of hearing loss prevention and the costs of follow-up tests, counseling, referrals, and potential job changes.

STS and PTS in Military Hearing Conservation Programs Chapter 3 reviewed the percentages of STS and permanent threshold shift (PTS) in the personnel enrolled in the military services' hearing conservation programs over recent years. In those data, a PTS is defined as an STS that is either (1) confirmed upon retest following at least 40 hours of quiet or (2) not resolved through follow-up testing. When the threshold shift is designated as a PTS (either through confirmation or lack of follow-up), the newest thresholds should, but may not always, become the baseline audiogram for future comparisons. As a result, PTS percentages may be artificially high from either a lack of follow-up testing or a failure to establish a new baseline following a previous PTS.

As shown in Figure 5-4, at 7–8 percent, the reported annual percentages of Air Force hearing conservation program enrollees with PTS for the period 1989–2004 are the lowest levels reported across the services. Rates for that period for the Army, Navy, and Marine Corps are approximately 15 percent. It is important to note, however, that these figures are based on data representing only about half of the enrollees in the hearing conservation programs. If this were a random sample of enrollees, it would be more than adequate for the evaluation of the programs. This is not the case, however, and the types of bias that might be introduced by unknown selection factors limit the utility of these data for the evaluation of the military's hearing conservation programs.

TABLE 5-4 Definitions of Significant Threshold Shift (STS) in the Military Services over Time

Year	DoD	Air Force	Army	Navy/Marine Corps
1956		For personnel with "Class A" hearing: ≥ 20 dB at any of 0.5, 1, 2, 3, 4, 6 kHz For personnel with "Class B" hearing: ≥ 10 dB at 2 kHz, or ≥ 15 dB at 3 kHz, or ≥ 20 dB at 4 kHz, and/or 6 kHz, one or both ears		
1976		Same as previously		
1978	≥ 20 dB in either ear, at any test frequency Test frequencies were 0.5, 1, 2, 3, 4, and 6 kHz	Same as previously		
1979		≥ 20 dB at 1, 2, 3, or 4 kHz	≥ 20 dB at 1, 2, 3, or 4 kHz and/or ≥ 10 dB average at 2, 3, and 4 kHz No age correction	≥ 15dB at any frequency 0.5–6 kHz in either ear
1982		≥ 20 dB at 1, 2, 3, or 4 kHz		
1983				≥ 15 dB at1, 2, 3, or 4 kHz and/or an average change of ≥ 10 dB at 2, 3, and 4 kHz

1987		≥ 20 dB at l, 2, 3, or 4 kHz and/or ≥ 10 dB average at 2, 3, and 4 kHz No age correction	≥ 20 dB at 1, 2, 3, or 4 kHz and/or ≥ 10 dB average at 2, 3, and 4 kHz With age correction	
1999 (DOEHRS-HC)	≥ 15 dB at 1, 2, 3, or 4 kHz and/or ≥ 10 dB average at 2, 3, and 4 kHz No age correction	≥ 15 dB at 1, 2, 3, or 4 kHz and/or ≥ 10 dB average at 2, 3, and 4 kHz No age correction	≥ 15 dB at 1, 2, 3, or 4 kHz and/or ≥ 10 dB average at 2, 3, and 4 kHz No age correction	≥ 15 dB at 1, 2, 3, or 4 kHz and/or ≥ 10 dB average at 2, 3, and 4 kHz No age correction
2004 (DOEHRS-HC)	≥ 10 dB average at 2, 3, and 4 kHz in either ear No age correction ≥ 15 dB in either ear at any test frequency from 1 to 4 kHz considered an early warning of potential future STS	≥ 10 dB average at 2, 3, and 4 kHz in either ear No age correction ≥ 15 dB in either ear at any test frequency from 1 to 4 kHz considered an early warning of potential future STS	≥ 10 dB average at 2, 3, and 4 kHz in either ear No age correction ≥ 15 dB in either ear at any test frequency from 1 to 4 kHz considered an early warning of potential future STS	≥ 10 dB average at 2, 3, and 4 kHz in either ear No age correction ≥ 15 dB in either ear at any test frequency from 1 to 4 kHz considered an early warning of potential future STS

NOTES: "Class A" and "Class B" hearing defined according to Department of the Air Force (1956).
SOURCES: Department of the Air Force (1956, 1976, 1982); Department of the Navy (1979); Schulz (1994, 2005a,b); Ohlin (2005b); Page (2005b).

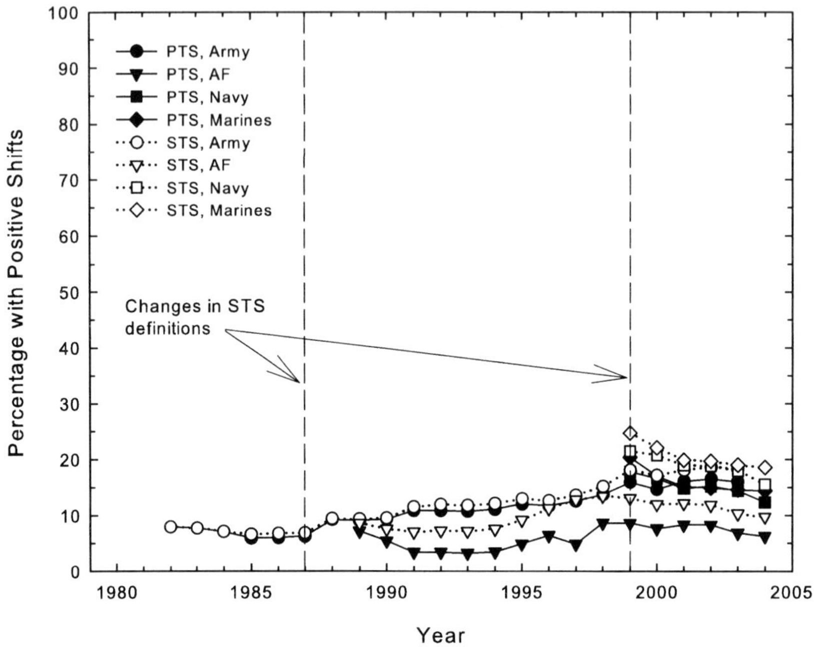

FIGURE 5-4 Percentage of personnel receiving audiometric tests who were identified as showing positive significant threshold shifts (STSs) and permanent threshold shifts (PTSs) (worse hearing) as reported by DOEHRS-HC for the Army, Air Force, Navy, and Marine Corps for 1982–2004. During this time the definition of STS changed, as detailed in Table 5-4.
SOURCES: Air Force Hearing Conservation Registry (2004b); Navy Environmental Health Center (2004a); U.S. Army Center for Health Promotion and Preventive Medicine (2004c).

PTS and STS values for each service are of interest, but far more useful for intervention and improvement are STS and PTS values reported by installation or by military occupational code, which permit a more focused assessment of subpopulations at greatest continuing risk. A review of data for 1991 from the Air Force hearing conservation data registry reported PTS rates of 1 percent and STS rates of 4 percent among pilots (Department of the Air Force, 1992). Among aerospace maintenance workers, 2 percent were reported to have a PTS and 4 percent an STS. Cases of PTS and STS were also reported among other personnel, such as those with financial and paralegal occupational specialties. Davis (1994) assessed the risk of PTS in the Air Force by Air Force Skill Code in 1992. The crude overall risk of PTS among those in the hearing conservation program was 2 percent, with risks ranging from 0 percent to 11 percent across skill codes for military and civilian participants combined.

A series of studies carried out in the mid-1990s assessed hearing thresholds among Navy personnel, as reviewed in Chapter 3. The studies documented continuing STS, not just among engineering and aviation personnel, but also in administrative and supply occupational categories where it would not be anticipated (Wolgemuth et al., 1995; Page et al., 2002).

In sum, the information available from audiometric monitoring carried out by the military services provides a complex picture of their hearing conservation programs. Most prominent, however, is the handicap posed by the poor compliance with requirements for reporting periodic audiometric data to a central repository, limiting the usefulness of the data registry as a surveillance and evaluation tool.

Program Evaluation

No single approach has been taken for program evaluation by the military services' hearing conservation programs. Except for the Air Force, a variety of ad hoc efforts to evaluate the effectiveness of the hearing conservation programs have been made (as drawn upon in Chapter 3 and above). Air Force regulations require annual reports concerning the overall state of the hearing conservation program (Department of the Air Force, 1991). Such reporting began in the mid-1970s, continued intermittently into the mid-1990s, and has recently resumed (Department of the Air Force, 1974, 1980, 1992, 1993, 1995; Meyer and Wirth, 1993; Pluta, 2003, 2005b).

Army regulations require reporting of effectiveness indicators at the installation level (Department of the Army, 1998). Data on cases of STS were reported for several years during the 1990s, but such data were unavailable from 1998 through late 2002 because of limitations in the reporting capability of DOEHRS-HC. Overall statistics for compliance with monitoring requirements, hearing profiles, and positive and negative STS cases are reported on the Army's hearing conservation program website, but no periodic servicewide evaluation is currently undertaken (Ohlin 2005c). However, the U.S. Army Center for Health Promotion and Preventive Medicine provides a program evaluation checklist for use throughout the Army (Ohlin, 1999; U.S. Army Center for Health Promotion and Preventive Medicine, 2005). In the Navy regulations, annual program performance evaluations are to be carried out at the local level but are not required or carried out servicewide (Navy Environmental Health Center, 2004b). According to DoD policy, components are to evaluate the effectiveness of their hearing conservation programs annually based on the prevalence of STS and the percentage of compliance with requirements for annual audiograms (DoD, 2004b).

Since the services' hearing conservation programs were introduced, individuals and groups with a particular commitment to hearing health have worked hard to draw attention to problems and challenges and to evaluate and improve the programs. The committee heard presentations from, and reviewed reports by, people who demonstrated a commitment to the assessment and improvement of their service's hearing conservation program. They are using the tools available to them (e.g., DOEHRS-HC, surveys, specific self-audit software, and epidemiological studies) to assess the effectiveness of their programs. DOEHRS-HC appears to have as yet unrealized potential to improve evaluation of the hearing conservation programs. Reported compliance with requirements for annual audiograms is low, limiting the data available for review and analysis, and reporting functions of the system are limited.

Conclusions About Program Adequacy

The effectiveness of the military hearing conservation programs is difficult to evaluate because of the disjointed and limited information available. The military services must contend with substantial challenges beyond their control, including the mobility and high turnover of their workforce and most significantly, the extreme and frequently unpredictable exposure to hazardous noise in combat.

Several important aspects of the hearing conservation programs, however, are largely within the discretion of DoD and the military services. These include the degree of funding; number of staff; extent of training; command emphasis on the importance of hearing protection; implementation of noise controls; degree of compliance with requirements for entrance, periodic, and termination audiograms; and reporting of audiometric data to a central repository. Although the committee was not able to systematically review each of these categories, the available information, taken together, is sufficient to conclude that the services' hearing conservation programs have been and remain inadequate to protect the hearing of service members. This does not suggest that there are not strong and effective efforts at local levels within the services, or even in leadership roles, but that the sum of these efforts is not yet sufficient.

FINDING: The evidence reviewed by the committee—including information on the effectiveness of available hearing protection devices and indicators regarding use of hearing protection, the completeness of audiometric monitoring, and compliance with requirements for entrance and separation audiograms—was sufficient to conclude that hearing conservation programs in the military are currently not adequate to protect the hearing of military service members, and have not been adequate for the period since World War II. This has important human health, personnel readiness, and financial implications.

REFERENCES

Acoustical Society of America. 2002. *Evaluating the Effectiveness of Hearing Conservation Programs Through Audiometric Data Base Analysis.* ANSI Technical Report: ANSI S12.13 TR-2002. New York: Acoustical Society of America.

Acton WI. 1987. History and development of hearing protection devices. *Journal of the Acoustical Society of America* 81(Suppl 1):S4.

Adera T, Donahue AM, Malit BD, Gaydos JC. 1993. Assessment of the proposed Draft American National Standard method for evaluating the effectiveness of hearing conservation programs. *Journal of Occupational Medicine* 35(6):568–573.

Adera T, Gullickson GM, Helfer T, Wang L, Gardner JW. 1995. Should the audiometric database analysis method (draft ANSI S12.13-1991) for evaluating the effectiveness of hearing conservation programs be accepted as a US national standard? *Journal of the American Academy of Audiology* 6(4):302–310.

Adera T, Amir C, Anderson L. 2000a. Use of comparison populations for evaluating the effectiveness of hearing loss prevention programs. *American Industrial Hygiene Association Journal* 61(1):11–15.

Adera T, Amir C, Anderson L. 2000b. Time trends analysis of hearing loss: An alternative approach to evaluating hearing loss prevention programs. *American Industrial Hygiene Association Journal* 61(2):161–165.

Air Force Hearing Conservation Registry. 2004a. DOEHRS Data Repository: Hearing Conservation Compliance Report, 2000–2003. Data provided to the Institute of Medicine Committee on Noise-Induced Hearing Loss and Tinnitus Associated with Military Service from World War II to the Present, Washington, DC.

Air Force Hearing Conservation Registry. 2004b. DOEHRS Data Repository: Positive Significant Threshold Shift, 1999–2004. Data provided to the Institute of Medicine Committee on Noise-Induced Hearing Loss and Tinnitus Associated with Military Service from World War II to the Present, Washington, DC.

AMSA (Army Medical Surveillance Activity). 1997. U.S. Army Hearing Conservation Program (HCP): Assessment of long term performance. *Medical Surveillance Monthly Report* 3(9):16–18.

AMSARA (Accession Medical Standards Analysis and Research Activity). 2002. *Annual Report 2002.* Washington, DC: Walter Reed Army Institute of Research.

Anderson BW, Garinther GR. 1997. Effects of active noise reduction in armor crew headsets. In: *Audio Effectiveness in Aviation.* NATO AGARD-CP-596. Neuilly-sur-Seine, France: NATO Advisory Group for Aerospace Research and Development. Pp. 20-1–20-6.

Bailey RC, Walker TJ. 1979. *Report on Use of Foam Ear Plugs at Reese AFB, Texas, June 1977–July 1979.* Reese Air Force Base, TX: United States Air Force Hospital Reese.

Berger EH. 1993. *The Naked Truth About NRRs.* EARLog No. 20. Indianapolis, IN: Aearo Co.

Berger EH. 2000a. Hearing protection devices. In: Berger EH, Royster LH, Royster JD, Driscoll DP, Layne M, eds. *The Noise Manual.* 5th ed. Fairfax, VA: American Industrial Hygiene Association. Pp. 379–454.

Berger EH. 2000b. Hearing protection device utilization around the world. *Spectrum* 17 (Suppl 1):18.

Berger EH, Franks JR, Lindgren F. 1996. International review of field studies of hearing protector attenuation. In: Axelsson A, Borchgrevink HM, Hamernik RP, Hellstrom PA, Henderson D, Salvi RJ, eds. *Scientific Basis of Noise-Induced Hearing Loss.* New York: Thieme Medical Publishers. Pp. 361–377.

Berger EH, Royster LH, Royster JD, Driscoll DO, Layne M. 2000. *The Noise Manual.* 5th ed. Fairfax, VA: American Industrial Hygiene Association.

Bergman M. 2002. *On the Origins of Audiology: American Wartime Military Audiology.* Audiology Today. Monograph No. 1. McLean, VA: American Academy of Audiology.

Bjorn VS, Albery CB, Shilling R. 2004a. *U.S. Navy Flight Deck Hearing Protection Use Trends: Survey Results* [abstract]. Presentation at the HFM Symposium on New Directions for Improving Audio Effectiveness, NATO Research and Technology Organization, April 11–13, 2005.

Bjorn VS, Shilling R, McKinley RL. 2004b. *Naval Aviation: Advanced Hearing Protection 201*. Slide set provided to Institute of Medicine Committee on Noise-Induced Hearing Loss and Tinnitus Associated with Military Service from World War II to the Present.

Blackstock DR, Von Gierke HE. 1956. *Development of an Extra Small and Extra Large Size for the V-51R Earplug*. Wright-Patterson Air Force Base, OH: Wright Air Development Center.

Bolger A. 1956. Ryan's hearing conservation program. *Industrial Hygiene Quarterly* 17(March): 522–554.

Byrne D, Monk W. 1993. Evaluating a hearing conservation program: A comparison of the USAEHA method and the ANSI S12.13 method. *Spectrum* 10(Suppl 1):19.

Camp RT, Mozo BT, Kuc LF, Schott GA. 1972. *Real Ear Attenuation Characteristics of Hearing Protective Devices Available Through Federal Supply Channels*. Fort Rucker, AL: U.S. Army Aeromedical Research Laboratory.

Chandler D. 2004. Department of Defense Hearing Conservation Program. Presentation to the Institute of Medicine Committee on Noise-Induced Hearing Loss and Tinnitus Associated with Military Service from World War II to the Present, Meeting I, Washington, DC. May 12.

Chandler DW, Sedge RK, Erdman SA. 1986. Accuracy of pre-induction audiograms. *Military Medicine* 151(5):264–267.

Dancer A, Buck K, Hamery P, Parmentier G. 1999. Hearing protection in the military environment. *Noise Health* 2(5):1–16.

Davis JA. 1994. *A Retrospective Study of a 1992 Subcohort of Air Force Personnel Exposed to Hazardous Noise by Air Force Skill Code*. M.P.H. thesis, University of Texas Health Science Center, Houston.

Department of the Air Force. 1948. *AF Regulation 160-3: Precautionary Measures Against Noise Hazards*. Washington, DC: Department of the Air Force.

Department of the Air Force. 1949. *AF Regulation 160-3: Precautionary Measures Against Noise Hazards*. Washington, DC: Department of the Air Force.

Department of the Air Force. 1956. *AF Regulation 160-3: Hazardous Noise Exposure*. Washington, DC: Department of the Air Force.

Department of the Air Force. 1973. *AF Regulation 161-35: Hazardous Noise Exposure*. Washington, DC: Department of the Air Force.

Department of the Air Force. 1974. *Hearing Conservation Program Status Report (Jan–Jun 1974)*. Brooks Air Force Base, TX: United States Air Force School of Aerospace Medicine.

Department of the Air Force. 1976. *AF Regulation 161-35: Hazardous Noise Exposure*. Washington, DC: Department of the Air Force.

Department of the Air Force. 1980. *Hearing Conservation Program Status Report (1980)*. Brooks Air Force Base, TX: United States Air Force School of Aerospace Medicine.

Department of the Air Force. 1982. *AF Regulation 161-35: Hazardous Noise Exposure*. Washington, DC: Department of the Air Force.

Department of the Air Force. 1991. *AFOSH Standard 161-20: Hearing Conservation Program*. Washington, DC: Department of the Air Force.

Department of the Air Force. 1992. *Hearing Conservation Program Status Report (1992)*. Brooks Air Force Base, TX: Armstrong Laboratory.

Department of the Air Force. 1993. *Hearing Conservation Program Status Report (1993)*. Brooks Air Force Base, TX: Armstrong Laboratory.

Department of the Air Force. 1994. *AFOSH 48-19: Hazardous Noise Program.* Washington, DC: Department of the Air Force.

Department of the Air Force. 1995. *Hearing Conservation Program Status Report (Oct 1993–Sept 1995).* Brooks Air Force Base, TX: Armstrong Laboratory.

Department of the Air Force. 2000. *Integration Guidance (Interim) Between DOHRS-HC and AFR 161-20 Pending AFI 48-20 Publication.* Washington, DC: Department of the Air Force.

Department of the Army. 1956. *TB MED 251: Noise and Conservation of Hearing.* Washington, DC: Department of the Army.

Department of the Army. 1965. *TB MED 251: Noise and Conservation of Hearing.* Washington, DC: Department of the Army.

Department of the Army. 1972. *TB MED 251: Noise and Conservation of Hearing.* Washington, DC: Department of the Army.

Department of the Army. 1980. *TB MED 501: Hearing Conservation.* Washington, DC: Department of the Army.

Department of the Army. 1998. *DA PAM 40-501: Hearing Conservation Program.* Washington, DC: Department of the Army.

Department of the Navy. 1955. *BUMED 6260.6: Hearing Conservation Program.* Washington, DC: Bureau of Medicine and Surgery.

Department of the Navy. 1970. *BUMED 6260.6B: Hearing Conservation Program.* Washington, DC: Bureau of Medicine and Surgery.

Department of the Navy. 1979. *OPNAVINST 6260.2: Hearing Conservation Program.* Washington, DC: Office of the Chief of Naval Operations.

Department of the Navy. 1984. *NAVMEDCOMINST 6260.5 CH-1: Occupational noise control and hearing conservation.* Washington, DC.

Department of the Navy. 1998. *Marine Corps Order 5100.8F: Marine Corps Occupational Safety and Health Program Manual.* Washington, DC: Department of the Navy.

Department of the Navy. 2000. *Marine Corps Order 6260.1E: Marine Corps Hearing Conservation Program Procedures.* Washington, DC: Department of the Navy.

Department of the Navy. 2001. *OPNAVINST 5100.19D: Navy Occupational Safety and Health: Program Manual for Forces Afloat.* Washington, DC: Department of the Navy.

Department of the Navy. 2002. *OPNAVINST 5100.23F: Navy Occupational Safety and Health Program Manual.* Washington, DC: Department of the Navy.

Department of Veterans Affairs. 2004. *Handbook of Standard Procedures and Best Practices for Audiology Compensation and Pension Examinations.* Dennis KC, Beck LB, Chandler DW, Schafer J, eds. Washington, DC: Department of Veterans Affairs.

Department of Veterans Affairs. 2005. *America's Wars.* [Online]. Available: http://www1.va.gov/opa/fact/amwars.html [accessed May 2005].

Dobie RA. 1995. Prevention of noise-induced hearing loss. *Archives of Otolaryngology—Head and Neck Surgery* 121(4):385–391.

Dobie RA. 2005. Audiometric threshold shift definitions: Simulations and suggestions. *Ear and Hearing* 26(1):62–77.

DoD (Department of Defense). 1978. *Department of Defense Instruction 6055.3: Hearing Conservation.* Washington, DC: Department of Defense.

DoD. 1987. *Department of Defense Instruction 6055.12: Hearing Conservation.* Washington, DC: Department of Defense.

DoD. 1996. *Department of Defense Instruction 6055.12: DoD Hearing Conservation Program.* Washington, DC: U.S. Department of Defense.

DoD. 1997. *MIL-STD-1474D: Department of Defense Design Criteria Standard—Noise Limits.* Washington, DC: Department of Defense.

DoD. 2004a. *Active Duty Military Personnel by Rank/Grade. December 2004.* [Online]. Available: http://www.dior.whs.mil/mmid/military/miltop.htm [accessed May 9, 2005].

DoD. 2004b. *Department of Defense Instruction 6055.12: DoD Hearing Conservation Program.* Washington, DC: Department of Defense.

DoD. 2005. *Active Duty Military Strength Report for March 31, 2005. Military Personnel Statistics.* [Online]. Available: http://www.dior.whs.mil/mmid/military/miltop.htm [accessed May 9, 2005].

Donahue AM, Ohlin DW. 1993. Noise and the impairment of hearing. In: Deeter DP, Gaydos JC, eds. *Occupational Health: The Soldier and the Industrial Base.* Falls Church, VA: Office of the Surgeon General.

Franks JR, Davis RR, Krieg EF. 1989. Analysis of a hearing conservation program database: Factors other than workplace noise. *Ear and Hearing* 10(5):273–280.

Frost AW. 2004. DoD Hearing Conservation Program Management Tools. Presentation at the 43rd Navy Occupational Health and Preventive Medicine Workshop. Navy Environmental Health Center, Portsmouth, Virginia.

GAO (General Accounting Office). 1977. *Hearing Protection: Problems in the Department of Defense.* GAO 03495. Washington, DC: GAO.

GAO. 1998. *Military Attrition: Better Data, Coupled With Policy Changes, Could Help the Services Reduce Early Separations.* GAO/NSIAD-98-213. Washington, DC: GAO.

GAO (Government Accountability Office). 2004. *Efforts to Coordinate a Single Physical Exam Process for Servicemembers Leaving the Military.* GAO-05-64. Washington, DC: GAO.

Gardner R, Berger EH. 1994. *History and Development of the EAR Foam Earplug.* EAR Tech. Rept. 94-10/HP. Indianapolis, IN: Aearo Corporation.

Gasaway DC. 1985. Significant milestones in the evolution of hearing conservation programs. In: Gasaway D. *Hearing Conservation: A Practical Manual and Guide.* Englewood Cliffs, NJ: Prentice-Hall. Pp. 1–11.

Gasaway DC. 1988. Occupational hearing conservation in the military. In: Lipscomb DM, ed. *Hearing Conservation in Industry, Schools, and the Military.* Boston, MA: Little, Brown. Pp. 243–262.

Gates K. 2005. *72CINVENTORY.xls* [Excel file]. Washington, DC: Department of the Army.

Guild E. 1966. Personal protection. In: *Industrial Noise Manual.* Akron, OH: American Industrial Hygiene Association. Pp. 84–109.

Gwin LP, Lacroix PG. 1985. A prevalence study of hearing conservation in the United States Naval Submarine Force. *Military Medicine* 150(12):652–656.

Hall MD. 2001. *US Coast Guard Hearing Conservation Program.* [Online]. Available: http://www.uscg.mil/hq/g-w/g-wk/wkh/omsep/lectures/Mod9/Mod_9_HCP.doc.doc [accessed February 4, 2005].

Haluska FP. 1964. Hearing conservation in a metal stamping and forging plant. *National Safety Congress Transactions* 3:11–15.

Hatton JF. 1956. Lockheed's ear protection program. *Industrial Hygiene Quarterly* 17(March):48–49.

Hetu R, Quoc HT, Duguay P. 1990. The likelihood of detecting a significant hearing threshold shift among noise-exposed workers subjected to annual audiometric testing. *Annals of Occupational Hygiene* 34(4):361–370.

Holland HH. 1967. Attenuation provided by fingers, palms, tragi, and V51R ear plugs. Letter to the editor. *Journal of the Acoustical Society of America* 41(6):1545.

Lee SE. 1998. *Role of Driver Hearing in Commercial Motor Vehicle Operation: An Evaluation of the FHWA Hearing Requirement.* Ph.D. dissertation, Virginia Polytechnic Institute and State University, Blacksburg.

Loeb M, Luz GA, Sheidler D, Vanderhei S. 1973. *Assessment of Auditory Acuity, Ear Protection, and Prediction of Hearing Loss in Military Personnel.* Louisville, KY: University of Louisville Psychology Research Department and Performance Research Laboratory.

McConnell W. 2004. Hearing Conservation in the Coast Guard. Presentation to the Institute of Medicine Committee on Noise-Induced Hearing Loss and Tinnitus Associated with Military Service from World War II to the Present, Meeting I, Washington, DC. May 12.

McConnell W (U.S. Coast Guard). 2005. Technical Review of Draft Section on Coast Guard. Comments provided to the Institute of Medicine Committee on Noise-Induced Hearing Loss and Tinnitus Associated with Military Service from World War II to the Present, Washington, DC.

McKinley RL, Nixon CW. 1993. Active noise reduction headsets. In: Vallet M, ed. *Noise & Man 93: Noise as a Public Health Problem.* Vol. 2. Proceedings of the 6th International Congress. Nice, France, July 1993. Institut National de Recherche sur les Transport et leur Sécurité. Pp. 83–86.

Melnick W. 1984. Evaluation of industrial hearing conservation programs: A review and analysis. *American Industrial Hygiene Association Journal* 45(7):459–467.

Meyer GD, Wirth DB. 1993. An evaluation of the U.S. Air Force's detailed follow-up audiometric examination program. *Military Medicine* 158(9):603–605.

Moritz CT, Bruce RD. 1994. A brief history of hearing protection in the United States. *Journal of the Acoustical Society of America* 95(5):2913.

Morrill JC, Sterrett ML. 1981. Quality controls for audiometric testing. *Occupational Health & Safety* 50(8):26–33.

Mozo BT, Murphy BA. 1997. *The Assessment of Sound Attenuation and Speech Intelligibility of Selected Active Noise Reduction Devices and the Communications Earplug When Used with the HGU-56/P Aviator Helmet.* USAARL Rept. No. 97-08. Fort Rucker, AL: U.S. Army Aeromedical Research Laboratory.

Mozo BT, Murphy BA. 1998. *Novel Communications and Hearing Protection for Helmet Systems: Communications Earplug.* USAARL Rept. No. 98-23. Fort Rucker, AL: U.S. Army Aeromedical Research Lab.

Narrigan JJ. 2004. U.S. Air Force: Hearing Conservation, a Historical Overview. Presentation to the Institute of Medicine Committee on Noise-Induced Hearing Loss and Tinnitus Associated with Military Service from World War II to the Present, Meeting I, Washington, DC. May 12.

Navy Environmental Health Center. 2004a. DOEHRS Data Repository: Positive Significant Threshold Shift, 1999–2004. Data provided to the Institute of Medicine Committee on Noise-Induced Hearing Loss and Tinnitus Associated with Military Service from World War II to the Present, Washington, DC.

Navy Environmental Health Center. 2004b. *Navy Medical Department Hearing Conservation Program Procedures TM 6260.51.99-2.* Portsmouth, VA: Bureau of Medicine and Surgery.

Neitzel R, Seixas N. 2005. The effectiveness of hearing protection among construction workers. *Journal of Occupational and Environmental Hygiene* 2(4):227–238.

Niebuhr DW. 2003. *A Comparison of the Medical Examination Processing Station Hearing Screen to the Baseline Audiogram: Fort Sill, OK.* Washington, DC: Walter Reed Army Institute of Research, Accession Medical Standards Analysis and Research Activity.

Niebuhr DW, Chandler DW, Helfer TM, Completo J. 2000. *A Comparison of the Medical Entrance Processing Station (MEPS) Screening Audiogram with the Defense Occupational and Environmental Health Readiness (DOEHRS) Baseline Audiogram.* Washington, DC: Walter Reed Army Institute of Research, Accession Medical Standards Analysis and Research Activity.

NIOSH (National Institute for Occupational Safety and Health). 1972. *NIOSH Criteria for a Recommended Standard: Occupational Exposure to Noise*. DHEW (NIOSH) Pub. No. HSM 73-1001. Cincinnati, OH: NIOSH.

NIOSH. 1996. *Preventing Occupational Hearing Loss—A Practical Guide*. Franks JR, Stephenson MR, Merry CJ, eds. DHHS (NIOSH) Pub. No. 96-110. Cincinnati, OH: NIOSH.

NIOSH. 1998. *NIOSH Criteria for a Recommended Standard: Occupational Noise Exposure*. DHHS (NIOSH) Pub. No. 98-126. Cincinnati, OH: NIOSH.

NIOSH. 2005. *Noise and Hearing Loss Prevention: Hearing Conservation Program Evaluation Checklist*. [Online]. Available: http://www.cdc.gov/niosh/topics/noise/workplace solutions/ hearingchecklist.html [accessed May 17, 2005].

Nixon CW. 1998. *A Glimpse of History: The Origin of Hearing Conservation Was in the Military?* Wright-Patterson Air Force Base, OH: U.S. Air Force Research Laboratory.

Ohlin D. 1999. Web-based self-audit protocol and effectiveness measures in the military. In: *Proceedings: Best Practices in Hearing Loss Prevention*. DHHS (NIOSH) Pub. No. 2001-157. Cincinnati, OH: NIOSH. Pp. 13–15.

Ohlin D (U.S. Army Center for Health Promotion and Preventive Medicine). 2004a. Materials distributed to the Institute of Medicine Committee on Noise-Induced Hearing Loss and Tinnitus Associated with Military Service from World War II to the Present, Meeting II, Washington, DC. July 19.

Ohlin D (U.S. Army Center for Health Promotion and Preventive Medicine). 2004b. Requested DOEHRS-HC reports for the Institute of Medicine Committee on Noise-Induced Hearing Loss and Tinnitus Associated with Military Service. E-mail to J. Durch, Institute of Medicine, July 23.

Ohlin D (U.S. Army Center for Health Promotion and Preventive Medicine). 2005a. Technical Review of Draft Section on Army Hearing Conservation Program. Comments provided to the Institute of Medicine Committee on Noise-Induced Hearing Loss and Tinnitus Associated with Military Service from World War II to the Present, Washington, DC.

Ohlin D (U.S. Army Center for Health Promotion and Preventive Medicine). 2005b. RE: Seek Help With STS Definitions. E-mail to L. Joellenbeck, Institute of Medicine, March 25.

Ohlin D (U.S. Army Center for Health Promotion and Preventive Medicine). 2005c. Re: Seeking Assistance with Questions Posed by Reviewers. E-mail to L. Joellenbeck, Institute of Medicine, July 29.

Ohlin D. 2005d. Sound identification training: Auditory armament for the battlefield. *Update (CAOHC)* 17(2):1,9.

OSHA (Occupational Safety and Health Administration). 1971. Occupational noise exposure. 29 C.F.R. 1910.95. *Federal Register* 36(105):10518.

OSHA. 1983. Occupational Noise Exposure; Hearing Conservation Amendment. Final Rule. 29 C.F.R. 1910.95. *Federal Register* 48(46):9738–9785.

Page JC. 2004a. U.S. Navy: Assessment of Acoustic Trauma Associated With Military Service from WW II to the Present. Presentation to the Institute of Medicine Committee on Noise-Induced Hearing Loss and Tinnitus Associated With Military Service From World War II to the Present, Meeting I, Washington, D.C. May 12.

Page JC (Navy Environmental Health Center). 2004b. RE: DOEHRS-HC data for Navy. E-mail to L. Joellenbeck, Institute of Medicine, October 27.

Page JC (Navy Environmental Health Center). 2005a. RE: Census Count, DoD Audiologists. E-mail to R. Danielson, Institute of Medicine Committee on Noise-Induced Hearing Loss and Tinnitus Associated With Military Service From World War II to the Present, May 17.

Page JC (Navy Environmental Health Center). 2005b. RE: Request technical review of draft report sections. E-mail to L. Joellenbeck, Institute of Medicine, June 9.

Page JC, Bohnker BK, Rovig G, Betts LS, Sack DM. 2002. Navy hearing conservation program: Threshold shifts in enlisted personnel, 1995–1999. *Military Medicine* 167(1):48–52.

Pluta R. 2003. *HCDR Annual Report 2003*. Brooks City–Base, TX: U.S. Air Force.

Pluta R. 2004. The USAF Hearing Conservation Program—Sustaining Readiness Through Healthy Communities. Presentation to the Institute of Medicine Committee on Noise-Induced Hearing Loss and Tinnitus Associated with Military Service from World War II to the Present, Meeting IV, Washington, DC. December 7.

Pluta R (U.S. Air Force). 2005a. RE: Seek help with technical review. E-mail to L. Joellenbeck, Institute of Medicine, June 24.

Pluta R (U.S. Air Force). 2005b. 2003 Annual Report. E-mail to L. Joellenbeck, Institute of Medicine, June 17.

Price IR, Whitaker G. 1986. Noise exposure during shotblasting and the acoustic properties of air-fed helmets. In: Lotz R, ed. *Proceedings of Inter-Noise 86*. Poughkeepsie, NY: Noise Control Foundation. Pp. 559–564.

Ridgley CD, Wilkins JR. 1991. A comparison of median hearing thresholds from U.S. Navy and U.S. Army audiometric data bases. *Military Medicine* 156(7):343–345.

Robertson RM, Williams CE. 1984. *Development of a Navy Hearing Conservation Management Information System (HECMIS)*. Pensacola, FL: Naval Aerospace Medical Research Laboratory.

Rosenblith WA, Wheeler DE, Smedal H. 1952. *Problems of High-Intensity Noise: A Survey and Recommendations*. Cambridge, MA: Office of Naval Research.

Rovig GW, Bohnker BK, Page JC. 2004. Hearing health risk in a population of aircraft carrier flight deck personnel. *Military Medicine* 169(6):429–432.

Royster JD. 1992. *Evaluation of Different Criteria for Significant Threshold Shift in Occupational Hearing Conservation Programs*. Raleigh, NC: Environmental Noise Consultants, Inc.

Royster JD. 1996. *Evaluation of Additional Criteria for Significant Threshold Shift in Occupational Hearing Conservation Programs*. Raleigh, NC: Environmental Noise Consultants, Inc.

Royster LH, Royster JD. 1990. Important elements and characteristics of hearing conservation programs and determination of their effectiveness. *Environment International* 16: 339–352.

Royster JD, Royster LH. 2000. Evaluating hearing conservation program effectiveness. In: Berger EH, Royster LH, Royster JD, Driscoll DP, Layne M, eds. *The Noise Manual*. 5th ed. Fairfax, VA: American Industrial Hygiene Association. Pp. 518–546.

Schleifer J, Fankhauser C, Strom CG, Loyborg W. 1984. Questionnaire survey of hearing protector use. *U.S. Navy Medicine* 75(2):14–15.

Schulz T. 1994. *Monitoring Audiometry in Hearing Conservation Programs*. Ph.D. dissertation, Ohio State University, Columbus.

Schulz T (James H. Quillen VA Medical Center). 2005a. IOM Answers 8.7.05. E-mail to L. Joellenbeck, Institute of Medicine, August 8.

Schulz T (James H. Quillen VA Medical Center). 2005b. Clarification of Reviewer Comments. E-mail to L. Joellenbeck, Institute of Medicine, August 9.

Shaw EAG. 1979. Hearing protector attenuation: A perspective view. *Applied Acoustics* 12:139–157.

Shaw WA, Veneklasen PS. 1945. *The Development of Ear Wardens Type V-51R*. Cambridge, MA: Office of Scientific Research and Development, National Defense Research Committee.

Simpson TH, Stewart M, Kaltenbach JA. 1993. Effects of audiometric threshold step size on proposed ANSI S12.13 outcomes for characterizing hearing conservation program effectiveness. *Journal of the American Academy of Audiology* 4(4):258–263.

Simpson TH, Stewart M, Kaltenbach JA. 1994. Early indicators of hearing conservation program performance. *Journal of the American Academy of Audiology* 5:300–306.

Simpson TH, Amos N, Rintelmann WF. 1998. Effects of pre-existing hearing loss on proposed ANSI S12.13 outcomes for characterizing hearing conservation program effectiveness: Follow-up investigation. *Journal of the American Academy of Audiology* 9(2):112–120.

Smoorenburg GF, ten Raa BH, Mimpen AM. 1986. *Real-World Attenuation of Hearing Protectors.* Presentation at the 12th International Congress on Acoustics, Toronto, Canada.

Suter AH. 1986. Hearing conservation. In: Berger EH, Ward WD, Morrill JC, Royster LH, eds. *Noise & Hearing Conservation Manual.* 4th ed. Akron, OH: American Industrial Hygiene Association. Pp. 1–18.

Suter AH. 1988. The development of federal noise standards and damage risk criteria. In: Lipscomb DM, ed. *Hearing Conservation in Industry, Schools, and the Military.* Boston, MA: College-Hill Press. Pp. 45–66.

Suter AH. 2000. Standards and regulations. In: Berger EH, Royster LH, Royster JD, Driscoll DP, Layne M, eds. *The Noise Manual.* 5th ed. Fairfax, VA: American Industrial Hygiene Association. Pp. 639–668.

Suter AH, Franks JR. 1990. *A Practical Guide to Effective Hearing Conservation Programs in the Workplace.* DHHS (NIOSH) Pub. No. 90–120. Cincinnati, OH: National Institute for Occupational Safety and Health.

Thomas JW. 1995. *The Application of Audiometric Data Base Analysis to Selected Air Force Bases.* Master's thesis, University of North Carolina, Chapel Hill.

U.S. Army Center for Health Promotion and Preventive Medicine. 2004a. *Defense Occupational and Environmental Health Readiness System—Hearing Conservation.* [Online]. Available: http://chppm-www.apgea.army.mil/hcp/about.aspx [accessed May 4, 2004].

U.S. Army Center for Health Promotion and Preventive Medicine. 2004b. DOEHRS Data Repository: Army Hearing Conservation Program Compliance Reports, 1988–2003, Including HEARS Data Comparison for 1989–1998. Data provided to the Institute of Medicine Committee on Noise-Induced Hearing Loss and Tinnitus Associated with Military Service from World War II to the Present, Washington, DC.

U.S. Army Center for Health Promotion and Preventive Medicine. 2004c. DOEHRS Data Repository: Hearing Conservation Program Positive STS Report, 1983–2003. Data provided to the Institute of Medicine Committee on Noise-Induced Hearing Loss and Tinnitus Associated with Military Service from World War II to the Present, Washington, DC.

U.S. Army Center for Health Promotion and Preventive Medicine. 2004d. *Just the Facts: The Combat Arms Earplug.* Aberdeen Proving Ground, MD: U.S. Army Center for Health Promotion and Preventive Medicine.

U.S. Army Center for Health Promotion and Preventive Medicine. 2005. *Hearing Conservation Program Evaluation Profile (HCPEP).* [Online]. Available: http://chppm-www.apgea.army.mil/hcp//HCPEP/ [accessed January 28, 2005].

U.S. Coast Guard. 1982. *Recommendations on Control of Excessive Noise (Navigation and Vessel Inspection Circular No. 12-82).* Washington, DC: U.S. Coast Guard.

U.S. Coast Guard. 1990. *COMDTINST M6000.1B: Medical Manual.* Washington, DC: U.S. Coast Guard.

U.S. Coast Guard. 1998. *When Was the Coast Guard Established?* [Online]. Available: http://www.uscg.mil/hq/g-cp/history/faqs/when.html [accessed February 2, 2005].

U.S. Coast Guard. 2003. *COMDTINST M5100.47: Safety and Environmental Health Manual.* Washington, DC: U.S. Coast Guard.

U.S. Department of Labor. 1969. Occupational noise exposure. *Federal Register* 34:7948.

Walden BE, Worthington DW, McCurdy HW. 1971. *The Extent of Hearing Loss in the Army: A Survey Report.* Washington, DC: Walter Reed Army Medical Center.

Walden BE, Prosek RA, Worthington DW. 1975. *The Prevalence of Hearing Loss within Selected U.S. Army Branches.* Washington, DC: Walter Reed Army Medical Center.

Walpole RH. 1943. *Test of Acousti-Guard (Ear Protective Device).* Fort Knox, KY: Armored Medical Research Laboratory.

Wilkins R. 1956. One approach to hearing conservation. *Industrial Hygiene Quarterly* 17 (March):54–55.

Wolgemuth KS, Luttrell WE, Kamhi AG, Wark DJ. 1995. The effectiveness of the Navy's hearing conservation program. *Military Medicine* 160(5):219–222.

Yankaskas K. 2004. Shipboard Noise and Hearing Conservation Issues. Presentation to the Institute of Medicine Committee on Noise-Induced Hearing Loss and Tinnitus Associated with Military Service from World War II to the Present, Meeting IV, Washington, DC. December 7.

Yankaskas KD, Shaw MF. 1999. Landing on the roof: CVN noise. *Naval Engineers Journal* 111(4):23–34.

6

Reports of Audiometric Testing in Service Medical Records of Military Veterans

One of the tasks for this study was to review the service medical records of military veterans to examine the compliance by the military services with regulations requiring audiograms. This chapter describes and presents results from the study to evaluate service medical records for the presence of audiograms performed when service members entered and left active duty (referred to here as entrance and separation audiograms). The Army, Navy, and Marine Corps have required such testing for all service members since at least the early 1980s, but some audiometric testing was being done in all of the services as early as the 1940s.

The specific language of the Statement of Task called for records to be "examined for regulatory compliance regarding audiometric surveillance (including reference, periodic, and termination audiograms)." Service members enrolled in a hearing conservation program are to receive "reference, periodic, and termination" audiograms at the beginning, annually, and at the end of their enrollment in the program. As noted in Chapter 5, only service members who meet certain noise-exposure criteria are placed in a hearing conservation program (see Table 5-1).

As framed, the stated task presented two problems. First, the hearing conservation programs of the Army, Navy, and Marine Corps did not have mandatory testing until the late 1970s and 1980. As a result, there is no basis for "compliance" with testing requirements for the period from World War II through the 1970s. Second, the services do not have central registries of personnel enrolled in their hearing conservation programs. As a result, it was not feasible to draw study samples limited to personnel who had participated in those programs.

The study that was conducted was based on data from service medical records of individuals who had served in the military, without regard to their enrollment in a hearing conservation program. In the committee's view, this was a more appropriate basis for the study than considering only personnel who had been enrolled in hearing conservation programs. As has been noted throughout this report, it is critical that measures of hearing and tinnitus be obtained at entry and exit from military service for all personnel in order to document any changes in hearing or tinnitus that occur during military service. Although participants in hearing conservation programs may be exposed to known noise hazards, other military personnel may also be exposed to hazardous noise. Moreover, the Department of Veterans Affairs (VA) must consider claims regarding hearing loss or tinnitus from veterans, regardless of their enrollment in a service's hearing conservation program. The majority of veterans receiving compensation for hearing loss at the end of 2004, for example, served in the military before the widespread implementation of hearing conservation programs in the late 1970s (Department of Veterans Affairs, 2005).

As conducted, this study assessed the extent to which audiometric test results were present in the medical records of 3,570 randomly selected service members who had separated from military service during one of five eras spanning the period from World War II to 2002. These records were reviewed, and the dates of all reports of audiometric tests were abstracted. The information on these tests was used to assess the proportion of service members who received audiometric tests at the beginning and the end of their military service. The study was to have included a review of records for Coast Guard personnel, but the Coast Guard did not grant the Institute of Medicine permission to have access to Coast Guard service medical records. Therefore, Coast Guard records were not included in this study.

The study protocol was approved by the Institutional Review Board of the National Academies.

STUDY METHODS

Selection of Study Sample

Individual service medical records were selected for use in the study in the following manner. Random samples of service member identifiers were selected from listings of service members (referred to as rosters) available to the Medical Follow-up Agency (MFUA) of the Institute of Medicine. These rosters span the period from World War II to 2002. Five service eras were defined: 1940 to 1949, 1950 to 1969, 1970 to the year hearing conservation program regulations were implemented,[1] the year hearing conservation program regulations were implemented to 1993, and 1994 to 2002. Lists of service members were generated from the available rosters for each branch of service (Army, Navy, Air Force, Marine Corps) and for each of the five service eras used. Records were assigned to one of the five service eras based on the service member's date of separation from service (release from active duty). Additional detail about the sampling frames for these service eras is provided in Box 6-1.

Sample-size targets were set at 100 records for each service and era prior to 1970, and 200 records for each service and era after 1970. These targets were established based on a priori estimates of the percentages of files with audiograms and the degree of confidence sought in those numbers. On the basis of MFUA experience in obtaining service medical records, rosters were oversampled for each service branch–service era category to take into account files that were missing or otherwise unavailable. To reach the overall target of 3,200 records, 6,218 records were requested. When the desired sample size for a branch of service and service era had been met or exceeded, record review for that time period was discontinued and any remaining files were returned without review.

Record Access and Data Abstraction

It was possible for the service medical records that were sought for the study to be located at the National Personnel Records Center (NPRC) in St. Louis, Missouri, the VA Regional Management Center in St. Louis, or VA regional offices across the country (see Box 6-2). Lists of service members selected for the study were matched against VA's Beneficiary Identification and Records Locator Subsystem (BIRLS) to ascertain the location of the records. Records were requested from the appropriate source. Records held by VA regional offices were mailed to the National Research Council office at the VA Regional Office in Washington, D.C., where they were abstracted under the supervision of MFUA staff. Records stored in St. Louis, at either the VA Regional Management Center or NPRC, were abstracted on-site under the supervision of MFUA staff.

Presence in the service medical record of any of the following forms was noted and recorded: report of medical examination (SF 88, DD Form 2808); medical history (SF 89, DD Form 2807); reference audiogram (AF 1491, DD Form 2215); or monitoring or termination audiogram (AF 1490,

[1]Hearing conservation programs were implemented in 1980 by the Army and in 1979 by the Navy and Marine Corps (for the purposes of this study, however, 1983 was used as the year of regulation for the Navy and Marine Corps). The Air Force updated its hearing conservation program requirements in 1982.

BOX 6-1
Sampling Frames for Service Eras

No single comprehensive list exists of persons who served in the Armed Forces during the period from World War II to 2002, or of service members for whom a service medical record is available. Several different representative listings of service members were used to span this time period:

World War II through 1949: Medical Follow-up Agency's (MFUA) World War II database (Roster #500) is derived from 1 percent or 2 percent samples of National Service Life Insurance policyholders. This insurance program for service members and veterans issued roughly 22 million policies from October 1940 through April 1951, and most of the participants served during World War II. Because Roster #500 differs from the other listings in that it is not based on year of separation from military service, the potential existed for persons selected from this roster to also be selected from one of the other listings. The records selected from Roster #500 were reviewed to identify any duplication, and none was found. Where available, entrance and separation dates for service members identified through this roster were ascertained using dates provided through the Beneficiary Identification and Records Locator Subsystem (BIRLS).

1950 through 1969: Two MFUA rosters were used to span the time period from 1950 through 1969. Roster #552 is a 0.1 percent sample of Armed Forces separations and spans separation years from 1950 to 1959 for some of the services, but it is most useful for the period 1950–1954. Roster #588 is a 1 percent sample of Armed Forces separations from 1955 through 1969.

1970 through 2002: The Defense Manpower Data Center (DMDC) of the Department of Defense maintains a computerized database of service records covering the period from the early 1970s to the present. DMDC has provided MFUA with a 1 percent sample of all the separations in its database. This listing provided sampling frames for the time periods "1970 through year of regulation," "year of regulation through 1993," and "1994 through 2002."

BOX 6-2
Storage of Service Medical Records

Service medical records are either stored by the services at the National Personnel Records Center (NPRC) in St. Louis, Missouri, or held by the Department of Veterans Affairs (VA). For individuals who left military service before approximately 1994, VA has custody in one of its regional offices of those service medical records that have been used in support of a medical claim. Service medical records of those who left military service before 1994, but have not filed medical claims, are stored at NPRC. VA holds the service medical records of all service members who separated from military service since 1994, either at the VA Regional Management Center in St. Louis or at VA sites across the country, where they are in use in conjunction with claims.

DD Form 2216). From these forms, information was recorded about the date, duty occupation code, presence/absence of numerical data from an audiogram, type of audiometer, type of hearing protection issued/used, purpose of or reason for the test, and the presence/absence of a significant threshold shift (STS). Numerical reports of hearing thresholds were not abstracted.

Table 6-1 shows the number of medical records reviewed for each service and time period. As can be seen, except for the Air Force for the two earliest eras, sample sizes were at least 90 percent of the targeted size and, most often, exceeded the targets. The fact that the Air Force was established as a separate service only in 1947 may have influenced the availability of Air Force records for the earliest era.

RESULTS

Presence of Any Audiograms

Overall, 82 percent (95% CI 80–83 percent) of the records reviewed contained the report of at least one audiometric test.[2] As shown in Table 6-2, the percentage of records containing audiometric data was lowest for those who separated from service in the earliest time periods. The Air Force was an early leader in audiometric testing. By the 1970s, however, at least 93 percent, and typically 98–100 percent, of the medical records sampled from each of the services had at least one audiogram.

Presence of Entrance Audiograms

Table 6-3 shows the percentages of service medical records with audiograms obtained within 60 days before or after the service member's entrance into active duty. The committee decided on ±60 days as a reasonable, but arbitrary, window. Using this window reflects the committee's recognition that not all audiograms will have been obtained for all personnel exactly on the dates of entry into and separation from the military while still being in reasonably close proximity to the dates of entry and separation. There is no question that the duration of this time window will have an impact on the percentage of service medical records considered to contain entry and separation audiograms. The wider the window, the higher the resulting percentages are. The maximum possible

---

[2]Counts and frequencies were tabulated using Microsoft Access and Statistical Applications Software. Confidence intervals were calculated according to Gardner and Altman (1989) as $p - (N_{1-\alpha/2} \times SE)$ to $p + (N_{1-\alpha/2} \times SE)$, where p is the observed proportion, N is the number of records, and SE (the standard error) = $\sqrt{[p\,(1 - p) / n]}$.

TABLE 6-1 Number of Service Medical Records Reviewed and Abstracted

Branch	Before 1950	1950–1969	1970–Regulation Date	Regulation Date–1993	1994–2002	Total
Army	187	164	220	217	273	1,061
Air Force	54	81	202	203	214	754
Marine Corps	93	95	207	187	250	832
Navy	122	102	225	204	270	923
Total	456	442	854	811	1,007	3,570

NOTE: Regulation date = 1980 for Army, 1983 for Navy and Marine Corps. The Air Force date of regulation is discussed in the text.

percentages for an unlimited time window are captured by the data shown in Table 6-2. Again, to be of value in assessing whether hearing loss or tinnitus are service connected, measurements must be obtained near the beginning and end of military service, and for this analysis, the committee considered 60 days to be an acceptable time frame for obtaining these measurements.

In general, although the effect varied considerably with the military service branch and era, additional analyses of these data indicated that doubling of the time window from ±60 days to ±120 days increased the percentages shown in Tables 6-3 through 6-5 by approximately 9–12 percentage points. That is, if 30 percent of the service medical records for a given branch and era had entrance audiograms when using a time window of ±60 days, then about 39–42 percent of the records contained entrance

TABLE 6-2 Percentages of Service Medical Records (95% Confidence Intervals) with Reports Containing Any Numeric Data from an Audiogram (n = 3,570)

Branch	Before 1950	1950–1969	1970–Regulation Date	Regulation Date–1993	1994–2002
Army	8 (4–12)	32 (25–39)	99 (97–100)	98 (96–100)	99 (97–100)
Air Force	81 (71–92)	77 (67–86)	100 (98–100)	100 (99–100)	100 (99–100)
Marine Corps	9 (3–14)	53 (43–63)	99 (97–100)	100	100
Navy	9 (4–14)	34 (25–44)	93 (90–97)	100	100

NOTE: The time periods reflect the era of the service member's release from active duty.

TABLE 6-3 Percentages of Service Medical Records (95% Confidence Intervals) with Reports of Audiometric Examinations Within 60 Days of Entry into Active Duty (n = 3,212)

Branch	Before 1950	1950–1969	1970–Regulation Date	Regulation Date–1993	1994–2002
Army	1 (0–2)	7 (3–11)	36 (30–43)	26 (20–32)	30 (24–35)
Air Force	*	17 (9–26)	30 (23–37)	25 (19–31)	20 (15–25)
Marine Corps	0	13 (4–23)	37 (30–44)	51 (44–59)	69 (63–74)
Navy	0	6 (1–11)	35 (28–41)	56 (49–63)	70 (64–75)

*Fewer than 40 records in the denominator.
NOTE: The time periods reflect the era of the service member's release from active duty.

audiograms using a time window of ±120 days. Appendix E provides tables identical to Tables 6-3 through 6-5, but using the larger (±120-day) time window.

From Table 6-3, for personnel who left military service during the 1970s, the percentage of service medical records containing such audiograms was 30–37 percent across all branches of the military. For the two earlier periods, the percentages were appreciably lower for all branches. For both the Army and the Air Force personnel who left military service during the periods since the 1980s, the percentage of records with entrance audiograms was slightly lower than for the 1970s era, but the differences were not statistically significant. Percentages of Marine Corps and Navy records with audiograms, on the other hand, increased in each era since the 1970s and for the most recent time period were 69 percent and 70 percent, respectively.

Presence of Separation Audiograms

Separation audiograms were defined as audiograms recorded within 60 days of a service member's release from active duty. As shown in Table 6-4, the percentage of records with a separation audiogram has ranged from 0 to 54 percent over the time periods considered. The percentage of records with separation audiograms has declined for all the services in the most recent period. The general trend across branches is for the percentage of files with separation audiograms to increase steadily or remain the same up to the 1970s, to remain fairly constant from the 1970s through the 1980s, and to decrease during the most recent era. From 1970 to 2002, however, the percentages for the Navy and Marine Corps are about twice as high as those for the Army or Air Force.

TABLE 6-4 Percentages of Service Medical Records (95% Confidence Intervals) with Reports of Audiometric Examinations Within 60 Days of Release from Active Duty (n = 3,226)

Branch	Before 1950	1950–1969	1970–Regulation Date	Regulation Date–1993	1994–2002
Army	1 (0–3)	12 (7–17)	27 (21–33)	29 (23–35)	14 (9–18)
Air Force	*	30 (20–40)	25 (19–31)	23 (17–28)	7 (4–10)
Marine Corps	0	2 (0–6)	49 (42–57)	53 (46–61)	36 (30–42)
Navy	0	11 (5–17)	54 (47–61)	54 (48–61)	44 (39–50)

*Fewer than 40 records in the denominator.

Presence of Entrance and Separation Audiograms

The percentages of records containing both an audiogram obtained within 60 days of entrance and an audiogram obtained within 60 days of separation from military service are shown in Table 6-5. As noted with regard to either of these audiograms alone, there is an increase in the percentages from the earliest era to the 1970s. After that, the percentages for the Army and Air Force decreased in the more recent periods, whereas the percentages for the Navy and Marine Corps increased slightly or held steady in each successive time period. Although the general trends over time are similar to those observed for each type of audiogram (at entry and at separation) alone (see Tables 6-3 and 6-4), the overall percentages of files having both audiograms are considerably lower, as expected.

TABLE 6-5 Percentages of Service Medical Records (95% Confidence Intervals) with Reports of Audiometric Examinations Within 60 Days of Entrance into and Release from Active Duty (n = 3,210)

Branch	Before 1950	1950–1969	1970–Regulation Date	Regulation Date–1993	1994–2002
Army	0	4 (1–7)	13 (8–17)	12 (7–16)	5 (2–7)
Air Force	*	10 (3–16)	12 (7–17)	5 (2–9)	1 (0–3)
Marine Corps	0	0	25 (19–32)	29 (22–35)	31 (25–37)
Navy	0	1 (0–3)	24 (18–30)	33 (27–40)	34 (28–39)

*Fewer than 40 records in the denominator.

NOTE: The time periods reflect the era of the service member's release from active duty.

COMPLIANCE WITH REGULATIONS

As noted in Chapter 5, the Army, Navy, and Marine Corps issued regulations by 1980 requiring that audiograms be performed at entrance into and separation from active duty (regulations issued in 1980 for the Army and 1979 for the Navy and Marine Corps). These regulations do not specify time windows for performing the audiograms, so, as noted, a ±60-day window was selected for this analysis.

A 1956 Air Force regulation required audiograms in conjunction with routine physical examinations at entrance into and separation from active duty, but by 1973 audiograms were no longer required immediately upon entrance. Standard practice, as of 1982, was to obtain an audiogram within 90 days of service members' assignment to their first permanent duty stations (Department of the Air Force, 1982), which might typically take place up to 6 months after entry into service (Pluta, 2005). Separation audiograms were required only of those Air Force personnel routinely exposed to hazardous noise (Department of the Air Force, 1982). As of this writing, it is not clear that any Air Force regulation requires entrance or exit audiograms for all Air Force military personnel. As of 1996, however, the Department of Defense established a requirement that all military personnel receive a reference audiogram at basic training prior to noise exposure (DoD, 1996). The percentages in Table 6-3 for the era 1994–2002 suggest that this testing occurred for about 20–30 percent of the personnel in the Army or Air Force and about 70 percent of those in the Navy or Marine Corps. At present, the Department of Defense does not require an audiogram upon separation from service.

Thus, with regard to the Army, Navy, and Marine Corps, it appears that compliance with the regulations in place since the early 1980s has been and is incomplete, particularly with regard to audiograms obtained at the time of separation from active duty. Navy and Marine Corps percentages are consistently higher than those for the Army, however. The percentages of Air Force records with entrance and separation audiograms declined over the three later time periods and were the lowest of the four services for the most recent time period. However, because Air Force regulations for the two later periods did not call for entrance or separation audiograms for all personnel, the service's testing practices may have been in better compliance with existing regulations than is reflected under the criteria used in the analysis reported here. Nevertheless, as noted, since 1996, the Department of Defense has required entrance audiograms before noise exposure at basic training for all military personnel.

Several factors should be taken into account in interpreting these data. One is that the data on the presence of an audiogram within 60 days of entrance into active duty are reported on the basis of the era in which the service member was released from active duty. Thus, individuals who served for many years may have begun their military service during a previous era with requirements for entrance testing that were different from those in place at the time they left military service. Because the Air Force has had higher reenlistment rates (DoD, 1997) and a longer median time-in-service than the other services, this potential discrepancy was more likely for Air Force study participants. For example, for the current study, whereas the median length of service for the Army, Navy, and Marine Corps across all five time periods ranged from 2 to 4 years, the corresponding values for the Air Force were 7, 7, and 10 years in the three most recent eras.

When a ±1-year window was used for analysis of "separation" audiograms, the percentages of personnel receiving an audiogram within the 1994–2002 time period were 38, 65, 83, and 86 percent in the Air Force, Army, Marine Corps, and Navy, respectively.[3] All these percentages are clearly much higher than the values derived for the same time period using a testing window of ±60 days (Table 6-4), as expected. Even here, however, the percentage of personnel tested is appreciably lower for the Air Force than for the other military services.

It should be noted that this analysis did not evaluate the extent to which the services obtained reference or termination audiograms for personnel entering and leaving hearing conservation programs. The reasons for not focusing exclusively on those personnel in hearing conservation programs were noted previously.

The following findings are based on the data in Tables 6-2 through 6-5, each of which makes use of a ±60-day time window. Although the specific percentages cited in some of the findings are dependent upon the time window used in the analysis, the general features of the data are the same for time windows of ±60 days (Tables 6-3 through 6-5) or ±120 days (Appendix E).

FINDING: Review of a sample of service medical records of military veterans indicates that compliance with requirements for audiometric testing at entrance into service has been limited, even in the most recent eras, and did not exceed 70 percent in any branch or era when using a ±60-day window for analysis.

FINDING: Review of a sample of service medical records of military veterans indicates that audiometric testing at separation from service has

---

[3]Percentages of personnel receiving an audiogram within a ±120-day window of release from active duty during the 1994–2002 time period were 17, 34, 60, and 66 in the Air Force, Army, Marine Corps, and Navy, respectively.

been limited, even in the most recent eras, and did not exceed 54 percent in any branch or era when using a ±60-day window for analysis.

FINDING: Review of a sample of service medical records of military veterans indicates that audiometric testing at both entrance into and separation from service has been extremely limited, even in the most recent eras, and did not exceed 34 percent in any branch or era when using a ±60-day window for analysis.

REFERENCES

Department of the Air Force. 1982. *AF Regulation 161-35: Hazardous Noise Exposure.* Washington, DC: Department of the Air Force.

Department of Veterans Affairs. 2005. Service-Connected Compensation—Veterans with Hearing Impairment as of December 21, 2004. Spreadsheets provided to J. Durch, Institute of Medicine, March 21.

DoD (Department of Defense). 1996. *Department of Defense Instruction 6055.12: DoD Hearing Conservation Program.* Washington, DC: Department of Defense.

DoD. 1997. *Selected Manpower Statistics: Fiscal Year 1997.* DTIC/NTIS No. DIOR/M01-9. Washington, DC: Department of Defense.

Gardner MJ, Altman DG. 1989. *Statistics with Confidence: Confidence Intervals and Statistical Guidelines.* London, United Kingdom: British Medical Journal.

Pluta R (U.S. Air Force). 2005. RE: Seek Help with Technical Review. E-mail to L. Joellenbeck, Institute of Medicine, June 24.

7

Conclusions and Comments

In this chapter the committee draws on the evidence it has reviewed to respond more directly to the specific points in its charge. The chapter also discusses potential opportunities identified by the committee to improve the effectiveness of the military services' hearing conservation programs designed to prevent noise-induced hearing loss and tinnitus. Finally, research needs and opportunities suggested by the committee's review are described. They cover the science of noise and hearing loss and tinnitus, as well as the protection of hearing and the course of hearing loss and tinnitus among military personnel.

RESPONDING TO THE ELEMENTS OF THE CHARGE

1. What sources of potentially damaging noise have been present in military settings since the beginning of World War II?

Many sources of potentially damaging noise have long existed in military settings. For the period addressed by this report—World War II to the present—some of these sources include weapons systems (e.g., hand guns, rifles, artillery pieces, rockets), wheeled and tracked vehicles, fixed- and rotary-wing aircraft, ships, and communications devices (Chapter 3). Service members may encounter these noise sources through training, standard military operations, and combat. Exposure to combat-related noise may be unpredictable in onset and duration. In addition, service members may be exposed to hazardous noise through activities that are not unique to the military environment, including various engineering, industrial, construction, or maintenance tasks.

2. What levels of noise exposure are necessary to cause hearing loss or tinnitus?

The specific noise levels that cause noise-induced hearing loss vary with the duration of the exposure, the type of noise, and the frequency content of the noise, as well as the susceptibility of the exposed individual (Chapters 1 and 2). Time-weighted average noise exposures of approximately 85 dBA for 8 hours per day for a 40-hour work week, or the equivalent, are considered to be hazardous, but a person must be so exposed for a number of years before developing noise-induced hearing loss. On the other hand, impulse noise with peak levels exceeding approximately 140 dB SPL may be hazardous even for a single exposure. These guidelines for safe noise exposures are designed to protect the majority of individuals from noise-induced hearing loss, but not to ensure that every individual is protected. With regard to noise-induced tinnitus, specific parameters of hazardous noise exposure have not been defined, but noise levels associated with hearing loss are also likely to be associated with tinnitus (Chapter 4).

3. What is the evidence that hearing loss or tinnitus has been incurred by members of the armed services as a result of noise exposure during military service since World War II?

Patterns of hearing loss consistent with noise exposure can be seen in cross-sectional studies of military personnel (Chapter 3). Because large numbers of people have served in the military since World War II, the total number who experienced noise-induced hearing loss by the time their military service ended may be substantial, but the available data provide no basis for a valid estimate of the number. Neither was it possible to estimate the proportion of a given military population that developed noise-induced hearing loss or tinnitus during military service, the amount of hearing loss incurred, or the relative risk of noise-induced hearing loss or tinnitus for a given individual, based on his or her branch of military service, occupational specialty, or service era.

With regard to hearing loss, the majority of the data available are average group hearing thresholds from cross-sectional studies. These average data indicate that hearing thresholds are worse in those groups with more years of military service. However, these cross-sectional data are not a sufficient basis for attributing greater hearing loss solely to a longer exposure to noise while in the military. The timing of exposure to noise and the noise doses received (or other factors that may affect hearing loss) may have differed among personnel who entered military service in different years by virtue of such factors as the timing of periods of combat or differences in use of hearing protection or recreational use of firearms.

Cross-sectional data can be used to identify associations between military service and noise-induced hearing loss, but are not sufficient to show causal relationships. In contrast, longitudinal data on hearing thresholds at the beginning and end of military service provide a basis for establishing that hearing loss occurred after exposure to noise during military service, a temporal ordering necessary for a causal relationship and estimation of risks.

With regard to tinnitus, even less information is available than for hearing loss (Chapter 4). The committee identified no epidemiological studies of tinnitus among U.S. military personnel, and the services' hearing conservation programs do not include surveillance for tinnitus. Limited tinnitus surveillance was introduced in 2003 with post-deployment health assessments.

4. What is the evidence that the effects of noise exposure at younger ages can lead to delayed onset of noise-induced hearing loss later in life?

There is little evidence available with which to address this question (Chapter 2). No longitudinal studies have examined patterns of hearing loss over time in noise-exposed humans or laboratory animals who did not develop hearing loss at the time of the noise exposure. The committee's understanding of the mechanisms and processes involved in the recovery from noise exposure suggests that a delay of many years in the onset of noise-induced hearing loss following an earlier noise exposure is extremely unlikely.

When hearing loss is known to have occurred as a result of a noise exposure, it has generally been thought that hearing loss for pure tones does not worsen following the cessation of a given noise exposure. However, there are no longitudinal data from humans who developed noise-induced hearing loss in early adulthood and were followed into their 60s, 70s, or 80s. Data from a few longitudinal studies of older adults, which differed in the way prior noise exposure was documented, have not produced conclusive results. To the committee's knowledge, only one longitudinal study has examined changes in hearing in laboratory animals after a noise-induced hearing loss. In middle-aged gerbils that sustained a slight noise-induced hearing loss and were followed for most of their remaining lifetimes, no change in the amount of noise-induced hearing loss was seen over time.

It is possible, however, that an individual's *awareness* of the effects of noise on hearing may be delayed considerably after the noise exposure. As illustrated in Chapter 2, young adults with a slight noise-induced high-frequency hearing loss (e.g., 15–30 dB HL at 6000 Hz), one not likely to cause much difficulty with communication if present at the time a young adult might be discharged from military service, will likely exhibit greater hearing loss as they age than young adults with normal hearing (0 dB HL) at discharge. As demonstrated previously in Figures 2-6 and 2-7, a slight noise-induced hearing loss of 20–30 dB HL incurred as a young adult, when combined with a similar amount of hearing loss associated with aging alone, can become a moderate hearing loss of 40–50 dB HL at an age of 50 or 60 years. This amount of hearing loss is often sufficient to interfere with everyday communication, and it may make the individual more aware of the effects of the earlier noise-induced hearing loss, especially in comparison to same-aged peers without prior noise-induced hearing loss (who have approximately half as much hearing loss).

5. What additional risk factors for noise-induced hearing loss or tinnitus are supported by a good level of evidence?

In humans, no specific exogenous or endogenous factors were identified that correlated with increased susceptibility to noise-induced hearing loss or tinnitus (Chapters 2 and 4). Exogenous factors for hearing loss considered by the committee included exposure to aminoglycoside antibiotics, cisplatin, diuretics, salicylates, solvents, carbon disulfide, carbon monoxide, cigarette smoking, whole-body vibration, body temperature, exercise, and electromagnetic fields. Some of these agents (e.g., aminoglycoside antibiotics and cisplatin) are known to be ototoxins that may induce hearing loss unrelated to noise exposure. Studies in humans of the effect on hearing of exposure to any of these agents in combination with noise either have not been done or have not produced conclusive results. Endogenous factors considered by the committee included (old) age, gender, race, eye color, and prior hearing loss, but these factors did not correlate with increased susceptibility to noise-induced hearing loss.

The committee identified only one study in humans that had investigated the association between tinnitus and combined exposures to noise and other factors. Tinnitus risk factors, independent of noise exposure, include hearing loss, head injury, middle ear disease, and certain medications (e.g., salicylates, aminoglycoside antibiotics).

6. When were the military services' hearing conservation measures adequate to protect the hearing of service members?

Data analyzed by the committee led to the conclusion that military hearing conservation programs, dating from the late 1970s, cannot be considered adequate to protect the hearing of service members. The committee concluded that hearing conservation activities from World War II through the 1970s would have been even less adequate to protect the hearing of service members than programs in place since the 1980s, because only early hearing protection devices of limited effectiveness were available and mandatory hearing conservation measures were in place only in the Air Force (Chapter 5).

Given that engineering measures to reduce noise levels and administrative measures to reduce noise exposures may not be compatible with military operations, use of hearing protection devices is often the primary defense against noise-induced hearing loss for military personnel. The effectiveness of these devices has increased substantially since World War II, but still depends on how well and how often they are used. Data on the use of hearing protection by military personnel are limited, but a handful of reports over the past 30 years suggest that in some settings, only about half of those who should have been using hearing protection devices were doing so.

The services' hearing conservation programs require annual measurements of hearing thresholds for military personnel who are considered to be exposed to hazardous noise. This surveillance effort alone will not prevent noise-induced hearing loss, but it may serve to limit the loss if the detection of temporary hearing losses or small permanent losses results in increased use of hearing protection or reassignment of the individual to lower noise environments. Available data show that records of hearing tests are being collected for only about half of the Army and Air Force personnel in the hearing conservation programs. Some personnel may not be receiving the required tests, and some test results may not be reaching the central hearing conservation registry system. The Navy and the Marine Corps do not report the proportion of enrollees who are being tested each year. The percentage of military service members tested each year who have a significant threshold shift (STS) has been approximately 10 percent in the Air Force over the past 3 years, and close to 18 percent in the Army, Navy, and Marine Corps, which is two to five times higher than rates considered appropriate in industrial hearing conservation programs.

7. When did the audiometric measures used by the military services become adequate to evaluate individual changes in hearing associated with military service?

A review of service medical records for veterans who left military service during the period from World War II to 2002 suggests that documented audiometric testing at entrance into and separation from service has been and remains limited, even in the most recent eras (Chapter 6). As argued repeatedly in this report, it is critical to obtain an audiogram at entry into and exit from military service to clearly establish whether noise-induced hearing loss developed during military service. The service medical records audited revealed that about 30 percent of personnel who left the Navy and Marine Corps during the period from the early 1980s to 2002 had both an entry and separation audiogram within ±60 days of entry or separation, whereas the percentages were even lower, typically less than 12 percent, for personnel who had served in either the Army or the Air Force. As expected, the percentage of service medical records containing audiograms of any type was lowest for the period before 1950, except for the Air Force, an early leader in requiring the collection of audiograms. The results of the review of service medical records indicate that audiometric testing by the military services has not been adequate, throughout the period from World War II to the present, to evaluate changes in hearing associated with military service for the majority of service members.

OPERATIONAL NEEDS SUGGESTED BY THE REPORT

The current irreversibility of noise-induced hearing loss and tinnitus means that preventing these problems, or limiting their progression, is especially important. From the review of information on noise exposure in military settings, hearing loss and tinnitus experienced by some service members, and the hearing conservation activities of the military services, the committee identified several steps that may enhance hearing protection for service members and improve the effectiveness of the services' hearing conservation programs. Although this report was prepared for the Department of Veterans Affairs (VA), it is the Department of Defense and the individual military services that can take these important steps to minimize the adverse effects of noise exposure on military personnel and better document hearing loss or tinnitus when either occurs during military service. The committee strongly recommends the following practices be implemented:

1. Work to achieve more extensive and consistent use of hearing protection by military personnel.
2. Include questions about the presence and severity of tinnitus in each ear on all audiometric records obtained from enlistment through the end of military service. (In the remaining suggestions, audiograms and audiometric records are assumed to include responses to questions about the presence and severity of tinnitus.)
3. Enforce requirements for audiograms prior to noise exposure for all new military service members at *all* basic training sites.
4. Enforce, and establish where they do not presently exist, require-

ments for audiograms at the completion of military service to ensure that any hearing loss or tinnitus arising during military service is adequately documented. The Department of Defense and the Department of Veterans Affairs should explore whether resources are available within the VA system to aid the military services in conducting audiometric tests and tinnitus assessments for personnel completing their military service.

5. Given the likely occurrence of maximum noise-induced hearing loss at 6000 Hz, include the measurement of hearing thresholds at 8000 Hz in all audiograms to allow for detection of the noise-notch pattern of hearing loss associated with noise exposure.

6. Enforce hearing conservation requirements for annual monitoring audiograms, as well as for follow-up audiograms if significant threshold shift is detected in annual monitoring audiograms.

7. Continue to develop the Defense Occupational and Environmental Health Readiness System (DOEHRS) to improve its reporting capabilities to match and exceed those available with the services' previous systems. Further development of this system should include modification of the hearing conservation component (DOEHRS-HC) to track reports of tinnitus. It should also include implementation of the industrial hygiene component (DOEHRS-IH) to provide information on exposures to hazardous noise and other chemical, physical, biological, and ergonomic hazards.

8. Develop mechanisms to provide VA personnel access to records from DOEHRS-HC for review of disability claims for hearing loss or tinnitus that are not otherwise supported by audiometric records in the service medical record.

RESEARCH NEEDS SUGGESTED BY THE REPORT

The committee also saw areas where further research would be valuable for improving understanding of broad scientific questions concerning the relationship between noise exposure and hearing loss and tinnitus. Research could also address more targeted questions concerning noise exposure, hearing loss, tinnitus, and hearing conservation measures related to military service.

Two broad scientific areas were of interest to the committee:

1. Further investigate, both in laboratory animals and humans, exposures to fluctuating noise, impulse/impact noise, and combinations of noise, as well as intermittent exposures to steady-state noise, to determine the acoustic parameters associated with noise-induced hearing loss and tinnitus.

2. Further investigate the mechanism, natural history, epidemiology, measurement, and treatment of noise-induced hearing loss and tinnitus.

Several avenues of research specifically related to military settings and military personnel could be considered. Many are offered as a means to fill the void for prospective, longitudinal, epidemiological data on noise-induced hearing loss and tinnitus in military personnel.

1. Obtain valid estimates of the incidence, prevalence, and severity of noise-induced hearing loss and tinnitus among military personnel, including gender-specific estimates. If the reporting ability and completeness of existing databases, such as DOEHRS-HC, improve, greater use might be made of their data for analyses for personnel enrolled in hearing conservation programs.
2. Establish cohorts of military veterans with various documented noise exposures, immediately upon discharge, and survey them periodically for ototoxic exposures, subsequent nonmilitary noise exposures, and hearing function, as well as presence and severity of tinnitus, in order to determine whether there is a delay in the effects of military noise exposure. These cohorts will need to be followed through the remainder of members' lifetimes, but this longitudinal study will reveal elements of the natural history of noise-induced hearing loss and tinnitus that otherwise will not be determined. The Millennium Cohort Study, which is designed to evaluate the long-term health of people who have served in the military, might provide a mechanism for conducting a longitudinal investigation of hearing health.
3. Conduct randomized trials of interventions within each military branch to determine with greater certainty which approaches to hearing conservation—including efforts to increase the use and effectiveness of hearing protection devices, compliance with requirements for audiometric testing, and the use of otoprotective medications—lead to lower incidence of noise-induced hearing loss and tinnitus.
4. On a sample basis, determine noise levels for modern military activities and also determine, with standard industrial hygiene methods, the noise dose experienced by individual military personnel where dosimetry has not been done.
5. Conduct real-world studies in military settings, including field and garrison conditions, to assess the noise attenuation and utilization rates of hearing protection devices, including the recently introduced earplugs that provide level-dependent sound attenuation.

Appendix A

Legislative Language from P.L. 107-330

(2) by striking "total deafness" the second place it appears and inserting "deafness".

Deadlines.

SEC. 104. ASSESSMENT OF ACOUSTIC TRAUMA ASSOCIATED WITH MILITARY SERVICE FROM WORLD WAR II TO PRESENT.

Contracts.

(a) ASSESSMENT BY NATIONAL ACADEMY OF SCIENCES.—The Secretary of Veterans Affairs shall seek to enter into an agreement with the National Academy of Sciences for the Academy to perform the activities specified in this section. The Secretary shall seek to enter into the agreement not later than 60 days after the date of the enactment of this Act.

(b) DUTIES UNDER AGREEMENT.—Under the agreement under subsection (a), the National Academy of Sciences shall do the following:

(1) Review and assess available data on hearing loss that could reasonably be expected to have been incurred by members of the Armed Forces during the period from the beginning of World War II to the date of the enactment of this Act.

(2) Identify the different sources of acoustic trauma that members of the Armed Forces could reasonably be expected to have been exposed to during the period from the beginning of World War II to the date of the enactment of this Act

(3) Determine how much exposure to each source of acoustic trauma identified under paragraph (2) is required to cause or contribute to hearing loss, hearing threshold shift, or tinnitus, as the case may be, and at what noise level.

(4) Determine whether or not such hearing loss, hearing threshold shift, or tinnitus, as the case may be, is—

(A) immediate or delayed onset;

(B) cumulative;

(C) progressive; or

(D) any combination of subparagraph (A), (B), and (C).

(5) Identify age, occupational history, and other factors which contribute to an individual's noise-induced hearing loss.

(6) Identify—

(A) the period of time at which audiometric measures used by the Armed Forces became adequate to evaluate individual hearing threshold shift; and

(B) the period of time at which hearing conservation measures to prevent individual hearing threshold shift were available to members of the Armed Forces, shown separately for each of the Army, Navy, Air Force, Marine Corps, and Coast Guard, and, for each such service, shown separately for members exposed to different sources of acoustic trauma identified under paragraph (2).

(c) REPORT.—Not later than 180 days after the date of the entry into the agreement referred to in subsection (a), the National Academy of Sciences shall submit to the Secretary a report on the activities of the National Academy of Sciences under the agreement, including the results of the activities required by subsection (b).

(d) REPORT ON ADMINISTRATION OF BENEFITS FOR HEARING LOSS AND TINNITUS.—(1) Not later than 180 days after the date of the enactment of this Act, the Secretary of Veterans Affairs shall submit to the Committees on Veterans' Affairs of the Senate and the House of Representatives a report on the claims submitted to the Secretary for disability compensation or health care for hearing loss or tinnitus.

(2) The report under paragraph (1) shall include the following:

(A) The number of decisions issued by the Secretary in each of fiscal years 2000, 2001, and 2002 on claims for disability compensation for hearing loss, tinnitus, or both.

(B) Of the decisions referred to in subparagraph (A)—

(i) the number in which compensation was awarded, and the number in which compensation was denied, set forth by fiscal year; and

(ii) the total amount of disability compensation paid on such claims during each such fiscal year.

(C) The total cost to the Department of Veterans Affairs of adjudicating the claims referred to in subparagraph (A), set forth in terms of full-time employee equivalents (FTEEs).

(D) The total number of veterans who sought treatment in Department of Veterans Affairs health care facilities during fiscal years specified in subparagraph (A) for hearing-related disorders, set forth by the number of veterans per year.

(E) The health care furnished to veterans referred to in subparagraph (D) for hearing-related disorders, including the number of veterans furnished hearing aids and the cost of furnishing such hearing aids.

TITLE II—MEMORIAL AFFAIRS

SEC. 201. PROHIBITION ON CERTAIN ADDITIONAL BENEFITS FOR PERSONS COMMITTING CAPITAL CRIMES.

(a) PRESIDENTIAL MEMORIAL CERTIFICATE.—Section 112 is amended by adding at the end the following new subsection:

"(c) A certificate may not be furnished under the program under subsection (a) on behalf of a deceased person described in section 2411(b) of this title.".

(b) FLAG TO DRAPE CASKET.—Section 2301 is amended—

(1) by redesignating subsection (g) as subsection (h); and

(2) by inserting after subsection (f) the following new subsection (g):

"(g) A flag may not be furnished under this section in the case of a person described in section 2411(b) of this title.".

(c) HEADSTONE OR MARKER FOR GRAVE.—Section 2306 is amended by adding at the end the following new subsection:

"(g)(1) A headstone or marker may not be furnished under subsection (a) for the unmarked grave of a person described in section 2411(b) of this title.

"(2) A memorial headstone or marker may not be furnished under subsection (b) for the purpose of commemorating a person described in section 2411(b) of this title.

"(3) A marker may not be furnished under subsection (d) for the grave of a person described in section 2411(b) of this title.".

(d) EFFECTIVE DATE.—The amendments made by this section shall apply with respect to deaths occurring on or after the date of the enactment of this Act. Applicability. 38 USC 112 note.

Appendix B

Agendas for Information-Gathering Meetings

MEETING I
MAY 12–13, 2004

The Keck Center of the National Academies
500 Fifth Street, N.W.
Washington, D.C.

Wednesday, May 12, 2004

11:30 a.m. Introductory Remarks
Larry Humes, Ph.D.
Chair, Committee on Noise-Induced Hearing Loss and Tinnitus Associated with Military Service from World War II to the Present

Introductions by Committee Members and Meeting Attendees

11:45 Study Context and Goals, Sponsor Perspective
Michael Hodgson, M.D., M.P.H.
Director, Occupational Health Program
Veterans Health Administration
Department of Veterans Affairs

Discussion

12:30 p.m. Lunch in meeting room

1:15 Study Origins and Intent, Congressional Perspective
Mary Schoelen, J.D.
Deputy Staff Director (Benefits) and General Counsel
Senate Veterans Affairs Committee

Discussion

Military Services' Brief Perspectives on Noise-Induced Hearing Loss Issues and Hearing Conservation Programs and History

2:00 *Colonel David W. Chandler, Ph.D.*
Director, Army Audiology and Speech Center
Consultant to the Army Surgeon General for Audiology and Hearing Conservation
Walter Reed Army Medical Center

2:20 *John Page*
Occupational Audiology Team Leader
Navy Environmental Health Center

2:40 *Major Joseph J. Narrigan, Au.D.*
Deputy Commander
Air Force Medical Element
Andrews Air Force Base, MD

3:00 *Commander Wade McConnell* [by telephone]
Chief, Environmental Health Division
Office of Safety and Environmental Health
U.S. Coast Guard

3:20 Discussion

3:45 Adjourn Open Session

Thursday, May 13, 2004

10:00 a.m. Begin Open Session

Military Services' Brief Perspectives on Noise-Induced Hearing Loss Issues and Hearing Conservation Programs and History (continued)
Al Lillibridge
Occupational Safety and Health Manager
U.S. Marine Corps

Discussion

10:30 Adjourn Open Session

MEETING II
JULY 19–20, 2004

The Keck Center of the National Academies
500 Fifth Street, N.W.
Washington, D.C.

Monday, July 19, 2004

1:30 p.m. Introductory Remarks
Larry Humes, Ph.D.
Chair, Committee on Noise-Induced Hearing Loss and Tinnitus Associated with Military Service

Introductions by Committee Members and Meeting Attendees

1:45 Military Services' Data and Databases on Hearing, Hearing Conservation Programs, and Noise

Department of Defense and Army Data and Databases
Douglas Ohlin, Ph.D.
Program Manager, Hearing Conservation
U.S. Army Center for Health Promotion and Preventive Medicine

Navy Data and Databases
John Page (by telephone)
Occupational Audiology Team Leader
Navy Environmental Health Center

Discussion

4:00 Adjourn Open Session

Tuesday, July 20, 2004

9:15 a.m. Introductory Remarks
Larry Humes, Ph.D.
Chair, Committee on Noise-Induced Hearing Loss and Tinnitus Associated with Military Service

Introductions by Committee Members and Meeting Attendees

9:20 Department of Veterans Affairs Claims Review and Examinations for Hearing Loss and Tinnitus

Veterans Health Administration:
Kyle Dennis, Ph.D.
Audiology and Speech Pathology National Program Office

Judy Schafer, Ph.D.
Audiologist, VA Medical Center, Washington, D.C.

Veterans Benefits Administration:
Bradley Flohr
Chief, Judicial/Advisory Review, Compensation and Pension

Discussion

11:30 Adjourn Open Session

MEETING IV
DECEMBER 6–7, 2004

The Keck Center of the National Academies
500 Fifth Street, N.W.
Washington, D.C.

Tuesday, December 7, 2004

8:00 a.m. Continental Breakfast

8:30 Introductory Remarks
Larry Humes, Ph.D.
Chair, Committee on Noise-Induced Hearing Loss and Tinnitus Associated with Military Service

Introductions by Committee Members and Meeting Attendees

8:40 Shipboard Noise and Hearing Conservation Issues
Kurt Yankaskas
Branch Manager, Human Systems Integration Programs
Naval Sea Systems Command

9:20 Air Force Hearing Conservation Program
Major Robert Pluta, USAF
Chief, Hearing Conservation Data Registry

10:00 Break

10:15	Presentations by Veterans
	Arnold Mathias
	Ansell (Mac) MacMillan
	Carroll Williams
	John Warwick
	Rick Weidman
	David Kayal
	Discussion
12:00 p.m.	Adjourn Open Session

Appendix C

Definitions

Audiogram: Graph of hearing threshold level as a function of frequency (ANSI, 1995).

Baseline audiogram: The initial audiogram to which subsequent audiograms are compared for the calculation of significant threshold shift. The Occupational Safety and Health Administration requires that a baseline audiogram be obtained from an examination administered before employment or within the first 30 days of employment and that is preceded by a period of at least 12 hours of quiet.

Continuous noise: Noise with negligibly small fluctuations of level within the period of observation (ANSI, 1995).

Cross-sectional study: A study that examines the relationship between diseases (or other health-related characteristics) and other variables of interest as they exist in a defined population at one particular time (Last, 1995).

Decibel (dB): The unit used to express the level of sound. The decibel is a logarithm of a ratio of two quantities, the denominator of which has been specified such that 0 dB approximates the threshold of hearing in the middle frequencies for young adults. The reference quantity in the denominator of the ratio is either a sound pressure of 20 micropascals (μPa) or a sound intensity of 10^{-12} watts/m^2.

Exchange rate: An increment of decibels that requires the halving of exposure time, or a decrement of decibels that allows the doubling of exposure time for a given noise dose. For example, a 3-dB exchange rate requires that noise exposure time be halved for each 3-dB increase in noise level. With a 5-dB exchange rate, exposure time may be doubled for each 5-dB decrease in noise level.

Frequency: The number of times that a periodic process, such as a sound wave, repeats each second.

Hearing threshold level (HTL): For a specified signal, the amount in decibels by which the hearing threshold for a listener exceeds a specified reference equivalent threshold level (ANSI, 1994).

Hertz (Hz): A unit of frequency equal to one cycle per second.

Impact noise: Impact noise is generated by the collision of one mass in motion with a second mass that may be in motion or at rest (ANSI, 1994).

Impulsive noise: Impulsive noise is characterized by a sharp rise and rapid decay in sound levels and is less than 1 second in duration, with 1 second between successive stimuli. For the purpose of this document, it refers to impact or impulse noise.

Intermittent noise: Noise levels that are interrupted by intervals of relatively low sound levels.

Inverse square law: A law in physics stating that the magnitude of a physical quantity varies inversely with the square of its distance from its source. For sound, sound level will diminish by 6 dB from a point source for every doubling of distance.

Longitudinal study: An epidemiological study in which subsets of a defined population can be identified who are, have been, or in the future may be exposed or not exposed, or exposed in different degrees, to a factor or factors hypothesized to influence the probability of occurrence of a given disease or other outcome. The main feature of longitudinal studies is observation of large numbers of people over a long period with comparison of incidence rates in groups that differ in exposure levels (after Last, 1995).

Noise: (1) Undesired sound. By extension, noise is any unwarranted disturbance within a useful frequency band, such as undesired electric waves in a transmission channel or device. (2) Erratic, intermittent, or statistically random oscillation (ANSI, 1994).

Noise dose: The noise exposure expressed as a percentage of the allowable daily exposure. For the Occupational Safety and Health Administration, which uses a 5-dB exchange rate (per 29 C.F.R. 1910.95) and a permissible exposure limit of 90 dBA, a 100-percent dose equals an 8-hour exposure to a continuous 90 dBA noise. A 50-percent dose (requiring enrollment in a hearing conservation program) corresponds to an 8-hour exposure to 85 dBA or a 4-hour exposure to 90 dBA. Different permissible exposure limits and exchange rates will change the dose computations accordingly (Berger et al., 2000).

Noise reduction rating (NRR): A single value in decibels that indicates a hearing protector's noise reduction capabilities, averaged across the range of audible frequencies, as measured under optimum laboratory conditions. By law, the NRR must appear on the label of all devices sold as personal hearing protectors in the United States.

Permanent threshold shift (PTS): Permanent increase, measured in decibels, in the threshold of audibility for an ear (ANSI, 1995).

Prevalence: The number of events (e.g., instances of a given disease or other condition) in a given population at a designated time (Last, 1995).

Prevalence rate (a proportion): The total number of individuals who have an attribute or disease at a particular time (or during a particular period) divided by the population at risk of having the attribute or disease at this point in time or midway through the period (Last, 1995).

Significant threshold shift: A shift in hearing threshold, outside the range of audiometric testing variability (±5 dB), that warrants follow-up action to prevent further hearing loss. The National Institute for Occupational Safety and Health defines significant threshold shift as an increase in the HTL of 15 dB or more at any frequency (500, 1000, 2000, 3000, 4000, or 6000 Hz) in either ear, which is confirmed for the same ear and frequency by a second test within 30 days of the first test (NIOSH, 1998).

Sound: Auditory sensation evoked by an oscillation in pressure, stress, particle displacement, particle velocity, and so on, in a medium with internal forces (e.g., elastic or viscous) (after ANSI, 1994).

Sound pressure: Root-mean-square instantaneous sound pressure at a point, during a given time interval. Sound pressure is measured in pascals (Pa) (ANSI, 1994).

Sound pressure level (SPL): Ten times the logarithm to the base 10 of the ratio of the time-mean-square pressure of a sound, in a stated frequency band, to the square of the reference sound pressure in gases of 20 Pa. Sound pressure levels are measured in decibels (ANSI, 1994).

Steady-state noise: Ongoing noise, the intensity of which remains at a measurable level without interruptions over an indefinite or specified period of time.

Temporary threshold shift (TTS): Temporary increase, measured in decibels, in the threshold of audibility for an ear (ANSI, 1995).

Threshold: The minimum sound pressure level of a pure tone that can be heard by an individual at least 50 percent of the time.

Time-weighted average (TWA): An A-weighted average sound level normalized to 8 hours, meaning that the average level over the time period that is observed is adjusted (i.e., normalized) to correspond to the sound level, which if present during a continuous 8-hour period would provide the same noise dose as the measured average level, for the specified exchange rate. For example, an average sound level of 100 dBA measured during a period of 4 hours using a 5-dB exchange rate corresponds to a TWA of 95 dBA (Berger et al., 2000).

Varying noise: Noise for which the level varies substantially during the period of observation (ANSI, 1995).

Weighted sound pressure levels: Two weighting curves are commonly applied to measures of sound levels to account for the way the ear responds to sound (ANSI, 1994):

> **A-weighting (dBA):** A measurement scale that approximates the loudness of sound relative to a 40 dB SPL 1000-Hertz (Hz) reference tone. A-weighting emphasizes the frequencies between 1000 and 4000 Hz and reduces the contributions from frequencies below 500 Hz.
>
> **C-weighting (dBC):** A measurement scale that approximates the loudness of tones relative to a 100 dB SPL 1000-Hz reference tone. C-

weighting allows for relatively flat sound pressure measurements by including lower frequency sounds than does the A-weighting scale.

REFERENCES

ANSI (American National Standards Institute). 1994. *American National Standard: Acoustical Terminology, ANSI S1.1-1994 (R 1999)*. New York: ANSI.

ANSI. 1995. *American National Standard: Bioacoustical Terminology, ANSI S3.20-1995*. New York: ANSI.

Berger EH, Royster LH, Royster JD, Driscoll DP, Layne M, eds. 2000. *The Noise Manual*. 5th ed. Fairfax, VA: American Industrial Hygiene Association.

Last JM. 1995. *A Dictionary of Epidemiology*. New York: Oxford University Press.

NIOSH (National Institute for Occupational Safety and Health). 1998. *Criteria for a Recommended Standard—Occupational Noise Exposure Revised Criteria*. DHHS (NIOSH) Pub. No. 98-126. Cincinnati, OH: NIOSH.

Appendix D

Summary Tables on Epidemiological Studies

TABLE D-1 Toluene Exposure as a Risk Factor for Noise-Induced Hearing Loss

Citation	Design	Population	Exposures and Source of Exposure Data
a. *Human Studies*			
Schaper et al. (2003)	Longitudinal study, 1996–2001	192 employees from 14 German rotogravure printing plants with 4 examinations Subjects at each examination: Exam 1: 333 Exam 2: 278 Exam 3: 241 Exam 4: 216 Stratification Toluene exposure: low vs high (based on worksite) Job tenure: short vs long Noise exposure: low (< 82 dBA) vs high (≥ 82 dBA)	Medical, psychological examinations Toluene and noise exposure measured 2 times per yr for each subject Historical records for past exposure estimates Toluene: Mean study exposure: High: 26 ppm Low: 3 ppm Lifetime weighted average daily exposure (for current exposure groups) High: 45 ppm Low: 10 ppm Biomarkers of exposure: hippuric acid, *o*-cresol Noise: lifetime average daily exposure Current high noise: 82 dBA Current low noise: 81 dBA
Morata et al. (1997)	Cross-sectional	124 male rotogravure printing workers, Sao Paulo, Brazil Mean age: 34 yrs (range 21–58 yrs) Employed at least 1 yr Mean tenure: 7 yrs (range 1–25 yrs)	Solvent exposure: TWA exposure evaluation for toluene, ethanol, and ethyl acetate Toluene levels (air): 0.14 to 919 mg/m^3 109 workers monitored for hippuric acid and creatinine in their urine Noise exposure: Continuous noise, 71 to 93 dBA; dosimetry for individual workers

Outcome Measure	Results	Comments
Hearing thresholds Hearing loss: thresholds > 25 dB Tested at 0.125–12 kHz Age adjustment based on ISO 7029 (1984) before repeated measures analysis	No sig effect on auditory thresholds for toluene intensity, exposure duration, or interactions Sig effect of current noise intensity ($F = 4.5$, $p = .04$)	Subjects were volunteers; some loss to follow-up No unexposed control group Little difference in noise exposure for high and low toluene exposure
High-freq hearing loss: notch in a freq b/t 3 and 6 kHz or thresholds poorest in this freq range Normal hearing: no single threshold > 25 dB Pure-tone audiometry: 0.5–8 kHz	No statistical interactions between noise and toluene Concentration of toluene in air was not sig associated w/ hearing loss Level of biological marker for toluene exposure (urinary hippuric acid) sig associated w/ hearing loss (OR = 1.76, 95% CI 1.00–2.98)	93% of subjects reported no exposure to major sources of nonoccupational noise (e.g., firearms, motorcycles, etc.) 11% of those exposed to noise > 85 dBA used hearing protection

continued

TABLE D-1 continued

Citation	Design	Population	Exposures and Source of Exposure Data
Morata et al. (1993)	Cross-sectional	190 male printing and paint manufacturing workers, Brazil Employed at least 1 yr Mean employment: Printing: 8 to 13 yrs Paint mfg: 6 yrs Exposure groups Noise only (printing): 50 Noise and toluene (printing): 51 Mixed solvents, no excess noise (paint mfg): 39 Unexposed (printing): 50	Noise only group: 88–97 dBA (continuous); dose 209–335% (5-dB exchange rate) Noise and toluene group: 88–98 dBA; dose 140–350% (5-dB exchange rate); toluene TWA 75–600 ppm Mixed solvents group: no dose data; toluene concentration 10–70 ppm (11 samples) Interviewed for work, exposure, and medical histories
b. *Animal Studies*			
Davis et al. (2002)	Experimental	33 chinchillas, in 6 exposure groups 6 adult rats as control group	10-day exposures Toluene: 2000 ppm Noise: 500 Hz octave band noise, 97.5 dB SPL Background noise < 60 dBA 22 chinchillas (monaural) Group 1: 8 h toluene, background noise only Group 2: no toluene, 8 h noise Group 3: 8 h toluene, 8 h noise Group 4: control group

Outcome Measure	Results	Comments
Normal hearing Worst threshold at 3–8 kHz (avg left and right ears): 0–25 dB; average of 0.5, 1, 2 kHz ≤ 25 dB High-freq hearing loss Categories based on worst threshold at 3–8 kHz (avg left and right ears) and average of 0.5, 1, 2 kHz ≤ 25 dB: (I) 30–40 dB, (II) 45–55 dB, (III) ≥ 60 dB (IV) average of 0.5, 1, 2 kHz > 25 dB Other hearing loss Unilateral, conductive Pure-tone audiometry: 0.5–8 kHz Otoscopy, immittance audiometry	Prevalence of high-freq hearing loss: 8% unexposed 26% noise 53% noise and toluene 18% mixture of solvents Relative risk of high-freq hearing loss: Noise only: 4.1 (95% CI 1.4–12.2) Noise and toluene: 10.9 (95% CI 4.1–28.9) Solvents only: 5.0 (95% CI 1.5–17.5)	Noise and solvent exposures in the different groups were not equivalent Without a group exposed to only toluene, could not assess whether effect of combined exposure was additive or multiplicative
		Differences in liver metabolism of toluene suggest that rats and mice are better models for human ototoxicity than chinchillas
Chinchillas ABR threshold shifts (pre- vs postexposure) Tested at 0.5, 1, 2, 4, 8, 16 kHz Postexposure testing on days 1, 3, 7, 14, and 30	Chinchillas Noise effects, but no ototoxicity Noise: 12 dB permanent threshold shift at 2 and 4 kHz Analysis of variance: no sig main effect for toluene alone or interaction of toluene w/ noise	

continued

TABLE D-1 continued

Citation	Design	Population	Exposures and Source of Exposure Data
			11 chinchillas (binaural) Group 5: 12 h toluene, 8 h noise; noise from 2 h after start to 2 h before end of toluene exposure Group 6: 12 h toluene, background noise only Rat comparison group 6 adult rats exposed to toluene at 2000 ppm, 8 h / day for 5 days Only background noise
Johnson et al. (1990)	Experimental	49 young male rats Exposure groups: Controls: 10 Noise: 10 Toluene: 10 Noise followed by toluene: 10 Noise, rest, toluene: 9	Toluene: 1000 ppm, 16 h/d, 7 d/w (11:00 am to 3:00 am) Noise: 10 h/d, 7 d/w Continuously varying signal: 2 kHz wide noise band, sweeping from 3 to 30 kHz at freq of 0.5 Hz Equivalent to sound level of 100 dB Controls: no noise or toluene Noise: 4 wks Toluene: 2 wks Noise followed by toluene: 4 wks noise, 2 wks toluene Noise, rest, toluene: 4 wks noise, 4 wks rest, 2 wks toluene

Outcome Measure	Results	Comments
Rats ABR threshold shifts (pre- vs postexposure) Tested at 8, 16, 32 kHz Postexposure testing on day 30	Rats Sig permanent threshold shift w/ shorter toluene exposure Threshold shifts of 20 and 15 dB at 16 and 32 kHz, respectively	
ABR thresholds at 1.6, 3.15, 6.3, 12.5, 20.0 kHz Measured at 1–5 wks after termination of exposure	Effect of noise followed by toluene, w/ or w/o rest, was larger than exposure to noise or toluene alone at 6.3, 12.5, and 20.0 kHz Noise: higher thresholds than controls at 6.3 (9 dB, $p < .05$), 12.5 (26 dB, $p < .001$), and 20.0 kHz (18 dB, $p < .001$) Toluene: higher thresholds than controls at all freq (1.6 to 12.5 kHz, 15–32 dB, $p < .001$; 20.0 kHz, 15 dB, $p < .01$) Noise followed by toluene: higher thresholds than controls at all freq (1.6 kHz, 8 dB, $p < .05$; 3.15 to 20.0 kHz, 34–45 dB, $p < .001$)	Additive effects from noise and toluene Exposure to toluene after noise may produce smaller losses than exposure to noise after toluene

continued

TABLE D-1 continued

Citation	Design	Population	Exposures and Source of Exposure Data
Johnson et al. (1988)	Experimental	39 male rats Exposure groups: Controls: 10 Group T (toluene only) 12 Group N (noise only): 8 Group T+N (toluene followed by noise): 9	Toluene: 1000 ppm, 16 h/d, 5 d/wk (3 pm to 7 am) Noise: 10 h/d, 7 d/wk Continuously varying signal: 2 kHz wide noise band, sweeping from 3 to 30 kHz at a freq of 0.5 Hz Equivalent sound level 100 dB Controls: no toluene or noise Group T: 2 wks toluene Group N: 4 wks noise Group T+N: 2 wks toluene; 4 wks noise

NOTES: ABR, auditory brainstem response; CI, confidence interval; OR, odds ratio; TWA, time-weighted average.

Outcome Measure	Results	Comments
ABR threshold shifts at 1.6, 3.15, 6.3, 12.5, 20.0 kHz Controls: at age 5 mos Group T: 2–5 days after termination of exposure; repeated 1 and 6 mos later Group N: 2–5 days after termination of exposure Group T+N: 2–5 days after termination of noise exposure, repeated 6 mos later	Group T: Higher thresholds than controls at all freq ($p < .001$); greatest difference (40 dB) at 12.5 kHz; improvement at most frequencies at 1 mo (5–10 dB) and 6 mo (5 dB) Group N: Higher thresholds than controls; greatest difference at highest freqs (6.3, 12.5, and 20.0 kHz; $p < .001$); maximum difference (50 dB) at 12.5 kHz Group T+N: Higher thresholds than controls at all freq ($p < .001$); for most animals, threshold exceeded maximum stimulus intensity at 12.5 and 20.0 kHz For any exposure, threshold shift greatest at 6.3 and 12.5 kHz; after 6 mos, greatest at 12.5 kHz Combined exposure (toluene followed by noise) produced greater shifts at 3.15 ($p < .01$) and 6.3 kHz ($p < .0001$) than summed losses for single exposures	

TABLE D-2 Carbon Monoxide as a Risk Factor for Noise-Induced Hearing Loss in Animals

Citation	Design	Population	Exposures
Rao and Fechter (2000)	Experimental	Rats	
	A: Combined exposure	8 exposure groups: in each, *n* = 6 Random assignment	CO: 1200 ppm Octave band noise w/ center frequency of 13.6 kHz (9.6–19.2 kHz) Background noise: ~ 50 dBA CO only Noise only: 95 dB for 4 h 100 dB for 2 h 105 dB for 1 hr Noise plus CO: 95 dB for 4 h, 100 dB for 2 h 105 dB for 1 hr Control: Air only
	B: Increased duration of exposure	2 exposure groups: in each, *n* = 6	CO: 1200 ppm Noise: 105 dB Noise only: 4 h Noise plus CO: 4 h

Outcome Measure	Results	Comments
CAP thresholds for 11 pure tones between 2 and 40 kHz 4-wk recovery period	Exposure groups significantly different from each other ($F(8, 45) = 13.04$, $p < .05$) Sig interaction b/t freq and exposure ($F(80, 450) = 4.35$, $p < .05$) Potentiation of threshold elevation for combined exposure does not increase beyond 100 dB for 2h 95 dB for 4h Combined exposure: 6 dB threshold elevation above noise only; difference not sig at any frequency 100 dB for 2h Combined exposure: sig elevation over noise only at all frequencies ($p < .05$) 105 dB for 1h Combined exposure: sig elevation over noise only at all frequencies ($p < .05$); greater dysfunction at lower freq than w/ other exposures Noise only: sig elevation over thresholds for 95 dB (4h), 100 dB (2h) ($p < .05$) CO only No sig difference from exposure to air only $F(1,10) = 1.72$, $p > .05$	Rats more resistant to CO than humans (30 min LD_{50} = 5000 ppm for rats; 1500 ppm immediately dangerous for humans)
CAP thresholds for 11 pure tones between 2 and 40 kHz 4-wk recovery period	Combined exposure: threshold elevations not sig diff from noise only Combined exposure: threshold elevations 15 dB greater than for 105 dB + CO for 1h	

continued

TABLE D-2 continued

Citation	Design	Population	Exposures
Fechter et al. (2000)	Experimental	Rats	
	A: Variation of CO dose	8 exposure groups: in each, *n* = 8 Random assignment	CO: varied exposures Noise: 100 dB octave band noise w/ center frequency of 13.6 kHz Background noise: < 35 dB CO only: 1200 ppm, 8 h Noise only: 8 h Noise plus CO (5 exposure groups): 8 h noise and CO at 300, 500, 700, 1200, or 1500 ppm Control: air only
	B: Variation of noise dose	8 exposure groups: in each, *n* = 6 Random assignment	CO: 1200 ppm Noise: Octave band noise w/ center frequency of 13.6 kHz CO only: 4 h Noise only: 95 dB: 2 h 100 dB: 2 h 100 dB: 4 h Noise plus CO: 95 dB: 2 h 100 dB: 2 h 100 dB: 4 h Control: Air only

Outcome Measure	Results	Comments
CAP thresholds for 11 pure tones between 2 and 40 kHz 4-wk recovery period	Noise plus CO Potentiation of noise effects by CO exposure emerges at CO exposures of 500 ppm and increases as CO level increases Sig elevation of thresholds over exposure to noise only with exposures to CO levels > 300 ppm Thresholds at lower frequencies affected only w/ CO levels ≥ 1200 ppm; CO only No effect on auditory function	
CAP thresholds for 11 pure tones between 2 and 40 kHz 4-wk recovery period	Nonlinear relationship b/t noise severity and potentiation of threshold elevation by CO Potentiation of noise effects by CO exposure greatest w/ noise exposure of 100 dB for 2 h For noise exposure of 100 dB for 4 h, no additional effect at higher freq from CO exposure; sig differences at some lower freq Sig difference from controls for noise exposure of 100 dB for 2 or 4 h, w/ or w/o CO	

continued

TABLE D-2 continued

Citation	Design	Population	Exposures
	C: Repeated exposures	4 exposure groups: in each, *n* = 8 Random assignment	CO level: 1200 ppm Noise exposure: Octave band noise w/ center frequency of 13.6 kHz Exposure duration: 5 successive days: CO only Noise only: 95 dB, 2 h Noise plus CO Control: Air only
Chen and Fechter (1999)	Experimental	Rats	
	A: Response to CO and high-frequency vs low-frequency noise	Exposure groups: High-freq noise CO: *n* = 4 Noise: *n* = 7 Noise plus CO: *n* = 7 Air: *n* = 7 Low-freq noise CO: *n* = 4 Noise: *n* = 4 Noise plus CO: *n* = 4 Air: *n* = 7 Random assignment	CO level: 1200 ppm Noise exposure: 8h, octave band noise High freq: 9.6–19.2 kHz at 100 dB (Ln) Low freq: 2.4–4.8 kHz at 115 dB (Ln)

Outcome Measure	Results	Comments
CAP thresholds for 11 pure tones between 2 and 40 kHz 4-wk recovery period	Sig elevation of thresholds for noise plus CO exposure compared with noise only No effect of repeated exposure to CO only	
CAP and CM threshold shifts at freq from 2 to 40 kHz 4-wk recovery period	High-freq noise Noise plus CO produced sig greater CAP threshold shifts than noise alone ($p < .05$); greater potentiation at higher freq Noise alone produced sig CAP threshold shifts from air alone at freq > 8 kHz CO alone produced no CAP threshold shifts from air alone CM elevations for noise and noise plus CO, at all freq, w/ greater elevations at high freq and for combined exposure Low-freq noise Noise plus CO produced sig greater threshold shifts than noise alone ($p < .05$); somewhat greater potentiation at low freq Noise alone produced sig CAP threshold shifts at all freq CO alone produced no CAP threshold shifts CM elevations for noise plus CO over noise alone at all freq, w/ greater differences at low freq; sig differences at only three freq	

continued

TABLE D-2 continued

Citation	Design	Population	Exposures
	B: Response to CO w/ varied noise exposure	Exposure groups: Noise Noise plus CO	CO: 1200 ppm Noise: octave band noises at 100 dB (Ln) 9.6–19.2 kHz 4.8–9.6 kHz 2.4–4.8 kHz 1.2–2.4 kHz
	C: Comparison of potentiation measured by CAP and CM	Exposure groups: Noise Noise plus CO	CO levels: 300, 500, 700, 1200, 1500 ppm High frequency noise: 9.6–19.2 kHz Low frequency noise: 2.4–4.8 kHz
	D: Recovery of auditory thresholds over time	Exposure groups for each recovery period: 1 wk Noise: $n = 4$ Noise plus CO: $n = 3$ 4 wks Noise: $n = 7$ Noise plus CO: $n = 7$ Air: $n = 7$	CO level: 1200 ppm Noise: 9.6–19.2 kHz at 100 dB (Ln)
	E: Hearing loss potentiation and CO concentration		CO level: 0–1500 ppm Noise: 8 h, 9.6–19.2 kHz at 100 dB (Ln)

Outcome Measure	Results	Comments
CAP and CM threshold shifts in three freq ranges: Low: 2–8 kHz Mid: 12–20 kHz High: 24–40 kHz	Noise plus CO: CAP threshold shifts and CM elevations greater than noise alone, especially at high and mid frequencies, for 9.6–19.2 kHz and 4.8–9.6 kHz bands Noise alone: only 9.6–19.2 kHz noise band caused significant CAP threshold shifts and CM elevations	
Difference in mean CAP or CM between noise plus CO and noise alone High-frequency noise: measured at freq > 8kHz Low-frequency noise: measured at 2–6 kHz	Potentiations shown by CAP and CM are correlated and similar	
CAP thresholds measured at 1 wk, 2 wks, and 4 wks after exposure	Thresholds sig higher than controls for noise plus CO and noise alone at freq > 8 kHz Noise plus CO: No sig difference b/t 1 wk and 4 wks Noise alone: Lower thresholds at 4 wks than at 1 wk; sig differences at 12, 16, 30, 35 kHz; remain sig higher than controls at > 8 kHz	
Average CAP thresholds measured at 2–8 kHz, 12–20 kHz, and 24–40 kHz 4-wk recovery period	Greater potentiation with higher CO levels; CO effect varies across freq ranges Potentiation of hearing loss begins at CO level of 300–500 ppm for freq > 8 kHz	

continued

TABLE D-2 continued

Citation	Design	Population	Exposures
Young et al. (1987)	Experimental	16 male rats 4 exposure groups: in each, n = 4	CO: 210 min, 1200 ppm Noise: 120 min, 110 dBA; peak intensity at 4–8 kHz w/ roll-off of 16 dB/octave CO only Noise only Noise plus CO: 90 min of CO before noise onset Control: air only

NOTES: CAP, compound action potential; CM, cochlear monophonic; CO, carbon monoxide; LD_{50}, lethal dose 50 (for 50% of a population).

Outcome Measure	Results	Comments
Reflex modulation audiometry: detection sensitivity thresholds at 10 and 40 kHz Thresholds measured before exposure and at 1 wk and 3 wks postexposure	CO only: no evidence of worse auditory functioning after exposure Noise only: worse thresholds after exposure ($p < .01$); 10 kHz worse than 40 kHz at 1 wk Noise plus CO: thresholds worse than noise alone; greater shift at 40 kHz than 10 kHz; sig interaction effect at 1 week ($p < .05$); not sig at 3 wks Control: thresholds unchanged	CO levels higher than those likely in occupational settings

TABLE D-3 Smoking as a Risk Factor for Noise-Induced Hearing Loss

Citation	Design	Population	Exposures and Source of Exposure Data
Ferrite and Santana (2005)	Cross-sectional	535 male metal plant workers participating in hearing screening and enrolled in health promotion program Northeast Brazil Exclusions: women; age > 55 yrs; missing audiometric data; hearing loss inconsistent with noise damage	Questionnaire: sociodemographics, lifestyle, occupational and health-related data, and smoking Noise exposure Based on job-noise matrix: Exposed: jobs w/ 81–93 dBA Nonexposed: jobs w/ < 81 dBA Pre-employment noise exposure assessed Total duration of exposure: 0 < 4 yrs and ≥ 4 yrs Smoking Nonsmokers (never smoked or < 6 months) Ever-smokers (current or past smokers) Age categories 20–40 yrs, 41–55 yrs
Palmer et al. (2004)	Cross-sectional	12,907 men and women 10,418 with responses on hearing Britain, 1997–1998 Age: 16–64 yrs Postal survey of randomly selected members of armed forces and persons from age-sex registers of 163 general practitioners in Britain	Survey questionnaire: exposure to vibration, time spent working in noisy jobs, smoking history, and hearing aid use Noise Time spent working in noisy places (a need to shout to be heard; considered equivalent to at least 85–90 dBA) Smoking history Smoker: smoked at least once/day for at least 1 month Current smokers Former smokers Lifelong nonsmokers

Outcome Measure	Results	Comments
Hearing loss: hearing threshold > 25 dB HL at 3, 4, 6, or 8 kHz Pure-tone hearing thresholds measured at 0.25, 0.5, 1, 2, 3, 4, 6, 8 kHz	Smoking w/ noise exposure, was sig. associated w/ greater hearing loss compared w/ younger nonsmokers not exposed to noise Younger workers: PR = 4.85 (90% CI 2.49–9.46) Older workers: PR = 7.65 (90% CI 4.43–13.23) Smoking alone not sig associated with greater hearing loss in younger workers (PR = 1.27, 90% CI: 0.37–4.32) Prevalence of hearing loss: Older noise-exposed smokers: 46% Younger noise-exposed smokers: 29% Older nonsmokers not exposed to noise: 24% Younger nonsmokers not exposed to noise: 6%	No other known ototoxic agents in the workplace Smoking, noise exposure, and age examined as dichotomous variables No information on use of hearing protection Possibility of selection bias from exclusion of workers without existing audiometric test data Nonoccupational noise exposure not considered
Self-reported hearing difficulties assessed by response to— "How well can you hear a person who is talking to you when he is sitting on your right [left] side in a quiet room?" Severe difficulty in worse ear, or wore hearing aid Moderate difficulty in worse ear Normal: no or slight difficulty	Combined exposure to noise and smoking was consistent with an additive effect Current vs nonsmokers with moderate to severe hearing difficulty No work noise: PR = 1.5 (95% CI 1.1–2.1) 1–5 yrs work in noise: PR = 3.3 (95% CI 2.4 to 4.5) > 5 yrs work in noise: PR = 5.7 (95% CI 4.4 to 7.1) (PRs age- and sex-adjusted)	Response rate was 58%; response rate higher for women and older subjects No dose-response effect tested for smoking or noise Use of hearing protection not determined

continued

TABLE D-3 continued

Citation	Design	Population	Exposures and Source of Exposure Data
Mizoue et al. (2003)	Cross-sectional	4,624 male steel company workers Japan, 1999 Age: < 61 yrs Current smokers: 56% Working in potentially high noise levels: 29% Exclusions: no auditory examination, incomplete smoking history, ex-smoker	Noise exposure Company worksite records for workers who had hearing tests Worksite noise levels measured twice/yr Smoking history As reported at periodic audiometric tests Nonsmoker: never smoked Ex-smoker Current smoker Cigarettes per day: < 15, 15–24, ≥ 25 Age groups < 40, 40–49, 50–60 yrs
Starck et al. (1999)	Cross-sectional	199 professional forest workers 171 shipyard workers Finland Mean age: Forest workers: 43 yrs Shipyard workers: 38 yrs Exclusions: hearing loss from ear diseases or severe head injuries	Questions to establish work history and use of hearing protection, smoking history Medical records reviewed, overall health status, otological examination Noise exposure A-weighted noise level for average working day for both groups: 100 dB Effective exposure with hearing protectors (measured for each worker) Forest workers: 95 dB Shipyard workers: 85 dB Smoking Nonsmokers: never smoked or quit > 10 yrs ago Smokers

Outcome Measure	Results	Comments
Hearing loss (worse ear) Low frequency: pure-tone threshold > 25 dB HL at 1 kHz High frequency: pure-tone threshold > 40 dB HL at 4 kHz Pure-tone air conduction audiometric tests: Workers w/o significant noise exposure: 1, 4 kHz Workers in noisy environments (> 85 dBA): 0.5, 1, 2, 4, 8 kHz	Combined effect of noise and smoking on high-frequency hearing loss is comparable to sum of their independent effects 4 kHz comparisons w/ non-smokers not exposed to noise Smokers exposed to occ noise: PRR = 2.56 (95% CI 2.12 to 3.07) Smokers not exposed to noise: PRR = 1.57 (95% CI 1.31 to 1.89) Nonsmokers exposed to noise: PRR= 1.77 (95% CI 1.36 to 2.30) Smoking not associated with low-frequency hearing loss	No control for past occupational noise exposure or leisure time noise exposure No information on ear disease or injuries that might have affected hearing PRR calculated by Cochran-Mantel-Haenszel method, w/ age stratification
Hearing level at 4 kHz Measured hearing level compared with expected level (ISO-1999), calculated on the basis of age, noise level (A-wt), and duration of exposure (50% estimate) Age correction of hearing levels with ISO 1999 (1990) model	Variation in hearing loss explained (linear regression) Forest workers Age: 26% Noise exposure: 10% Smoking: 1% (p=ns) Shipyard workers Age: 48% Noise exposure: 15% Smoking: 3% (p < .05) Hearing levels for smokers and nonsmokers not significantly different	All workers exposed to noise; differed in noise levels and duration of exposure

continued

TABLE D-3 continued

Citation	Design	Population	Exposures and Source of Exposure Data
Cruickshanks et al. (1998)	Population-based, cross-sectional	3,753 adults, ages 48–92 yrs Beaver Dam, Wisconsin, 1993–1995 Mean age: 66 yrs 58% women	Questionnaire and examination <u>Noise exposure</u> History of occupational noise exposure: having to speak in a loud voice to be heard; farmer driving tractor w/o cab, or military service with noise (pilot, aircraft or tank crew, ship engine room, use of grenades, mortars, multiperson weapons systems) <u>Smoking status at examination</u> Nonsmoker: < 100 cigarettes (lifetime) Ex-smoker Current smoker <u>Amount smoked</u> Pack-years
Virokannas and Anttonen (1995)	Cross-sectional	433 reindeer herders Northern Finland, 1988 Mean age: 43 yrs (range 18–64 yrs) Exclusions: accidental exposure to explosion; suffered ear disease; abnormal findings for tympanic membrane	Clinical examination and questions on exposure to noise, smoking history, use of ear protectors Principal noise sources: Snowmobiles: 92–104 dBA Chainsaws: 96–103 dBA Gunshots: under 80 dBA (annual equivalent) <u>Noise exposure</u> (based on cumulative hrs of use of noisy tools and vehicles) Mild: 0–3,700 h Moderate: 3,701–8,700 h Heavy: 8,701–15,000 h

Outcome Measure	Results	Comments
Hearing loss: Average hearing threshold at 0.5, 1, 2, and 4 kHz > 25 dB HL (worse ear) Mild: > 25, ≤ 40 dB HL Moderate: > 40, ≤ 60 dB HL Marked: > 60 dB HL	Current smokers had sig increased risk of hearing loss compared w/ nonsmokers among those w/ and w/o occupational noise exposure W/ occupational noise: OR = 1.85 (95% CI 1.33–2.57) W/o occupational noise: OR = 1.53 (95% CI 1.03–2.29)	Self-reported smoking may have been understated
Pure-tone hearing thresholds(audiometers calibrated to ISO 389 standard) Frequencies tested: 0.5, 1, 2, 3, 4, 6, 8 kHz Age adjustment: measured hearing thresholds transformed to values at age 18 (ISO 7029)	Analysis of Covariance Covariant: exposure time to noise Significant effect of heavy or very heavy lifetime smoking on age-adjusted hearing thresholds: 3 kHz (right p = .044, left p = .001) 4 kHz (right p = .055, left p = .086) Not significant at other frequencies	Very heavy and heavy smokers tended to use noisier tools than the moderate and nonsmokers Non-, moderate, and heavy smokers used hearing protectors more often (61–64%) than very heavy smokers (47%)

continued

TABLE D-3 continued

Citation	Design	Population	Exposures and Source of Exposure Data
			Lifetime smoking (number of cigarettes) Moderate: > 0 to < 36,000 Heavy: 36,000 to < 144,000 Very heavy: 144,000 or more
Pyykko et al. (1988)	Cross-sectional	199 professional forest workers Northeastern Finland (1970s ?) Exclusions: bilateral ear disease	Survey: history of working habits, use of ear protectors, history of vibration syndrome, smoking history Complete medical evaluation Noise exposure Time-weighted median inside earmuffs: 99 dBA (91 dBA average inside muffs, 103 dBA outside muffs); measured for sample of 6 workers Smoking Smoker: smoked w/in past 10 yrs Nonsmoker: never smoked or no smoking in past 10 yrs

NOTES: CI, confidence interval; OR, odds ratio; PR, prevalence ratio; PRR, prevalence rate ratio.

Outcome Measure	Results	Comments
Mean hearing threshold of both ears at 4 kHz Age correction based on A-weighted noise levels and duration of exposure	Smoking not sig correlated with sensorineural hearing loss at 4 kHz	

TABLE D-4 Progression of Hearing Loss After Noise-Induced Hearing Loss

Citation	Design	Population	Exposures and Source of Exposure Data
Lee et al. (2005)	Longitudinal cohort	188 participants (91 women, 97 men); 376 ears Participation for at least 3 yrs (mean 6.4 yrs) Ages at entry: 60–81 yrs (mean 68 yrs) Recruitment through advertisements and referral, began in 1987 Conventional thresholds tested annually Exclusions: conductive hearing loss, active otologic/neurologic disease Longitudinal study of presbyacusis, Medical University of South Carolina	Questionnaire: noise history, medication use, self-evaluation of hearing handicap History of noise exposure: 56 of 85 men 18 of 78 women

Outcome Measure	Results	Comments
Rate of change in pure-tone hearing thresholds (0.25, 0.5, 1, 2, 3, 4, 6, 8 kHz) (slope from linear regression of changes for each ear) Testing also done for extended high frequencies	No significant difference in rate of threshold change at 1–2 kHz for noise-exposed and unexposed Rate of change at 6–8 kHz lower for noise-exposed than unexposed males; lower at 2 kHz for noise-exposed females Hearing thresholds at 2–8 kHz were significantly higher (7.7 to 12.1 dB, $p < 0.05$) for noise-exposed men Noise-exposed women had smaller threshold elevations (2.2 to 7.6 dB) Overall, men had significantly poorer initial thresholds than women at 2–8 kHz Rates of change at 0.25–2 kHz had positive correlations w/ initial thresholds at higher frequencies Rates of change at 3–8 kHz had negative correlations w/ initial thresholds at those frequencies	

continued

TABLE D-4 continued

Citation	Design	Population	Exposures and Source of Exposure Data
Cruickshanks et al. (2003)	Longitudinal cohort	Baseline (1993) 3,753 participants, ages 48 to 92 years 1,925 w/o hearing loss [PTA (0.5, 1,2,4 kHz) in both ears ≤ 25 dB] Mean age: 61 yrs 1,631 w/ hearing loss [PTA (0.5, 1,2,4 kHz) in either ear > 25 dB and < 100 dB in worse ear] Mean age: 71 yrs 5-yr follow-up (1998) 2,800 participants Losses to death, refusal, loss to follow-up; losses greater among persons w/ hearing loss Epidemiology of Hearing Loss Study, Beaver Dam, WI	Questionnaires: occupation and exposure to occupational noise, leisure noise, military service
Rosenhall (2003)	Mixed longitudinal and cross-sectional cohort	Gerontological and geriatric population study, Gothenburg, Sweden Total population: 616 men 869 women Cohorts 1. 1971, 70 yr olds; seven exams, to age 90 yrs; cohort supplemented beginning at age 85 2. 1976, 70 yr olds; second exam at 75 yrs 4. 1990–1991, 75 yr olds 5. 1992–1993, 70 yr olds	Questionnaires: occupation and exposure to occupational noise

Outcome Measure	Results	Comments
Incident hearing loss PTA (0.5, 1, 2, 4 kHz) in either ear > 25 dB at follow-up	Incident hearing loss All: 21% (95% CI 19.4–23.4) Men (age-adj): 31% (CI 26.8–34.6) Current or former occupation (age-, sex-adj): Mgmt/profess: OR = 1.00 Production: OR = 1.92 (95% CI 1.10 – 3.37) Operations/fabricators: OR = 1.92 (95% CI 1.26–2.93) Self-reported occupational noise exposure: not significant	
Progression of hearing loss PTA (0.5, 1, 2, 4 kHz) > 5 dB increase over baseline level	Progression of hearing loss All: 53.5% (CI 50.2–56.4) Men (age-adj): 51.5% (CI 47.0–55.2) Occupation: no significant assoc between likely noise exposure and progression	
Change in pure-tone hearing thresholds (0.25, 0.5, 1, 2, 4, 6, 8 kHz) from age 70 to age 75 yrs	Greater increases in hearing thresholds at 1, 2, 8 kHz for men exposed to noise than men not exposed Increases at 4 kHz similar for exposed and non-exposed Less change between ages 75 and 79; similar changes for exposed and non-exposed	No measures of statistical significance reported

continued

TABLE D-4 continued

Citation	Design	Population	Exposures and Source of Exposure Data
Gates et al. (2000)	Longitudinal cohort	203 male members of the initial Framingham Heart Study cohort with audiometric tests from examinations E15 and E22 E15 No notch (N0): 75 right ears (RE), 68 left ears (LE) Small notch (N1): 50 RE, 47 LE Large notch (N2): 78 RE, 88 LE Mean age at E15: 64 yrs Exclusions: men w/ 20 dB difference in PTA b/t right and left ears	Represented by presence and characteristics of audiometric notch at Examination E15 Notch determined from two-part function for linear pattern at lower frequencies and vertically oriented, concave parabolic form at higher frequencies (2–8 kHz) Depth-of-notch categories: N0: < 15 dB N1: 15 dB–35 dB N2: ≥ 35 dB N1: possible noise damage N2: probable noise damage

NOTES: CI, confidence interval; LE, left ear; OR, odds ratio; PTA, pure-tone average; RE, right ear.

Outcome Measure	Results	Comments
15-year change (E22–E15) in audiometric threshold for 8 frequencies: 0.25–8 kHz (each ear assessed separately) Secondary analysis included as predictor variables cardiovascular disease events (lifetime), smoking history (E15–E22), number of prescription medications used regularly (E15–E22)	0.25, 0.5, 1 kHz: threshold shifts not significantly different across notch groups 2 kHz: notch groups are significantly different; (e.g., LE shifts: N0 = 12.4; N1 = 16.0; N2 = 18.8; p = 0.0018) 4, 6 kHz: N2 shifts significantly smaller than N0, N1 8 kHz: N1 shift significantly greater than N0 or N2; N0 and N2 not significantly different With E15 thresholds as covariates, notch category was significant only at 2 kHz ($p < .001$) Similar patterns for right and left ears Results adjusted for age, smoking, medications	Actual noise exposures during E15–E22 interval not documented; presumed no additional occupational noise exposure for most subjects and little recreational exposure Past noise exposure inferred from presence of audiometric notch Possible survival bias

TABLE D-5 Features of Studies Included in Analysis of Hearing Loss Among Military Personnel

Citation	Design	Timing of Data Collection	Study Population	Stratification
a. *Air Force Studies*				
Sutherland and Gasaway (1976)	Cross-sectional	Jan–June 1975 Data from AF Form 1490 (annual hearing conservation audiogram), received Jan–Jun 1975	56,951 USAF personnel (men only?) 48,262 military 8,689 civilians Personnel tested as part of the hearing conservation program	Age (yrs) 18–24; 25–34; 35–44; 45–54; 55–64
Sutherland and Gasaway (1978)	Cross-sectional	1 June 1975–31 May 1976 Data from AF Form 1490 (annual hearing conservation audiogram) for test dates w/in study period	117,454 USAF personnel (men only?) 99,318 military 18,136 civilians Received annual hearing conservation audiogram during 1-yr study period (one record/person)	Age (yrs) 18–24; 25–34; 35–44; 45–54; 55–64
Thomas (1995)	Longitudinal and cross-sectional reporting	Not specified Data from Air Force hearing conservation program test results	6,655 persons in hearing conservation program (8 bases) 3,029 military 2,859 civilians 6,207 men 365 women Only personnel with at least 4 sequential audiograms	Tests 1–4 (mean age for personnel at each test)

Frequencies Tested	Audiometric Zero, Audiometer, and Earphones	Summary Measures Reported*	Comments, Conclusions
0.5, 1, 2, 3, 4, 6 kHz	ANSI, 1969 TDH-39 earphones w/ MX-41/AR cushions	Median hearing threshold levels (military and civilians reported separately) % distribution of HTLs for each frequency % STS	Hearing levels of military noise-exposed USAF personnel better than USAF civilians and U.S. population (except at ages 18–24, 0.5 kHz, left ear).
0.5, 1, 2, 3, 4, 6 kHz	ANSI, 1969 TDH-39 earphones w/ MX-41/AR cushions	Median hearing threshold levels (military and civilians reported separately) % distribution of HTLs for each frequency % STS	Noise-exposed USAF military personnel show better hearing than noise-exposed USAF civilians and general U.S. population
0.5, 1, 2, 3, 4, 6 kHz	Not specified	Mean hearing threshold levels (ears not reported separately) (military and civilians reported separately) % worse on subsequent test % better or worse on subsequent test	Total personnel in hearing conservation program was 14,166 Less variability in women's HTLs than men's (combined military and civilian) Analysis of variability of HTLs classifies USAF hearing conservation program as unacceptable to marginal

continued

TABLE D-5 continued

Citation	Design	Timing of Data Collection	Study Population	Stratification
b. *Army Studies*				
Walden et al. (1971)	Cross-sectional (pilot study)	1971 (Sept–Nov?)	2,726 men Active duty Officers and enlisted Included recruits, advanced trainees, regular personnel from infantry, armor, artillery, pilots Convenience sample from 6 bases	Age (yrs) 16–20; 21–25; 26–30; 31–35; 36–40; 41–45; 46–50; 51–55 LOS (yrs): 0–2; 2–4; 4–6; 6–8; 8–10; 10–15; 15–20; 20–25; 25–30; (w/in branches: < 4, 4–10, > 10 yrs)
Walden et al. (1975)	Cross-sectional	June 1974–Sept 1974	3,000 enlisted men (10 bases; 1,000 each infantry, armor, and artillery) 75% of active duty spent in designated branch 300 male inductees (3 bases) Random selection	Branch LOS (yrs): 1.5–2.4; 2.5–7.4; 7.5–12.4; 12.5–17.4; 17.5–22.4
Peters and Ford (1983)	Cross-sectional	Feb–Aug 1982	145 aviators (all men?) Ft. Rucker; officers and warrant officers Age range 24–45 yrs (mean 32 yrs) 54% w/ substantial exposure to small arms and artillery fire Random sample	Flight hours 50–400; 401–600; 601–800; 801–1000; 1001–2000; 2001–3000; 3001–4000; 4001–5000; 5001–6000; 6001–7000

Frequencies Tested	Audiometric Zero, Audiometer, and Earphones	Summary Measures Reported*	Comments, Conclusions
0.25, 0.5, 1, 2, 4, 6 kHz	ISO, 1964 Portable audiometers w/ minimum threshold of -10 dB HL	Mean hearing thresholds (standard deviation) % distribution by hearing profile	Higher % with hearing impairment w/ longer active duty service Higher thresholds seen as early as first 4–6 months of active duty Authors cite need for longitudinal study
0.25, 0.5, 1, 1.5, 2, 3, 4, 6, 8 kHz < 81 dBA exposure in 14 hours before test	ANSI, 1969 Manual testing w/ audiometers and test booths available at each base TDH-39 earphones w/ MX-41/AR cushions	Mean hearing thresholds (standard deviation) % distribution by hearing profile (standard error)	Hearing ability decreases as time-in-service increases Clinically significant losses: 20–30% at ≥ 2 yrs LOS; > 50% at ≥ 15 yrs LOS Middle to high frequencies most affected Assigned hearing profile category not correct for many
2, 3, 4, 6 kHz	ANSI, 1969 Grason-Stadler audiometer, Model 1701 w/ TDH 49 earphones or Grason-Stadler GSI 10 w/ TDH 50P earphones	Mean and median hearing thresholds (standard deviation, standard error)	Mean thresholds lower (better) than aviators in Walden et al. (1971) Thresholds higher for left ear at 2, 4 kHz, higher for right ear at 6 kHz Significant effect of flight hours (50–400; 401–3000; 3001–6000 hrs) on left-ear threshold differences between 4 and 2 kHz

continued

TABLE D-5 continued

Citation	Design	Timing of Data Collection	Study Population	Stratification
Chandler and Fletcher (1983)	Retrospective analysis of longitudinal data	Date not specified Reference and current audiograms (w/in past 12 mos)	209 men in six engineering MOS (1) Basic engineers; (2) carpenters, plumbers, electricians; (3) heavy equipment operator/ mechanic; (4) maintenance (wheeled vehicle); (5) truck driver; (6) petroleum supply and storage Exclusions: history of nonoccupational noise hazards, head or acoustic trauma, middle-ear pathology, or family history of hearing loss Age range: 18–50 yrs	MOS
Ohlin (1992)	Cross-sectional	1989 Hearing conservation program reports Last test in 1989 DD Form 2215: 1,284 tests DD Form 2216: 1,625 tests	2,903 enlisted men (10 bases: 985 Infantry; 963 Armor; 959 Artillery) Random selection from registry for hearing conservation program 3,534 inductees (1 base) Test results from feasibility study for reference audiograms No otologic examinations	LOS (yrs) 1.5–2.4; 2.5–7.4; 7.5–12.4; 12.5–17.4; 17.5–22.4 yrs Branch (w/o LOS)

Frequencies Tested	Audiometric Zero, Audiometer, and Earphones	Summary Measures Reported*	Comments, Conclusions
0.5, 1, 2, 3, 4, 6 kc/s	ANSI, 1969 No information on audiometer or earphones	Mean hearing threshold levels (current) Change from reference thresholds (graphed data only)	No remarkable differences among MOS groups Better current HTLs at 3, 4, and 6 kc/s by 23–34 dB, than 1954 data for comparably aged men working in industry may reflect exclusions for other causes of hearing loss in military population
0.5, 1, 3, 4, 6 kHz (no quiet period required before DD Form 2216 test)	Standard hearing conservation program calibration HEARS microprocessor audiometers; group testing Circumaural test earphones (Audiocups)	Mean hearing thresholds (average of left and right ears) (standard error) % distribution by hearing profile (standard error)	59% of enlisted combat arms personnel from sampled bases in HEARS in 1989 Entrance standards stricter as of 1 Aug 87 (AR 40-501) No data on race or nonoccupational noise Systematic increase in HTLs with increase in LOS, w/ approx 50% of hearing loss attributed to aging Differences among branches not clinically significant Improvements over 1974 levels for all LOSs

continued

TABLE D-5 continued

Citation	Design	Timing of Data Collection	Study Population	Stratification
Henselman et al. (1995)	Cross-sectional	1989 Records from hearing conservation program tests (high-noise MOS) and periodic physical examinations (low-noise MOS)	39,006 enlisted men from high-noise-exposure MOS (infantry, armor, artillery) 18,730 enlisted men from low-noise-exposure MOS (administration, supply and services, medical, visual information, electronic maintenance and calibration, public affairs, automated data processing, topography, intelligence) Age range: 17–56 yrs	Race (white black, other), LOS (yrs): by yr for 0–14.9 yrs; 15–19.9; 20–24.9 yrs
c. *Navy and Marine Corps Studies*				
Robertson et al. (1978)	Cross-sectional	"[N]early three years" (mid 1970s? dates not specified)	3,050 enlisted sailors: 1,561 in "experimental" ratings (high noise exposure expected: airman, fireman, equipment operator, machinist mate, engineman, boiler tech, aviation mechanic, aviation machinist mate, aviation boatswain mate, aviation ordnance man)	LOS (yrs) For ratings: 1–2; 2–3; 3–4; 4–5; 5–10; 10–15; 15–20; 20–25 yrs For apprentices: 1–2; 2–3; 3–4 yrs

Frequencies Tested	Audiometric Zero, Audiometer, and Earphones	Summary Measures Reported*	Comments, Conclusions
0.5, 1, 2, 3, 4, 6 kHz	Standard hearing conservation program calibration Tracor RA 600AM microprocessor audiometer TDH-39 earphones with ear cushions	Age-corrected threshold average (1, 2, 3, and 4 kHz; left and right ears averaged) Age correction from ISO-1999, database B	Analyzed records cover 25% of total personnel in high-noise MOS and 18% of personnel in low-noise MOS Exposure categories based on current MOS; no information on noise-exposure history No control for nonoccupational noise Significant difference in average thresholds between high- and low-noise groups, but differences were < 5 dB
0.5, 1, 2, 3, 4, 6, 8 kHz	ANSI, 1969 Manual audiometry No information on audiometer or earphones	"Average" hearing threshold levels (standard deviation) % of subjects with significant high- and low-frequency hearing losses (≥ 30 dB)	37% of experimental group and 23% of control group had significant high-freq hearing losses at ≥ 5 yrs of service USMC high-freq losses described as generally greater than Navy; only those for Navy EO rating similar to USMC

continued

TABLE D-5 continued

Citation	Design	Timing of Data Collection	Study Population	Stratification
			1,489 in "control" ratings (low noise exposure expected: hospitalman, dentalman, hospital corpsman, dental tech, mess management spec., yeoman, personnelman, disbursing clerk, training device tech, aviation maintenance admin) 361 apprentices (airman, fireman, hospitalman, dentalman) 121 recruits 298 women (9% overall: 18% of recruits: 16% of control; 1.7% of experimental) "Subjects were identified by computer" Exclusions: conductive hearing loss	
Goldenberg (1977)	Cross-sectional	13-month period (early 1970s? date not specified)	11,577 men (Marine Corps enlisted personnel and officers) Consecutive unique test subjects at one site No otologic examination	Age (yrs) < 18; 18–24; 25–34; 35–44; 45–54; 55–64

Frequencies Tested	Audiometric Zero, Audiometer, and Earphones	Summary Measures Reported*	Comments, Conclusions
0.5, 1, 2, 3, 4, 6 kHz	ISO (1964) Rudmose ARJ-4A audiometer, 10-man booth TDH-39 earphones If questionable self-recording audiogram, retested in 1-man booth w/ Beltone 15 CW audiometer, TDH-39 earphones	Median hearing thresholds (better ear only; graphed data) % w/ > 25 dB and % w/ > 45 dB loss at avg of speech frequencies (0.5, 1, 2 kHz) or high frequencies (3, 4, 6 kHz)	Marine Corps data similar to USPHS for civilians Markedly higher thresholds at ages 35–44 yrs and older

continued

TABLE D-5 continued

Citation	Design	Timing of Data Collection	Study Population	Stratification
Bohnker et al. (2002)	Cross-sectional	Tests between 1995–1999 Hearing conservation program reports Data entered beginning 1999	68,632 enlisted personnel w/ "monitoring" audiogram reports from hearing conservation program Navy men = 51,643 Marine men = 12,251 Navy women = 4,184 Marine women = 554 Analysis based on 20–25% sample of 152,590 records	Service (USN/USMC), gender, and age (yrs) 17–24; 25–29; 30–34; 35–39; 40–44; 45–49; ≥ 50
d. *U.S. Adult Population*				
Glorig and Roberts (1965)	Cross-sectional	Oct 1959–Dec 1962	6,672 persons examined Noninstitutionalized civilians, ages 18–79 yrs Nationally representative random sample	Sex and age (yrs) 18–24; 24–34; 35–44; 45–54; 55–64; 65–74; 75–79

*Left ears and right ears reported separately unless otherwise noted.
NOTE: EO, equipment operator; HEARS; Hearing Evaluation Automated Registry System [Army] ; HTL, hearing threshold level; LOS, length of service; MOS, military occupational specialty; NEHC, Navy Environmental Health Center; OSHA, Occupational Safety and Health Administration; STS, significant threshold shift; USPHS, U.S. Public Health Service.

Frequencies Tested	Audiometric Zero, Audiometer, and Earphones	Summary Measures Reported*	Comments, Conclusions
0.5, 1, 2, 3, 4, 6 kHz	Not reported; reference to testing according to NEHC guidance	Mean hearing thresholds (standard deviation)	Mean thresholds for USN and USMC were generally worse than OSHA age-adjusted values Women had lower thresholds than men w/in each service at corresponding ages
0.5, 1, 2, 3, 4, 6 kcps	ASA, 1951 Beltone audiometers, TDH-39 earphones w/ MX-41/AR cushions Minimum threshold: −10 dB HL	Median hearing thresholds (method for calculating 95% confidence intervals described) % distribution by hearing threshold categories	No noise-exposure history Hearing levels higher w/ age from youngest to oldest

TABLE D-6 Studies on Prevalence of Tinnitus and Prevalence of Tinnitus with Hearing Loss

Study	Design, Population	Definition of Tinnitus	Definition of Hearing Level or Hearing Loss
a. *Community-based Studies*			
Sindhusake et al. (2003a,b, 2004) Blue Mountains Hearing Study, Australia, 1997	Cross-sectional 2,015 persons, ages 55+ yrs (mean 70 yrs) Residents of 2 suburban postal code areas	Tinnitus Sounds persisting for 5 minutes or longer during the past yr	Hearing level Lower frequencies: PTA for 0.5, 1, 2, and 4 kHz Higher frequencies: PTA for 4, 6, and 8 kHz Hearing loss PTA for 0.5, 1, 2, and 4 kHz > 25 dB HL in better ear Pure-tone testing at 0.25, 0.5, 1, 2, 4, 6, 8 kHz (3 kHz if 20 dB difference b/t 2 and 4 kHz)
Hoffman and Reed (2004) Tambs et al. (2003) Nord-Trøndelag Hearing Loss Study, Norway, 1995–1997	Cross-sectional 51,975 persons, ages 20–101 yrs (mean: 50 yrs) Residents of Nord-Trøndelag County	Tinnitus Bothered by ringing in the ears	Hearing loss PTA for 0.5, 1, 2, and 4 kHz in worse ear: ≤ 25 dB HL > 25, ≤ 40 dB HL > 40 dB HL

Noise Exposure and Source of Exposure Data	Prevalence of Tinnitus	Prevalence of Tinnitus, by Hearing Level	Comments
Questionnaire on noise exposure Duration (yrs) and level of occupational noise (quiet; tolerable; "unable to hear speech") Noise during military service Leisure noise (e.g., gunfire)	30% w/ tinnitus Age- and sex-standardized to Australian population	Hearing loss 35% w/ tinnitus Normal hearing 27% w/ tinnitus	75% participation rate for eligibles
Questionnaire on noise exposure Occupational noise: difficult to have a conversation; type of work; exposed to staple gun, hammering, chain saw, blasting, etc; Other noise: impulse noise (y/n); brass band; personal stereo	15% w/ tinnitus	Odds ratio for tinnitus ≤ 25 dB HL Men: 1.0 Women: 1.0 > 25, ≤ 40 dB HL Men: 2.84 (95% CI 2.55–3.16) Women: 2.78 (95% CI 2.45–3.15) > 40 dB HL Men: 4.18 (95% CI 3.66–4.77) Women: 5.40 (95% CI (4.67–6.24)	Possible response bias Participation rates were 65% for men; 73% for women; < 50% for ages < 30 yrs; ≥ 75% for ages 50–80 yrs

continued

TABLE D-6 continued

Study	Design, Population	Definition of Tinnitus	Definition of Hearing Level or Hearing Loss
Nondahl et al. (2002) Epidemiology of Hearing Loss Study, Beaver Dam, WI 1993–2000	Cross-sectional and longitudinal Baseline 3,737 participants, ages 48 to 92 years 5-yr follow-up 2,558 participants (75% of those w/o significant tinnitus at baseline)	Significant tinnitus "buzzing, ringing, or noise" in the ears in the past year, rated as at least moderately severe or causing problems w/ sleep or both Excludes tinnitus rated mild or of unknown severity and not causing problems w/ sleep	Hearing loss PTA of thresholds at 0.5, 1, 2, and 4 kHz > 25 dB HL in worse ear Pure-tone testing at 0.25, 0.5, 1, 2, 3, 4, 6, 8 kHz
Palmer et al., (2002) United Kingdom, 1997–1998	Cross-sectional 12,907 responses from adults, ages 16–64 yrs Postal survey using random sample from age-sex registers for 34 general medical practices, plus members of the armed services	Tinnitus Noises lasting longer than 5 minutes during the past 12 months Persistent tinnitus Occurring most or all of the time	No hearing difficulty Report of no or slight difficulty hearing a person talking in a quiet room w/ better ear Severe hearing difficulty Use of a hearing aid or report of severe difficulty hearing or inability to hear a person talking in a quiet room Intermediate hearing difficulty All others No audiometric testing done

Noise Exposure and Source of Exposure Data	Prevalence of Tinnitus	Prevalence of Tinnitus, by Hearing Level	Comments
Occupational noise history - full-time job required using loud voice to be heard at 2 ft; - farmer who drove tractor w/out cab at least half of the time; or - military duties on aircraft; tracked vehicle, ship engine room; on weapons range ≥ 7 times per yr; used grenades, mortars, or shoulder-held grenade launchers; used weapons system requiring more than one operator	Baseline (significant tinnitus) 8% [8.2%; 95% CI 7.4–9.1] Severe: 1.5% Moderate or causing sleep problems: 6.7% 5-yr Incidence 6% [5.7%; 95% CI 4.8–6.6]	Baseline w/ hearing loss 12% w/ significant tinnitus w/o hearing loss 5% w/ significant tinnitus	
Questionnaire response on number of years working in noisy places (need to shout to be heard)	Persistent tinnitus Men: 6% Women: 3%	W/ severe hearing difficulty Men: 16.1% w/ persistent tinnitus (age-standardized) Women: 33% w/ persistent tinnitus W/ no hearing difficulty Men: 5.0% w/ persistent tinnitus Women: 3% w/ persistent tinnitus	Possible response bias (58% response rate)

continued

TABLE D-6 continued

Study	Design, Population	Definition of Tinnitus	Definition of Hearing Level or Hearing Loss
Adams et al. (1999) National Health Interview Survey, United States, 1996	Cross-sectional 63,402 persons, all ages Nationally representative random sample; noninstitutionalized population; excludes armed forces and nursing home residents	"Does [household member] now have tinnitus or ringing in the ears?"	
Hoffman and Reed (2004) Adams and Marano (1995) Disability Supplement, National Health Interview Survey, United States, 1994–1995	Cross-sectional, household interviews 99,435 adults, ages 20 yrs and older Subset of participants in national sample of noninstitutionalized population; excludes armed forces and nursing home residents Excludes proxy responses on tinnitus	"Does [household member] have ringing, roaring or buzzing in the ears or head now that has lasted for at least 3 months?"	Moderate or worse hearing loss (subjective assessment, criteria not specified) Uses hearing aid
Hoffman and Reed (2004) Adams and Benson (1991) Hearing Supplement, National Health Interview Survey, United States, 1990	Cross-sectional, household interviews 59,343 adults, ages 20 yrs and older Subset of participants in national sample of noninstitutionalized population; excludes armed forces and nursing home residents Excludes proxy responses on tinnitus	"At any time over the past 12 months, have you ever noticed ringing in the ears, or have you been bothered by other funny noises in your ears or head?" If yes: Occurs all the time/ every few days/ less often Bothered quite a bit/ just a little/ not at all	Moderate or worse hearing loss (subjective assessment, criteria not specified)

Noise Exposure and Source of Exposure Data	Prevalence of Tinnitus	Prevalence of Tinnitus, by Hearing Level	Comments
No exposure data	3% w/ tinnitus Age < 45 yrs: 1% 45–64 yrs: 6% 65+ yrs: 9% 18+ yrs: 4% 45+ yrs: 7%		94% participation rate
	4% w/ tinnitus	Moderate or worse hearing loss 32% w/ tinnitus Uses hearing aid 12% w/ tinnitus	Unknown effect of exclusion of proxy responses on representativeness of data (40,570 adult proxy responses)
No exposure data	8% w/ tinnitus	Moderate or worse hearing loss 42% w/ tinnitus	Unknown effect of exclusion of proxy responses on representativeness of data (27,364 proxy responses)

continued

TABLE D-6 continued

Study	Design, Population	Definition of Tinnitus	Definition of Hearing Level or Hearing Loss
Coles (1996) Tier B, National Study of Hearing, United Kingdom	Cross-sectional 3,234 people selected in stratified random sample from postal survey responses Stratification to ensure larger proportion of persons w/hearing disorders and tinnitus	Prolonged spontaneous tinnitus: lasting at least 5 minutes and not temporarily induced by noise, drugs, or ear or respiratory illness	Hearing threshold levels
Medical Research Council's Institute of Hearing Research, (1981) Coles (1984) Tier A, National Study of Hearing, United Kingdom, 1978–1982	Cross-sectional (multiple samples) Postal survey, responses from random samples of adults, age 17+ yrs, from 4 cities: Prepilot: 522 Pilot: 5,000 (74%) Phase I: 8,069 Phase II: 7,645	<u>Prolonged spontaneous tinnitus</u> lasting at least 5 minutes and not temporarily induced by noise, drugs, or ear or respiratory illness <u>Prepilot</u> Ever noticed noises in your head <u>Spontaneous tinnitus</u> <u>Pilot</u> Ringing or buzzing lasting 5 minutes or more, excluding those occurring only after exposure to loud noise <u>Phase I</u> Ever have noises lasting 5 minutes or more, excluding those occurring only after exposure to loud noise	

Noise Exposure and Source of Exposure Data	Prevalence of Tinnitus	Prevalence of Tinnitus, by Hearing Level	Comments
		Presence of moderately or severely annoying tinnitus <u>HTL 10–19 dB vs HTL < 10 dB</u> OR=2 <u>HTL >80 dB vs HTL < 10 dB</u> OR= 27	Controlling for hearing thresholds eliminates associations w/ age, noise exposure, or socioeconomic status
Little or no occupational noise exposure: 8% w/ tinnitus High lifetime noise dose: 21% w/ tinnitus	<u>Prolonged, spontaneous</u> 10% <u>Any tinnitus</u> 34% to 39% <u>Brief or nonspontaneous</u> 23% to 27% <u>Spontaneous</u> 11% to 18% Moderately or severely annoying: 5% Sleep-disturbing: 5% Severe effect on quality of life: 1% Severe effect on ability to lead a normal life: 0.5%		80% response rate 7% of all adults have sought a doctor's help for tinnitus

continued

TABLE D-6 continued

Study	Design, Population	Definition of Tinnitus	Definition of Hearing Level or Hearing Loss
		Phase II "Nowadays" noises lasting 5 minutes or more, excluding those occurring only after exposure to loud noise	
Parving et al. (1993) Copenhagen Male Study, Denmark, 1985–1986	Cross-sectional assessment of tinnitus 3,387 men, ages 53–75 in 1985–86 Participants in a prospective cohort study of cardiovascular health	Tinnitus of greater than 5 minutes duration	Self-assessed hearing ability: do you think your hearing is affected?
Rosenhall and Karlsson (1991) Gothenburg, Sweden, 1971–1976	Repeated cross-sectional assessments of defined cohorts Cohorts F01: 377 initial members; born 1901–02; first examined in 1971 F06: 297 initial members; born 1906–07; first examined in 1976 Gerontological and geriatric population study	Tinnitus (buzzing): none, occasional, continuous	Pure-tone thresholds at 0.25, 0.5, 1, 2, 4, and 8 kHz

Noise Exposure and Source of Exposure Data	Prevalence of Tinnitus	Prevalence of Tinnitus, by Hearing Level	Comments
	17% w/ tinnitus	Significantly higher prevalence of tinnitus if hearing affected (X^2, $p < 0.001$) Hearing affected: Yes: 27% w/ tinnitus No: 11% w/ tinnitus	
	<u>Continuous tinnitus</u> <u>F01</u> Age 70: 8% Age 75: 12% Age 79: 11% <u>F06</u> Age 70: 12% <u>Occasional tinnitus</u> <u>F01</u> Age 70: 20% Age 75: 17% Age 80: 30% <u>F06</u> Age 70: 19%		

continued

TABLE D-6 continued

Study	Design, Population	Definition of Tinnitus	Definition of Hearing Level or Hearing Loss
Axelsson and Ringdahl (1989) Gothenburg, Sweden, 1980s	Cross-sectional 2,378 responses from adults, ages 20–79 yrs Postal survey using age-stratified random sample from city population register	Do you suffer from tinnitus? (never/seldom/often/always) Additional characterization of tinnitus only for those responding "often" or "always"	Subjective hearing: - normal - some hearing loss - marked hearing loss - deaf
Roberts (1968) Health Examination Survey, United States, 1960–1962	Cross-sectional 6,672 adults, ages 18–79 yrs Nationally representative random sample; noninstitutionalized population; excludes armed forces and nursing home residents	"At any time over the past few years, have you ever noticed ringing (tinnitus) in your ears of have you been bothered by other funny noises in your ears?" If yes, how often (every few days / less often) If yes, do they bother you (quite a bit / just a little) Severe and mild tinnitus not explicitly defined	Average thresholds Better than normal: –5 dB HL or better Normal: –4 dB to 15 dB HL Some hearing impairment: 16 dB HL or worse Average of thresholds at 0.5, 1, and 2 kcps (audiometric zero: ASA, 1951) Testing at 0.5, 1, 2, 3, 4, 6 kcps

Noise Exposure and Source of Exposure Data	Prevalence of Tinnitus	Prevalence of Tinnitus, by Hearing Level	Comments
	Tinnitus Always: 6% Often: 8% Seldom/never: 86%		Response rate: 66% (usable responses)
No exposure data	32% w/ tinnitus Severe: 6% Mild: 27%	W/ better than normal hearing 3% w/ tinnitus W/ some hearing impairment 22% w/ tinnitus	

continued

TABLE D-6 continued

Study	Design, Population	Definition of Tinnitus	Definition of Hearing Level or Hearing Loss
b. *Industrial Workers*			
Sulkowski et al. (1999) Poland	Cross-sectional 261 male drop-forge operators Age range: 18–61 yrs (mean: 31 yrs) Employment: 1–28 yrs (mean: 10 yrs) 169 age-matched controls from low-noise factory areas (mean age: 35 yrs) Exclusions: ear disease or pre-employment noise exposure	No definition provided Interview to determine presence and characteristics of tinnitus	Pure-tone audiometric testing before the beginning of the workday
Griest and Bishop (1996) Oregon, 1971–1990, 1992	Longitudinal (retrospective cohort) 138 men in steel foundry hearing conservation program Initial ages: 18 to 41 yrs (mean 28 yrs) No exclusions for other tinnitus risk factors	Tinnitus reported at annual audiogram (never / 1–2 times / 3+ times) Tinnitus reported on 1992 questionnaire: Ever hear ringing noises or other sounds When first aware How often (rarely / several times a month / several times a week / several times a day / always there) How long (only a few minutes / several hours / several days / always there)	Annual audiograms starting in 1971 Threshold at 4 kHz for men 20–29 yrs in 1971

Noise Exposure and Source of Exposure Data	Prevalence of Tinnitus	Prevalence of Tinnitus, by Hearing Level	Comments
Noise survey; 0.34 impulses per second Hearing protectors not worn regularly	Noise-exposed 70% w/ tinnitus Controls 4% w/ tinnitus		
Noise levels: TWA of 85 to 101 dBA Exposure of ≥ 85 dBA for 10 yrs or more Sound surveys and dosimetry from 2 to 4 hour exposures Hearing protection use recorded	Tinnitus reports at audiograms Never: 62% 1–2 times: 17% 3+ times: 20% Tinnitus reports on questionnaire Never: 39% Rarely / several times a month: 43% Several times a week or more: 17% Significant association between frequency of tinnitus reports at audiogram and report of tinnitus several times a week or more in questionnaire (X^2, $p < 0.0001$)		Tinnitus at baseline not excluded For 20- to 29-yr olds at baseline, no significant differences between those reporting any tinnitus and no tinnitus in military, recreational, or other occupational noise exposure

continued

TABLE D-6 continued

Study	Design, Population	Definition of Tinnitus	Definition of Hearing Level or Hearing Loss
Gabriels et al. (1996) Western Australia	Cross-sectional 38,725 otologically normal, noise-exposed workers given baseline hearing tests required by workers compensation program Age range:16–55+ yrs Noise exposure: 0–25+ yrs Exclusions: indications of hearing loss from injury or disease	Tinnitus: Yes to "Do you ever have ringing noises . . . which last more than 5 minutes?" (yes/ no / maybe)	Noise-induced hearing loss Age-corrected "percentage loss of hearing" > 0% Percentage loss of hearing calculated from Australian National Acoustic Laboratory tables specific for age and sex Tested at 0.5, 1, 1.5, 2, 3, 4, 6, 8 kHz; 16 hrs of prior quiet
Phoon et al. (1993) Singapore, 1990	Cross-sectional 647 workers from 808 workers identified as having noise-induced hearing loss through annual monitoring audiograms Mean age: 39 yrs Exclusions: ear disease, ear drum abnormality, conductive hearing loss	Interview at follow-up examination: Any tinnitus w/in past 6 months? Frequency: all the time/ once a day/ ≥ once a week/ < once a week/ not sure	Hearing loss (average of thresholds at 1, 2, 3 kHz) Early: ≤ 25 dB HL Intermediate: > 25, < 50 dB HL Late: ≥ 50 dB HL

Noise Exposure and Source of Exposure Data	Prevalence of Tinnitus	Prevalence of Tinnitus, by Hearing Level	Comments
Workplace exposure of 8-hr TWA of 90 dBA, or peak 140 dB	18% w/ tinnitus 10% maybe tinnitus	>10% NIHL 37% w/ tinnitus 0% NIHL 16% w/ tinnitus	No significant differences for age or number of years of noise exposure
Assumed to be 8-hr time weighted average ≥ 85 dBA (basis for requirement for annual testing)	23% w/ tinnitus No significant differences by age, duration of noise exposure W/ tinnitus (n=151) Once a day or more: 34% < once a week: 39% Excluded cases 42% w/ tinnitus	Hearing loss Early: 20% w/ tinnitus Intermediate: 30% w/ tinnitus Late: 27% w/ tinnitus Prevalence significantly lower in early hearing loss cases than intermediate and late cases (X^2, p = 0.02)	Consulted a doctor: 14% Interfered w/ sleep: 14% Interfered w/ some daily activity: 30%

continued

TABLE D-6 continued

Study	Design, Population	Definition of Tinnitus	Definition of Hearing Level or Hearing Loss
Neuberger et al. (1992) Austria, 1984–1986	Cross-sectional 110,647 noise-exposed factory workers Documented noise-exposure history: > 4 hr daily, > 85 dBA for ≥ 6 months Ages 15–65 yrs (median: 38 yrs) Exclusions: insufficient or unspecified noise exposure	No definition provided	Roeser speech impairment index = 0.5 (HL 3 kHz) + HL 1 kHz – 15 Impairment: Index > 10% Pure-tone thresholds at 0.25 to 8 kHz (only results for left ears included in analysis)
Kamal et al. (1989) Egypt	Cross-sectional 88 forge hammering workers Age range: 30–60 yrs Noise exposure: 9–25 yrs Exclusions: ear infections	No definition provided Interview to obtain tinnitus reports	Pure-tone audiometric testing at 0.25, 0.5, 1, 2, 4, 6, 8 kHz Each worker tested before and after a work shift
Chung et al. (1984) British Columbia	Cross-sectional 33,168 workers in industrial hearing conservation programs Exclusions: history of ear disease, head injury, ear surgery, relative w/ hereditary hearing loss; current day-to-day changes in hearing, ruptured ear drum	From medical history interview: "do you now have ringing in your ears?" (instruction to interviewers: tinnitus present more than momentarily and at least recurring if not continuous)	Audiometric surveillance records for annual testing for hearing conservation programs

Noise Exposure and Source of Exposure Data	Prevalence of Tinnitus	Prevalence of Tinnitus, by Hearing Level	Comments
Median noise exposure: 90.5 dBA SPL Worksite records of A-weighted noise levels Self-report of previous noise exposure Self-reported hearing protector use: seldom (< 10% of work time), occasional (10–50%), or continuous (> 50%)	7% w/ tinnitus Rates described as higher w/ increasing hearing loss and w/ history of ear disease or head injury		93% exposed to steady-state noise Noise levels recorded w/ most recent audiogram attributed to prior periods of noise-exposed work (attribution justified by low workforce turnover)
Noise measured at workers' ears Background: 92–94 dBA Hammer: 120–135 dBA No hearing protection used	88% w/ tinnitus		
Assumed to be 8-hr TWA ≥ 85 dBA (basis for requirement for annual testing) Reports on shooting history	7% w/ tinnitus No association w/ age, smoking history (current or past), or shooting history after controlling for hearing thresholds	All thresholds ≤ 25 dB HL: 3% w/ tinnitus	

continued

TABLE D-6 continued

Study	Design, Population	Definition of Tinnitus	Definition of Hearing Level or Hearing Loss
c. Military Personnel			
Attias et al. (2002) Israel	Cross-sectional 2,200 male soldiers Random selection of noise-exposed personnel routinely screened for hearing problems Age: 22–50 yrs	Do you experience or hear sounds when no sound source appears to be present? Interview to determine tinnitus onset, variability, impact	Noise-induced hearing loss Average threshold at 2–8 kHz ≥ 25 dB HL Pure-tone audiometric testing
Ylikoski and Ylikoski (1994) Finland	Cross-sectional 699 male Finnish army officers Age-stratified random sample Mean age: 39.8 yrs (median 41.0, range 25–61) Exclusions: incomplete audiometric or questionnaire data; apparent ear disease	Continuous tinnitus: occurred practically always, steady-state character	Hearing loss, worse ear Slight: > 20 dB, ≤ 40 dB at any freq 3–8 kHz; and ≤ 20 dB at 0.5, 1, 2 kHz Moderate: 41 dB–64 dB at any freq 3–8 kHz; and ≤ 20 dB at 0.5, 1, 2 kHz Severe: ≥ 65 dB at any freq 3–8 kHz; and ≤ 20 dB at 0.5, 1, 2 kHz Disabling: > 20 dB at any freq 0.5, 1, 2 kHz
Christiansson and Wintzell (1993) Sweden, Nov 1986–Feb 1987	Cross-sectional 204 male infantry officers (entire unit) Providing training in use of small (rifles, machine guns) and heavy firearms (recoilless rifles, mortars) Exclusions: acute, chronic, secretory otitis	Questionnaire: occurrence of tinnitus No information on persistence (no questions or definitions provided)	Audiometric testing at 0.25 to 8 kHz

Noise Exposure and Source of Exposure Data	Prevalence of Tinnitus	Prevalence of Tinnitus, by Hearing Level	Comments
Interview to determine previous noise exposure	14% w/ tinnitus	Normal hearing 3% w/ tinnitus Hearing loss 19% w/ tinnitus	
Questionnaire to establish exposure to weapons fire	Continuous tinnitus: 9% Occasional tinnitus: 34% Correlations: Number of weapons impulses: $r = 0.27$, $p < 0.001$ Small-caliber weapons: $r = 0.26$, $p < 0.001$ Large-caliber weapons: $r = 0.1$, $p < 0.005$ Frequent use of hearing protectors: $r = -0.17$, $p < 0.001$	Normal hearing 2% continuous tinnitus 33% occasional tinnitus Slight/moderate loss 3% continuous 31% occasional Severe loss 20% continuous 32% occasional Disabling loss 26% continuous 43% occasional	Data on other noise exposure may be incomplete
Questionnaire on type of service, previous impulse noise exposure, use of hearing protectors Sound pressure levels for weapons measured at gunner's ear and likely position of instructor	17% w/ "annoyance of tinnitus" Ever exposed to heavy detonations: Yes: 26% w/ tinnitus No: 5% w/ tinnitus (X^2, $p < 0.001$) Age (yrs) < 30 11% 31–40 19% 41–50 16% > 50 24%		Possible recall bias in link between tinnitus and exposure to heavy detonations

continued

TABLE D-6 continued

Study	Design, Population	Definition of Tinnitus	Definition of Hearing Level or Hearing Loss
d. *Acoustic Trauma*			
Mrena et al. (2002) Finland, 1999	Longitudinal Former conscripts, Finnish Defense Forces 418 soldiers treated for acoustic trauma July 1984–April 1989, all w/ tinnitus 122 w/ persistent tinnitus at discharge, 1984–1989 101 reached in 1999 (83%), 66 still w/ tinnitus Mean age at exposure: 21 yrs (18.8–30.4) Age at follow-up 30–41 yrs Duration of tinnitus: 12 yrs (9.8–14.3) No prior tinnitus	Tinnitus Handicap Questionnaire Rating scales for intensity, level of annoyance (0–100, least to most)	Hearing loss: threshold > 20 dB HL at any frequency at discharge 6 cases w/ hearing loss on entering military service No indication of audiometric testing at follow-up
Temmel et al. (1999) Austrian military service, Jan 1995–June 1996	Cross-sectional 81 male acoustic trauma patients Mean age: 22 yrs Treated 3 days after exposure Exclusions: hearing threshold > 20 dB HL at any frequency at start of service; illnesses, conditions that might affect auditory system	No questions/definition provided	Acute acoustic trauma: acute acoustic exposure producing temporary or permanent pure-tone threshold shift Hearing loss: threshold > 20 dB HL Hearing thresholds at 0.125–8 kHz

Noise Exposure and Source of Exposure Data	Prevalence of Tinnitus	Prevalence of Tinnitus, by Hearing Level	Comments
Assault rifle: 42 cases Bazooka: 3 cases Single cases: handguns, cannons, grenades 73% had fired the weapon 2 cases wore hearing protection (ear plugs) Other exposure history from medical records	At discharge 29% (122 of 418) w/ tinnitus At follow-up (10–15 yrs) 66% (66 of 101) w/ tinnitus	Normal hearing 4 cases (at time of discharge) Greater hearing impairment associated w/ greater tinnitus disturbance Perceived problems 33% tinnitus worse than hearing loss 33% hearing loss worse than tinnitus	79% response (52 of 66)
80% not wearing hearing protection (accidental discharges, loss of protectors, etc.)	84% w/ tinnitus	Hearing loss 83% w/ tinnitus No hearing loss 100% w/ tinnitus 75% had hearing loss at frequencies above 2 kHz	No significant differences for: a. blank/live ammunition b. number of shots c. use of hearing protection

continued

TABLE D-6 continued

Study	Design, Population	Definition of Tinnitus	Definition of Hearing Level or Hearing Loss
Man and Naggan (1981) Israel	Cross-sectional 102 patients w/ acoustic trauma; 81 w/ tinnitus Age: 18–35 Selected for evidence of "cochlear trauma" (high-frequency hearing loss?) Exclusions: head injury, history of ear disease		Audiometric testing, ISO calibration Worst threshold: 6 kHz for 76% Ears pooled
Melinek et al. (1976) Israel, 1967–1970	Cross-sectional and longitudinal 433 soldiers treated for acute acoustic trauma 313 transferred to noncombat unit 120 continued in field units Inclusion criteria: Age 18 to 25 yrs; abrupt onset of tinnitus or hearing loss w/in 2 mos; audiometric grade of 2+ in at least one ear; diagnostic audiometry 1 wk or more after exposure; no history of prior industrial noise exposure; no history of ear disease; audiometric follow-up w/in 1 to 24 mos	No definition provided	Acute acoustic trauma: abrupt onset of symptoms (tinnitus or hearing loss) generally associated w/ unusually loud impact noise Severity grouping Normal: all thresholds except 8 kHz ≤ 15 dB Mild: thresholds 20–25 dB at 4 kHz and up to 30 dB or more at 6 kHz Moderate: thresholds ≥ 20 dB at 4 kHz w/ or w/out elevated thresholds at 3 kHz Severe: thresholds of ≥ 45 dB at 4 kHz w/ thresholds ≥ 35 dB at 2 kHz or thresholds of ≥ 25 dB at 1 kHz

Noise Exposure and Source of Exposure Data	Prevalence of Tinnitus	Prevalence of Tinnitus, by Hearing Level	Comments
Questionnaire on noise exposure and complaints about hearing, tinnitus	79% of subjects Always present: 70% (of ears) Sometimes present: 30% (of ears) Tinnitus matched at frequencies between 4 and 8 kHz (37% at 6; 23% at 4; 24% at 8)	Higher intensity tinnitus associated w/ greater hearing loss ($r = 0.71$, $p < 0.001$)	No statistical association between tinnitus level and disturbed sleep or concentration Potential selection effect from use of clinic population
No ear protection used	61% (of ears) w/ tinnitus Change in subjective symptoms (including tinnitus) at follow-up: Transferred Deterioration: 2% Improvement: 34% Continued field unit Deterioration: 15% Improvement: 22%	Normal 42% Mild AT 60% Moderate AT 66% Severe 66%	Some hearing loss may have existed before acoustic trauma Lower initial severity rating for group continuing in field units Statistical significance of differences not reported

continued

TABLE D-6 continued

Study	Design, Population	Definition of Tinnitus	Definition of Hearing Level or Hearing Loss
	Analysis based on ears because of differences in severity		Pure-tone audiometric testing at 0.5, 1, 2, 4, 8 kHz; at 3 and 6 kHz for some subjects
Salmivalli (1967) Finland, 1963	Cross-sectional 197 male infantry and artillery soldiers (officers and enlisted) exposed to gunfire or blast basis for selection not specified	No questions or definitions provided	Severity grades I: narrow dip, thresholds ≤ 30 dB HL II: thresholds 30–60 dB HL only at frequencies above 2 kcps III: thresholds > 60 dB HL above 2 kcps or elevated at 0.5 to 2 kcps IV: thresholds elevated at 0.5 to 1 kcps Pure-tone testing at 0.25, 0.5, 1, 1.5, 2, 3, 4, 6, 8, 12, kc/ sec

NOTE: AT, acoustic trauma; CI, confidence interval; NIHL, noise-induced hearing loss; OR, odds ratio; PTA, pure-tone average; TWA, time-weighted average.

Noise Exposure and Source of Exposure Data	Prevalence of Tinnitus	Prevalence of Tinnitus, by Hearing Level	Comments
Noise levels measured under field conditions No estimate of rounds fired by individuals	No overall prevalence reported	Prevalence of tinnitus Normal: 15.7% I: 33% II: 25% III: 35% IV: 56% Prevalence of tinnitus after firing Normal: 41% I: 49% II: 49% III: 57% IV: 64%	

REFERENCES

Table D-1

Davis RR, Murphy WJ, Snawder JE, Striley CA, Henderson D, Khan A, Krieg EF. 2002. Susceptibility to the ototoxic properties of toluene is species specific. *Hearing Research* 166(1–2): 24–32.

ISO (International Organization for Standardization). 1984. *ISO 7029: Acoustics—Threshold of Hearing by Air Conduction as a Function of Age and Sex for Otologically Normal Persons.* Geneva, Switzerland: ISO.

Johnson AC, Juntunen L, Nylen P, Borg E, Hoglund G. 1988. Effect of interaction between noise and toluene on auditory function in the rat. *Acta Oto-Laryngologica* 105(1–2): 56–63.

Johnson AC, Nylen P, Borg E, Hoglund G. 1990. Sequence of exposure to noise and toluene can determine loss of auditory sensitivity in the rat. *Acta Oto-Laryngologica* 109(1–2): 34–40.

Morata TC, Dunn DE, Kretschmer LW, Lemasters GK, Keith RW. 1993. Effects of occupational exposure to organic solvents and noise on hearing. *Scandinavian Journal of Work, Environment & Health* 19(4):245–254.

Morata TC, Fiorini AC, Fischer FM, Colacioppo S, Wallingford KM, Krieg EF, Dunn DE, Gozzoli L, Padrao MA, Cesar CL. 1997. Toluene-induced hearing loss among rotogravure printing workers. *Scandinavian Journal of Work, Environment & Health* 23(4):289–298.

Schaper M, Demes P, Zupanic M, Blaszkewicz M, Seeber A. 2003. Occupational toluene exposure and auditory function: Results from a follow-up study. *Annals of Occupational Hygiene* 47(6):493–502.

Table D-2

Chen GD, Fechter LD. 1999. Potentiation of octave-band noise induced auditory impairment by carbon monoxide. *Hearing Research* 132(1–2):149–159.

Fechter LD, Chen GD, Rao D, Larabee J. 2000. Predicting exposure conditions that facilitate the potentiation of noise-induced hearing loss by carbon monoxide. *Toxicological Sciences* 58(2):315–323.

Rao DB, Fechter LD. 2000. Increased noise severity limits potentiation of noise induced hearing loss by carbon monoxide. *Hearing Research* 150(1–2):206–214.

Young JS, Upchurch MB, Kaufman MJ, Fechter LD. 1987. Carbon monoxide exposure potentiates high-frequency auditory threshold shifts induced by noise. *Hearing Research* 26(1):37–43.

Table D-3

Cruickshanks KJ, Klein R, Klein BEK, Wiley TL, Nondahl DM, Tweed TS. 1998. Cigarette smoking and hearing loss: The Epidemiology of Hearing Loss Study. *Journal of the American Medical Association* 279(21):1715–1719.

Ferrite S, Santana V. 2005. Joint effects of smoking, noise exposure and age on hearing loss. *Occupational Medicine (London)* 55(1):48–53.

ISO (International Organization for Standardization). 1964. *ISO/R 389: Standard Reference Zero for the Calibration of Pure Tone Audiometers.* Geneva, Switzerland: ISO.

ISO. 1990. *ISO 1999: Acoustics—Determination of Occupational Noise Exposure and Estimation of Noise-Induced Hearing Impairment.* Geneva, Switzerland: ISO.

Mizoue T, Miyamoto T, Shimizu T. 2003. Combined effect of smoking and occupational exposure to noise on hearing loss in steel factory workers. *Occupational and Environmental Medicine* 60:56–59.

Palmer KT, Griffin MJ, Syddall HE, Coggon D. 2004. Cigarette smoking, occupational exposure to noise, and self reported hearing difficulties. *Occupational and Environmental Medicine* 61:340–344.

Pyykko I, Koskimies K, Starck J, Pekkarinen J, Inaba R. 1988. Evaluation of factors affecting sensory neural hearing loss. *Acta Oto-Laryngologica* (Suppl 449):155–158.

Starck J, Toppila E, Pyykko I. 1999. Smoking as a risk factor in sensory neural hearing loss among workers exposed to occupational noise. *Acta Oto-Laryngologica* 119(3):302–305.

Virokannas H, Anttonen H. 1995. Dose-response relationship between smoking and impairment of hearing acuity in workers exposed to noise. *Scandinavian Audiology* 24(4):211–216.

Table D-4

Cruickshanks KJ, Tweed TS, Wiley TL, Klein BEK, Klein R, Chappell R, Nondahl DM, Dalton DS. 2003. The 5-year incidence and progression of hearing loss: The Epidemiology of Hearing Loss Study. *Archives of Otolaryngology–Head and Neck Surgery* 129:1041–1046.

Gates GA, Schmid P, Kujawa SG, Nam B, D'Agostino R. 2000. Longitudinal threshold changes in older men with audiometric notches. *Hearing Research* 141(1–2):220–228.

Lee FS, Matthews LJ, Dubno JR, Mills JH. 2005. Longitudinal study of pure-tone thresholds in older persons. *Ear and Hearing* 26:1–11.

Rosenhall U. 2003. The influence of ageing on noise-induced hearing loss. *Noise Health* 5(20):47–53.

Table D-5

ANSI (American National Standards Institute). 1969. *ANSI S3.6: Specification for Audiometers.* New York: American National Standards Institute.

ASA (American Standards Association). 1951. *American Standard Specification for Audiometers for General Diagnostic Purposes.* Pub. No. Z24.5-1951. New York: ASA.

Bohnker BK, Page JC, Rovig G, Betts LS, Muller JG, Sack DM. 2002. U.S. Navy and Marine Corps hearing conservation program, 1995–1999: Mean hearing thresholds for enlisted personnel by gender and age groups. *Military Medicine* 167(2):132–135.

Chandler DW, Fletcher JL. 1983. Hearing levels in U.S. Army engineers. *Journal of Audiology Research* 23(1):23–32.

Glorig A, Roberts J. 1965. Hearing levels of adults by age and sex, United States, 1960–1962. *Vital and Health Statistics.* Series 11, No. 11. Washington, DC: Department of Health, Education, and Welfare.

Goldenberg RA. 1977. Hearing loss in the Marine Corps. *Transactions—American Academy of Ophthalmology and Otolaryngology* 84(6):991–996.

Henselman LW, Henderson D, Shadoan J, Subramaniam M, Saunders S, Ohlin D. 1995. Effects of noise exposure, race, and years of service on hearing in U.S. Army soldiers. *Ear and Hearing* 16(4):382–391.

ISO (International Organization for Standardization). 1964. *ISO/R 389: Standard Reference Zero for the Calibration of Pure Tone Audiometers.* Geneva, Switzerland: ISO.

ISO. 1990. *ISO 1999: Acoustics—Determination of Occupational Noise Exposure and Estimation of Noise-Induced Hearing Impairment.* Geneva, Switzerland: ISO.

Ohlin D. 1992. *15 Years Revisited: The Prevalence of Hearing Loss Among Selected U.S. Army Branches.* Hearing Conservation Special Study No. 51-01-PM82-93. Aberdeen Proving Ground, MD: U.S. Army Environmental Hygiene Agency.

Peters LJ, Ford H. 1983. *Extent of Hearing Loss Among Army Aviators at Fort Rucker, Alabama.* USAARL Rept. No. 83-12. Fort Rucker, AL: U.S. Army Aeormedical Research Laboratory.

Robertson RM, Page JC, Williams CE. 1978. *The Prevalence of Hearing Loss Among Selected Navy Enlisted Personnel.* NAMRL-1251. Pensacola, FL: Naval Aerospace Medical Research Laboratory, Naval Medical Research and Development Command.

Sutherland HC, Gasaway DC. 1976. *Hearing Levels of Noise-Exposed U.S. Air Force Personnel Compared to Those in the Total U.S. Population.* Report SAM-TR-76-27. Brooks Air Force Base, TX: United States Air Force School of Aerospace Medicine.

Sutherland HC, Gasaway DC. 1978. *Current Hearing Threshold Levels for Noise-Exposed U.S. Air Force Personnel. One Year's Reportings.* Report SAM-TR-78-39. Brooks Air Force Base, TX: United States Air Force School of Aerospace Medicine.

Thomas JW. 1995. *The Application of Audiometric Data Base Analysis to Selected Air Force Bases.* Master's thesis, University of North Carolina, Chapel Hill.

Walden BE, Worthington DW, McCurdy HW. 1971. *The Extent of Hearing Loss in the Army: A Survey Report.* Washington, DC: Walter Reed Army Medical Center.

Walden BE, Prosek RA, Worthington DW. 1975. *The Prevalence of Hearing Loss within Selected U.S. Army Branches.* Washington, DC: Walter Reed Army Medical Center.

Table D-6

Adams PF, Benson V. 1991. Current estimates from the National Health Interview Survey, 1990. *Vital and Health Statistics.* Series 10, No. 181. Hyattsville, MD: National Center for Health Statistics.

Adams PF, Marano MA. 1995. Current estimates from the National Health Interview Survey, 1994. *Vital and Health Statistics.* Series 10, No. 193. Hyattsville, MD: National Center for Health Statistics.

Adams PF, Hendershot GE, Marano MA. 1999. Current estimates from the National Health Interview Survey, 1996. *Vital and Health Statistics.* Series 10, No. 200. Hyattsville, MD: National Center for Health Statistics.

ASA (American Standards Association). 1951. *American Standard Specification for Audiometers for General Diagnostic Purposes.* Pub. No. Z24.5-1951. New York: ASA.

Attias J, Reshef I, Shemesh Z, Salomon G. 2002. Support for the central theory of tinnitus generation: A military epidemiological study. *International Journal of Audiology* 41(5):301–307.

Axelsson A, Ringdahl A. 1989. Tinnitus—a study of its prevalence and characteristics. *British Journal of Audiology* 23(1):53–62.

Christiansson BA, Wintzell KA. 1993. An audiological survey of officers at an infantry regiment. *Scandinavian Audiology* 22(3):147–152.

Chung DY, Gannon RP, Mason K. 1984. Factors affecting the prevalence of tinnitus. *Audiology* 23(5):441–452.

Coles RR. 1996. Epidemiology, aetiology, and classification. In: Reich GE, Vernon JA, eds. *Proceedings of the Fifth International Tinnitus Seminar 1995.* Portland, OR: American Tinnitus Association. Pp. 25–29.

Coles RR. 1984. Epidemiology of tinnitus: (1) Prevalence. *Journal of Laryngology and Otology Supplement* 9:7–15.

Gabriels P, Monley P, Guzeleva D. 1996. Noise-exposed workers: Is tinnitus being ignored? In: Reich GE, Vernon JA, eds. *Proceedings of the Fifth International Tinnitus Seminar 1995*. Portland, OR: American Tinnitus Association. Pp. 373–380.

Griest SE, Bishop PM. 1996. Evaluation of tinnitus and occupational hearing loss based on 20-year longitudinal data. In: Reich GE, Vernon JA, eds. *Proceedings of the Fifth International Tinnitus Seminar 1995*. Portland, OR: American Tinnitus Association. Pp. 381–394.

Hoffman HJ, Reed GW. 2004. Epidemiology of tinnitus. In: Snow JB, ed. *Tinnitus: Theory and Management.* Hamilton, Ontario, Canada: B.C. Decker. Pp. 16–41.

Kamal AA, Mikael RA, Faris R. 1989. Follow-up of hearing thresholds among forge hammering workers. *American Journal of Industrial Medicine* 16(6):645–658.

Man A, Naggan L. 1981. Characteristics of tinnitus in acoustic trauma. *Audiology* 20(1): 72–78.

Medical Research Council's Institute of Hearing Research. 1981. Epidemiology of tinnitus. In: Evered D, Lawrenson G, eds. *Tinnitus.* Ciba Foundation Symposium 85. London, England: Pitman Books Ltd. Pp. 16–34.

Melinek M, Naggan L, Altman M. 1976. Acute acoustic trauma—a clinical investigation and prognosis in 433 symptomatic soldiers. *Israel Journal of Medical Sciences* 12(6):560–569.

Mrena R, Savolainen S, Kuokkanen JT, Ylikoski J. 2002. Characteristics of tinnitus induced by acute acoustic trauma: A long-term follow-up. *Audiology and Neuro-Otology* 7(2):122–130.

Neuberger M, Korpert K, Raber A, Schwetz F, Bauer P. 1992. Hearing loss from industrial noise, head injury and ear disease. A multivariate analysis on audiometric examinations of 110,647 workers. *Audiology* 31(1):45–57.

Nondahl DM, Cruickshanks KJ, Wiley TL, Klein R, Klein BE, Tweed TS. 2002. Prevalence and 5-year incidence of tinnitus among older adults: The Epidemiology of Hearing Loss Study. *Journal of the American Academy of Audiology* 13(6):323–331.

Palmer KT, Griffin MJ, Syddall HE, Davis A, Pannett B, Coggon D. 2002. Occupational exposure to noise and the attributable burden of hearing difficulties in Great Britain. *Occupational and Environmental Medicine* 59(9):634–639.

Parving A, Hein HO, Suadicani P, Ostri B, Gyntelberg F. 1993. Epidemiology of hearing disorders. Some factors affecting hearing. The Copenhagen Male Study. *Scandinavian Audiology* 22(2):101–107.

Phoon WH, Lee HS, Chia SE. 1993. Tinnitus in noise-exposed workers. *Occupational Medicine (London)* 43(1):35–38.

Roberts J. 1968. Hearing status and ear examination: Findings among adults, United States, 1960–1962. *Vital and Health Statistics.* Series 11, No. 32. Washington, DC: Department of Health, Education, and Welfare.

Rosenhall U, Karlsson AK. 1991. Tinnitus in old age. *Scandinavian Audiology* 20(3):165–171.

Salmivalli A. 1967. Acoustic trauma in regular army personnel. Clinical audiologic study. *Acta Oto-Laryngologica* (Suppl 222):1–85.

Sindhusake D, Mitchell P, Newall P, Golding M, Rochtchina E, Rubin G. 2003a. Prevalence and characteristics of tinnitus in older adults: The Blue Mountains Hearing Study. *International Journal of Audiology* 42(5):289–294.

Sindhusake D, Golding M, Newall P, Rubin G, Jakobsen K, Mitchell P. 2003b. Risk factors for tinnitus in a population of older adults: The Blue Mountains Hearing Study. *Ear and Hearing* 24(6):501–507.

Sindhusake D, Golding M, Wigney D, Newall P, Jakobsen K, Mitchell P. 2004. Factors predicting severity of tinnitus: A population-based assessment. *Journal of the American Academy of Audiology* 15(4):269–280.

Sulkowski W, Kowalska S, Lipowczan A, Prasher D, Raglan E. 1999. Tinnitus and impulse noise-induced hearing loss in drop-forge operators. *International Journal of Occupational Medicine and Environmental Health* 12(2):177–182.

Tambs K, Hoffman HJ, Borchgrevink HM, Holmen J, Samuelsen SO. 2003. Hearing loss induced by noise, ear infections, and head injuries: Results from the Nord-Trøndelag Hearing Loss Study. *International Journal of Audiology* 42(2):89–105.

Temmel AF, Kierner AC, Steurer M, Riedl S, Innitzer J. 1999. Hearing loss and tinnitus in acute acoustic trauma. *Wiener Klinische Wochenschrift* 111(21):891–893.

Ylikoski ME, Ylikoski JS. 1994. Hearing loss and handicap of professional soldiers exposed to gunfire noise. *Scandinavian Journal of Work, Environment & Health* 20(2):93–100.

Appendix E

Results from Alternative Analyses of Data on Reports of Audiometric Testing in Service Medical Records

Chapter 6 reports the results of a review of service medical records to assess the availability of audiometric test records dating from a service member's entry into and separation from military service. The analysis reported in Chapter 6 used a ±60-day window around entry and separation dates to determine the percentage of service medical records in which these audiometric records were present. To assess the effect of the ±60-day window, an alternative analysis was conducted using a ±120-day window. The results of that alternative analysis are shown below.

TABLE E-1 Percentages of Service Medical Records (95% Confidence Intervals) with Reports of Audiometric Examinations Within 120 Days of Entry into Active Duty (n = 3,212)

Branch	Before 1950	1950–1969	1970–Regulation Date	Regulation Date–1993	1994–2002
Army	1 (0–2)	10 (5–14)	45 (39–52)	42 (36–49)	54 (48–60)
Air Force	*	27 (17–37)	43 (36–50)	43 (37–50)	43 (36–50)
Marine Corps	0	17 (7–28)	48 (41–56)	62 (55–69)	76 (71–82)
Navy	1 (0–4)	11 (5–17)	48 (41–55)	66 (59–72)	76 (71–81)

*Fewer than 40 records in denominator.

NOTE: The time periods reflect the era of the service member's release from active duty.

TABLE E-2 Percentages of Service Medical Records (95% Confidence Intervals) with Reports of Audiometric Examinations Within 120 Days of Release from Active Duty (n = 3,226)

Branch	Before 1950	1950–1969	1970–Regulation Date	Regulation Date–1993	1994–2002
Army	3 (0–6)	16 (10–21)	52 (45–59)	47 (40–54)	34 (28–40)
Air Force	*	41 (30–51)	49 (42–56)	38 (31–45)	17 (12–22)
Marine Corps	0	4 (0–9)	65 (57–72)	69 (62–76)	60 (54–66)
Navy	0	11 (5–17)	61 (54–68)	67 (61–74)	66 (61–72)

*Fewer than 40 records in the denominator.

TABLE E-3 Percentages of Service Medical Records (95% Confidence Intervals) with Reports of Audiometric Examinations Within 120 Days of Entrance into and Release from Active Duty (n = 3,210)

Branch	Before 1950	1950–1969	1970–Regulation Date	Regulation Date–1993	1994–2002
Army	1 (0–2)	7 (3–11)	28 (22–34)	22 (16–27)	19 (14–23)
Air Force	*	23 (14–33)	25 (19–31)	16 (11–21)	9 (5–13)
Marine Corps	0	2 (0–6)	38 (31–45)	44 (37–51)	51 (45–57)
Navy	0	2 (0–5)	33 (26–40)	46 (39–53)	52 (46–58)

*Fewer than 40 records in the denominator.

NOTE: The time periods reflect the era of the service member's release from active duty.

Appendix F

Selected Sources of Information on Sound Pressure Levels Measured in and Around Military Systems and Equipment

1. Ali AY. 1988. *Noise Hazard Assessment of the Enclosed Primary Cooling Tower System at Cheyenne Mountain Air Force Station, Colorado.* Brooks Air Force Base, TX: Air Force Occupational and Environmental Health Laboratory.
2. Averbuch A, Schomer P. 1978. *Noise Levels in U.S. Army Corps of Engineers Powerhouses.* Champaign, IL: Construction Engineering Research Laboratory.
3. Brouns AJ, Ely RA. 1978. *Analysis of Noise in U.S. Army Aircraft.* Dallas, TX: Vought Corp. Advanced Technology Center Inc.
4. Camp RT. 1965. *Noise Spectra of the Bell OH-13-T Helicopter.* USAARU Report No. 65-3. Fort Rucker, AL: U.S. Army Medical Research and Development Command.
5. Camp RT, Bailey RW. 1965. *Noise Spectra of the Turbo-Beaver.* USAARU Report No. 65-4. Fort Rucker, AL: U.S. Army Medical Research and Development Command.
6. Decker WH, Nixon CW. 1985. *B-52G Crew Noise Exposure Study.* Wright-Patterson Air Force Base, OH: Aerospace Medical Research Laboratory.
7. Ellis JC. 1988. *Impulse Noise Level Measurements in Defensive Fighting Position Shelters, AF Weapons Laboratory, Kirtland AFB New Mexico.* Brooks Air Force Base, TX: Air Force Occupational and Environmental Health Laboratory.
8. Garinther GR, Kalb JT. 1983. *An Acoustical Assessment of the Impulse Noise of Grenade Simulators Exploding in Enclosures.* Aberdeen Proving Ground, MD: Human Engineering Laboratory.

9. Gasaway DC. 1964. Characteristics of noise associated with the operation of military aircraft. *Aerospace Medicine* 35:327–336.
10. Gasaway DC. 1975. *A-Weighted Sound Levels in Cockpits of Fixed- and Rotary-Wing Aircraft*. Brooks Air Force Base, TX: School of Aerospace Medicine.
11. Gasaway DC. 1976. *Noise Levels Measured within Aircraft during Takeoff, Climb, and Cruise (Low, Normal, and High)*. Brooks Air Force Base, TX: School of Aerospace Medicine.
12. Gasaway DC. 1986. Noise levels in cockpits of aircraft during normal cruise and considerations of auditory risk. *Aviation, Space, and Environmental Medicine* 57(2):103–112.
13. Gasaway DC. 2002. *Donald C. Gasaway's Aircraft Noise Compendia*. Brooks City-Base, TX: U.S. Air Force.
14. Gasaway DC, Hatfield JL. 1963. *A Survey of Internal and External Noise Environments in US Army Aircraft*. USAARU Report No. 64-1. Fort Rucker, AL: U.S. Army Aeromedical Research Unit.
15. Goldstein J. 1975. *Noise Hazard Evaluation, Sound Level Data of Noise Sources*. Technical Guide TG-040. Aberdeen Proving Ground, MD: U.S. Army Environmental Hygiene Agency.
16. Goldstein J. 1975. *Soldier's Guide for Occupational Exposure Risk Assessment During Deployment*. Aberdeen Proving Ground, MD: U.S. Army Environmental Hygiene Agency.
17. Gordon E, Ahroon WA. 2000. *Noise Levels in the USAARL NUH-60 (Black Hawk) Aeromed Flight Simulator*. USAARL Report No. 2000-19. Fort Rucker, AL: U.S. Army Aeromedical Research Laboratory.
18. Gordon E, Murphy BA. 2000. *Impulse Noise Levels of the UH-60A/L Black Hawk Helicopter Cockpit Air Bag System*. USAARL Report No. 2000-24. Fort Rucker, AL: U.S. Army Aeromedical Research Laboratory.
19. Harris JD. 1971. *Hearing Conservation and Acoustic Habitability for Submariners*. NSMRL 682. Groton, CT: Naval Submarine Medical Research Laboratory.
20. Hatfield JL, Gasaway DC. 1963. *Noise Problems Associated with the Operation of U.S. Army Aircraft*. USAARU Report No. 63-1. Fort Rucker, AL: U.S. Army Aeromedical Research Unit.
21. Hille HK. 1979. *USAF Bioenvironmental Noise Data Handbook—Volume 127 F-15A In-Flight Crew Noise*. Wright-Patterson Air Force Base, OH: Aerospace Medical Research Laboratory.
22. Jones CM. 1983. *C-9A Interior Noise Evaluation*. Brooks Air Force Base, TX: Air Force Occupational and Environmental Health Laboratory.
23. Karta Technologies, Inc. 1999. *Lead-Free Small Arms Ammunition Study*. IERA-RS-BR-SR-2000-0001. San Antonio, TX: Air Force Institute for Environment, Safety and Occupational Health Risk Analysis.

24. Kugler AB, Hale ME, Rentz PE. 1977. *Occupational Noise Exposure on FF 1052 (KNOX) and DD 963 (SPRUANCE) Class Ships*. Canoga Park, CA: Bolt Beranek and Newman, Inc.
25. Lambert DR. 1980. *Airborne Noise Levels on U.S. Navy Ships*. NOSC TD 317. San Diego, CA: Naval Ocean Systems Center.
26. Lambert DR. 1981. *Setting Airborne Noise Limits for Navy Shipboard Compartments: Considerations Based on Data from Existing Ships*. NOSC TR 712. San Diego, CA: Naval Ocean Systems Center.
27. Luz GA. 1983. *Data Base for Assessing the Annoyance of the Noise of Small Arms*. Technical Guide TG-135. Aberdeen Proving Ground, MD: U.S. Army Environmental Hygiene Agency.
28. Magliozzi B, Metzger FB, Bausch W, King RJ. 1975. *A Comprehensive Review of Helicopter Noise Literature*. Washington, DC: Federal Aviation Administration.
29. McBryan JC, Schomer PD. 1976. *Noise in Dishwashing Rooms*. Champaign, IL: Construction Engineering Research Laboratory.
30. Mozo BT, Gordon E. 1994. *The Assessment of the AH-64D, Longbow, Mast-Mounted Assembly Noise Hazard for Maintenance Personnel*. USAARL Report No. 94-37. Fort Rucker, AL: U.S. Army Aeromedical Research Laboratory.
31. Newman JS, Rickley EJ. 1979. *Noise Levels and Flight Profiles of Eight Helicopters Using Proposed International Certification Procedures*. Report No. FAA-EE-79-03. Washington, DC: U.S. Department of Transportation.
32. Ostler DA, Cain PA, Gordon E, Fallon EW. 2002. *Noise Attenuation in HALO Vertical Wind Tunnel Training*. USAARL Report No. 2002-15. Fort Rucker, AL: U.S. Army Aeromedical Research Laboratory.
33. Patterson JH, Mozo BT, Marrow RH, McConnell JR, Lomba-Gautier I, Curd DL, Phillips YY, Henderson R. 1985. *Direct Determination of the Adequacy of Hearing Protective Devices for Use with the M198, 155mm Towed Howitzer*. USAARL Report No. 85-14. Fort Rucker, AL: U.S. Army Aeromedical Research Laboratory.
34. Powell RG, Rose JF. 1977. *USAF Bioenvironmental Noise Data Handbook: C130E Aircraft, Near and Far-field Noise*. Wright-Patterson Air Force Base, OH: Aerospace Medical Research Laboratory.
35. Ribera JE, Mozo BT, Mason KT, Murphy BA. 1995. *Communication and Noise Hazard Survey of CH-47D Crewmembers*. Fort Rucker, AL: U.S. Army Aeromedical Research Laboratory. [*Related publication*: Ribera JE, Mozo BT, Mason KT, Murphy BA. 1996. Communication survey of CH-47D crewmembers. *Military Medicine* 161(7):387–391.]
36. Rickley EJ, Jones KE, Keller AS, Fleming GG. 1993. *Noise Measurement Flight Test of Five Light Helicopters*. Cambridge, MA: John A. Volpe National Transportation Systems Center.

37. Rose JF. 1969. *Measurement and Analysis of Bioacoustic Environments Aboard AC-119G and AC-130A Gunships*. AMRL-TR-68-176. Wright-Patterson Air Force Base, OH: Aerospace Medical Research Laboratory.
38. Schomer PD, Averbuch A, Raspet R. 1982. *Operational Noise Data for UH-60A and CH-47C Army Helicopters*. Champaign, IL: U.S. Army Construction Engineering Research Laboratory.
39. Schomer PD, Averbuch A, Raspet R, Wolf RK. 1988. *Operational Noise Data for CH-47D and AH-64 Army Helicopters*. Champaign, IL: U.S. Army Construction Engineering Research Laboratory.
40. Schomer PD, Wagner LR, Benson LJ. 1995. *Human and Community Response to Military Noise—Results from Field-Laboratory Tests of Small Arms, 25 mm Cannons, Helicopters, and Blast Sounds*. Champaign, IL: U.S. Army Construction Engineering Research Laboratory.
41. Smith PF. 1997. *Development of a General Hearing Conservation Standard for Diving Operations: Research on Hearing Conservation for Exposure to Noise in Dry Hyperbaric Environments: I. Basic Considerations and Preliminary Experiments*. NSMRL 1204. Groton, CT: Naval Submarine Medical Research Laboratory.
42. Speakman JD, Powell RG, Lee RA. 1978. *Community Noise Exposure Resulting from Aircraft Operations: Volume 5. Acoustic Data on Air Force Propeller Aircraft*. Wright-Patterson Air Force Base, OH: Aerospace Medical Research Laboratory.
43. Sterba JA. 1987. *Evaluation of an Impulse Noise Producing Underwater Explosive Device on Hearing in Divers*. Report No. 10-87. Panama City, FL: Navy Experimental Diving Unit.
44. Thomas JW. 2003. *Noise Exposure and Risk of Hearing Loss for Air Force Welders*. Fort Collins, CO: Colorado State University.
45. True HC, Letty RM. 1977. *Helicopter Noise Measurements Data Report*. Washington, DC: U.S. Department of Transportation.
46. U.S. Army Center for Health Promotion and Preventive Medicine, Hearing Conservation Program. 2004. *Noise Levels of Common Army Equipment*. [Online]. Available: http://chppm-www.apgea.army.mil/hcp/NoiseLevelsPrint.htm [accessed April 21, 2004].
47. Webster JC. 1981. *Noise Levels on Aircraft Carrier Flight Decks, and Their Effects*. NELC 1762. San Diego, CA: Naval Electronics Laboratory Center.

Appendix G

Department of Defense Hearing Conservation Report Forms

NOISE SURVEY
(Sound Level Meter Survey)

1. DATE *(YYYYMMDD)*

2. TYPE SURVEY *(Enter code)* 1 - INITIAL SURVEY 2 - RE-SURVEY 3 - OTHER

3. SOUND LEVEL METER		4. MICROPHONE		5. CALIBRATOR	
a. MANUFACTURER		a. MANUFACTURER		a. MANUFACTURER	
b. MODEL	c. SERIAL NO.	b. MODEL	c. SERIAL NO.	b. MODEL	c. SERIAL NO.
d. LAST ELECTROACOUSTIC CALIB DATE *(YYYYMMDD)*		d. LAST ELECTROACOUSTIC CALIB DATE *(YYYYMMDD)*		d. LAST ELECTROACOUSTIC CALIB DATE *(YYYYMMDD)*	

6. WIND SCREEN *(X one)* USED NOT USED

7. MEASUREMENTS OBTAINED *(X one)* INDOORS OUTDOORS

8. DESCRIPTION OF AREAS/DUTIES WHERE NOISE SURVEY CONDUCTED *(Illustrate on additional sheet and attach to form)*

9. PRIMARY SOURCE OF NOISE

10. SECONDARY SOURCE OF NOISE

11. SOUND LEVEL DATA					12. PROTECTION REQUIRED *(re: dBA - Level)*			
a. LOCATION	b. METER ACTION	c. dBC	d. dBA	e. RISK ASSESSMENT CODE	a. NONE *(Less than 85)*	b. PLUG OR MUFF *(85-108)*	c. PLUG AND MUFF *(108-118)*	d. PLUG + MUFF + TIME LIMIT *(Greater than 118)*

NOTES: Range of levels noted by /; i.e., 102/109. At operator stations, measure at ear level.
METER ACTION: Enter F for fast meter action and S for slow meter action.

13. REMARKS *(i.e., Area and equipment posted, hearing protection in use, etc.)*

14. MORE DETAILED NOISE EVALUATION REQUIRED: YES NO *(If "YES," identify type evaluation needed.)*

15. NAME(S) OF PERSON(S) IDENTIFIED FOR AUDIOMETRIC MONITORING *(Use additional sheet if more space is needed and attach to form)*

16. SUPERVISOR OF NOISE-HAZARDOUS AREA OR OPERATION

a. NAME *(Last, First, Middle Initial)*	b. TELEPHONE *(Include area code)*	c. ORGANIZATION

17. SURVEY PERFORMED BY *(Last Name, First Name, MI)*

18. HEARING CONSERVATION MONITOR *(Last Name, First Name, MI)*

DD FORM 2214, JAN 2000 PREVIOUS EDITION MAY BE USED.

Reset

INSTRUCTIONS

(Refer to DoD Component Instructions for Additional Guidance)

PURPOSE: This form is intended to record noise survey results for the identification of potentially noise-hazardous environments.

GENERAL: Print all information in ink. Only medical, industrial hygiene, safety, or engineering personnel who meet training requirements specified by the DOD components will make sound level measurements.

1. Date - Enter date noise survey conducted (e.g., if Jan. 14, 1999, enter 19990114).

2. Type, Survey - Enter appropriate numeric code in box (e.g., enter "1" if area or operation not surveyed before or no available records of previous survey; enter "2" if resurvey conducted at regular intervals (such as once each 12 months); or enter "3" if noise being reevaluated to confirm validity of previously obtained measurements or for purposes other than indicated).

3. Sound Level Meter:
 a. Mfgr - Enter name of company that produced sound level meter.
 b. Model - Enter manufacturer's designation.
 c. Serial No. - Enter manufacturer's serial number.
 d. Last Electroacoustic Calib Date - Enter year, month, day (see Item 1) of last comprehensive calibration required by DOD component. Not to include calibration checks made with acoustical calibrator.

4. Microphone (Fill in this section if microphone is detachable from sound level meter)
 a. Manufacturer - Enter name of company that produced microphone.
 b. Model - Enter manufacturer's designation.
 c. Serial No. - Enter manufacturer's serial number.
 d. Last Electroacoustic Calib Date - Enter year, month, and day (see Item 1) of last comprehensive calibration as required by DOD component.

5. Calibrator:
 a. Manufacturer - Enter name of company that produced calibrator.
 b. Model - Enter manufacturer's designation.
 c. Serial Number. Enter manufacturer's serial number.
 d. Last Electroacoustic Calib Date. Enter year, month, and day (see Item 1) of last comprehensive calibration as required by DoD component.

6. Wind Screen - Check appropriate box indicating if manufacturer's device to reduce wind noise is mounted over microphone assembly.

7. Measurements Obtained - Check appropriate box indicating if measurements obtained indoors or outdoors.

8. Description of Areas/Duties Where Noise Survey Conducted - Include building number(s), name of activity and/or operation, identify specific microphone locations, performance conditions and descriptions of machinery (e.g., rpm, load, etc). Where applicable, include noise-hazard contours of area. On additional sheet make simple line drawing of area and identify noise sources and locations of measurement.

9. Primary Source of Noise - If possible, identify the location(s) of the highest dBA value.

10. Secondary Source of Noise - If possible, identify all other noise sources when the primary noise source is off (e.g. background noise sources and other noise sources that may or may not be noise hazardous).

11. Sound Level Data
 a. Location - Position where measurement is obtained should correspond with those identified, or illustrated on form.
 b. Meter Action - See Notes in Sound Level Data Sec. levels measured with weighting switch of meter in "C" position.
 c. dBC - If required by DOD component, enter sound levels measured with weighting switch of meter in "C" position.
 d. dBA - Enter sound levels measured with weighting switch of meter in "A" position. See NOTES in Sound Level Data Section.
 e. Risk Assessment Code - Enter expression of risk that combines elements of hazard severity and mishap probability. Hazard severity categories shall be assigned by roman numeral as follows:
 (1) Category I - Catastrophic: May cause death or loss of a facility (Code I).
 (2) Category II - Critical: May cause severe injury, e.g., severe occupational illness, or major property damage (Code II).
 (3) Category III - Marginal: May cause minor injury, e.g., minor occupational illness, or minor property damage (Code III).
 (4) Category IV - Negligible: Probably would not affect personnel safety or health, but is nevertheless in violation of specific criteria (Code IV). Mishap probability shall be assigned capital letter according to following criteria:
 (a) Subcategory A: Likely to occur immediately or within a short period of time (Code A).
 (b) Subcategory B: Probably will occur in time (Code B).
 (c) Subcategory C: May occur in time (Code C).
 (d) Subcategory D: Unlikely to occur (Code D).
Enter codes as IIB, IIIC, etc. Refer to DOD Instruction 6055.1/DOD component instructions for specific definitions and guidance.

12. Protection Required (re: dBA Level)
 a. None (less than 85: If dBA levels less than 85, check this column. No hearing protectors required.
 b. Plug or Muff (85 - 108): If dBA levels 85 - 108 inclusive, check this column. Earplugs, ear muffs, ear-canal caps, or noise-attenuating helmet required.
 c. Plug and Muff (108 - 118): If dBA levels over 108 to 118 inclusive, check this column. Earplugs worn in combination with ear muffs or noise-attenuating helmet required.
 d. Plug, Muff & Time: If dBA levels over 118, check this column. Earplugs worn in combination with ear muffs or noise-attenuating helmet and time limit (to be determined by DOD component) required.

13. Remarks - Enter type of hearing protection in use, whether area and equipment posted with appropriate caution signs, etc.

14. More Detailed Noise Evaluation Required - Check "yes" box if more detailed noise evaluation is required; check "no" box if not. Specify the type of evaluation needed (e.g., octave band analysis, etc.).

15. Name(s) of Persons Identified for Audiometric Monitoring - List names of individuals routinely exposed to noise in preceding locations.

16. Supervisor of Noise - Hazardous Area or Operation - Enter name (surname, given name, & middle initial) of the first-echelon (immediate) supervisor of the location (and personnel) surveyed.

17. Survey Performed by - Enter name (surname, given name & middle initial) of individual performing survey & signature.

18. Hearing Conservation Monitor - Enter name of individual reviewing survey results & signature. Usually local surgeon or designated representative.

DD FORM 2214 (BACK), JAN 2000

Reset

NOISE SURVEY *(Continuation Sheet)*
(Sound Level Meter Survey)

1. DATE *(YYYYMMDD)*	2. TYPE SURVEY *(Enter 1, 2, or 3 in box)* ☐ 1 - INITIAL SURVEY 2 - RE-SURVEY 3 - OTHER							
3. SOUND LEVEL DATA					4. PROTECTION REQUIRED *(re: dBA Level)*			
a. LOCATION	b. METER ACTION	c. dBC	d. dBA	e. RISK ASSESSMENT CODE	a. NONE *(Less than 85)*	b. PLUG OR MUFF *(85-108)*	c. PLUG AND MUFF *(108-118)*	d. PLUG + MUFF + TIME LIMIT *(Greater than 118)*

5. SURVEY PERFORMED BY *(Last Name, First Name, Middle Initial)*	6. HEARING CONSERVATION MONITOR *(Last Name, First Name, Middle Initial)*

DD FORM 2214C, JAN 2000 PREVIOUS EDITION MAY BE USED.

Reset

REFERENCE AUDIOGRAM

(This form is subject to the Privacy Act of 1974 - use Blanket PAS - DD Form 2005)

1. ZIP CODE/APO/FPO/PAS

2. DOD COMPONENT
A - ARMY F - AIR FORCE 1 - OTHER
N - NAVY M - MARINE CORPS

3. SERVICE COMPONENT
R - REGULAR G - NATIONAL GUARD
V - RESERVE 1 - OTHER

4. SOCIAL SECURITY NUMBER

5. NAME *(Last, First, Middle Initial)*

6. DATE OF BIRTH *(YYYYMMDD)*

7. SEX
M - MALE
F - FEMALE

8. PAY GRADE, UNIFORMED SERVICES

9. PAY GRADE, CIVILIAN

10. SERVICE DUTY OCCUPATION CODE

11. MAILING ADDRESS OF ASSIGNMENT

12. LOCATION - PLACE OF WORK

13. MAJOR COMMAND

14. DUTY TELEPHONE *(Include area code)*

AUDIOMETRY

15. REASON FOR CONDUCTING AUDIOGRAM
1 - REFERENCE ESTABLISHED PRIOR TO INITIAL DUTY IN HAZARDOUS NOISE AREAS
2 - REFERENCE ESTABLISHED FOLLOWING EXPOSURE IN NOISE DUTIES
3 - REFERENCE RE-ESTABLISHED AFTER FOLLOW-UP PROGRAM

16. AUDIOMETRIC DATA RE: ANSI S3.6 - 1989	LEFT						RIGHT					
	500	1000	2000	3000	4000	6000	500	1000	2000	3000	4000	6000
17. DATE OF AUDIOGRAM *(YYYYMMDD)*												

18. MEETS REFERRAL CRITERIA
1 - NO
2 - YES

19. MILITARY TIME OF DAY *(Optional)*

20. HOURS SINCE LAST NOISE EXPOSURE

21. EAR, NOSE, AND THROAT PROBLEM AT TIME OF TEST
1 - NO 2 - YES 3 - UNKNOWN

22. EXAMINER
a. NAME *(Last, First, Middle Initial)*
b. TRAINING CERTIFICATION NUMBER
c. SERVICE DUTY OCCUPATION CODE
d. OFFICE SYMBOL

23. AUDIOMETER
a. TYPE
1 - MANUAL
2 - SELF-RECORDING *(Automatic)*
3 - MICROPROCESSOR
b. MODEL
c. MANUFACTURER
d. SERIAL NUMBER
e. LAST ELECTROACOUSTIC CALIBRATION DATE *(YYYYMMDD)*

24. PERSONAL HEARING PROTECTION
a. TYPE ISSUED
1 - SINGLE FLANGE (VS1R) 4 - EAR CANAL CAPS
2 - TRIPLE FLANGE 5- NOISE MUFFS
3 - HAND FORMED EARPLUG 6 - OTHER
7 - NONE
b. SIZE EARPLUGS
L R 1 - XS 4 - L
2 - S 5 - XL
3 - M
c. DOUBLE PROTECTION USED
1 - NO
2 - YES
d. GLASSES WORN *(Including goggles)*
1 - NO
2 - YES
e. FREQUENCY GLASSES WORN
1 - ALWAYS
2 - SELDOM
3 - N/A

25. REMARKS *(Include exposure data)*

DD FORM 2215, JAN 2000 PREVIOUS EDITION MAY BE USED.

Reset

INSTRUCTIONS

(Refer to DoD Component Instructions for additional guidance)

PURPOSE: This form is used to record initial audiometric test results with which later audiometric test results can be compared (see DD Form 2216, "Hearing Conservation Data," to record periodic test results).

1. ZIP CODE/APO/FPO/PAS. Enter nine digit ZIP Code/APO/FPO/ PAS of where audiometric test is conducted.

2. DOD COMPONENT. Enter letter in box of major organizational subdivision of DoD to which military or civilian individual is assigned. Enter "1" if DoD component is not listed.

3. SERVICE COMPONENT. Enter letter in box corresponding to primary subdivision of separate military service in which military is assigned (e.g., Regular (R) - standing military component of armed forces in peace and war; Reserve (V) - component of ready trained personnel for military service when needed, etc.; National Guard (G) - component of National Guard personnel in full-time or part-time status). Enter "1" for all others, including civilians.

PERSONAL DATA OF INDIVIDUAL BEING TESTED:

4. SOCIAL SECURITY NUMBER. Enter nine digit social security number. If foreign national, enter "FN" in middle two blocks.

5. NAME. Enter surname, given name and middle initial.

6. DATE OF BIRTH. Enter year, month, day.

7. SEX. Enter "M" if male, "F" if female.

8. PAY GRADE, UNIFORMED SERVICES. For military personnel only, enter military personnel class and pay level serial number as follows:

O11 - General of the Army/General of the Air Force/Fleet Admiral
O10 - General/Admiral
O09 - Lieutenant General/Vice Admiral
O08 - Major General/Rear Admiral (Upper Half)
O07 - Brigadier General/Rear Admiral (Lower Half)/Commodore
O06 - Colonel (A,F,M)/Captain (N)
O05 - Lieutenant Colonel/Commander
O04 - Major/Lieutenant Commander
O03 - Captain (A,F,M)/Lieutenant (N)
O02 - First Lieutenant/Lieutenant Junior Grade
O01 - Second Lieutenant/Ensign
W05 - Chief Warrant Officer, W-5
W04 - Chief Warrant Officer, W-4
W03 - Chief Warrant Officer, W-3
W02 - Chief Warrant Officer, W-2
W01 - Warrant Officer, W-1
C00 - Cadet/Midshipman
E09 - Sergeant Major/Chief Master Sergeant/Master Chief Petty Officer
E08 - Master Sergeant (A,M)/Senior Chief Petty Officer/Senior Master Sergeant/First Sergeant(A)
E07 - Sergeant First Class/Gunnery Sergeant/Chief Petty Officer/ Master Sergeant (F)/Platoon Sergeant (A)/Specialist-7
E06 - Staff Sergeant/Technical Sergeant/Petty Officer First Class/ Specialist-6
E05 - Sergeant (A,M)/Staff Sergeant/Petty Officer Second Class/ Specialist-5
E04 - Corporal/Sergeant (F)/Petty Officer Third Class/Specialist-4
E03 - Private First Class (A)/Airman First Class/Lance Corporal/Seaman
E02 - Private (PV1)/Airman/Private First Class (M)/Seaman Apprentice
E01 - Private (PV2)/Private (M)/Airman Basic/Seaman Recruit

9. GRADE, CIVILIAN. Enter two letters and two numbers of Federal civilian employee rank (e.g., WG05, GS11, etc.). Letter entries will be WG, WL, WS, WN, WD or GS. Number entries will be 01 to 18. Enter "1111" if other (e.g., foreign national, contractor, etc.).

10. SERVICE DUTY OCCUPATION CODE. Enter code to which military member's duty occupation is assigned (e.g., MOS, SSI, NEC/Rating, NOBC or AFSC in which individual is actually working). Enter number code of civilian job series in which civilian member is actually working (e.g., for a carpenter enter "4607").

11. MAILING ADDRESS OF ASSIGNMENT. Enter installation name (and street address for Navy and Marines), unit, office symbol, and ZIP Code/APO/FPO/PAS of individual's current duty assignment.

12. LOCATION - PLACE OF WORK. Enter specific location where individual is routinely exposed to hazardous noise including building number (e.g., Corpus Christi, NAS, Building 1571, Carpenter Shop). For Air Force personnel, enter 12-digit Workplace Identifier Code per AFOSH Std. 161-17.

13. MAJOR COMMAND. Enter authorized abbreviation of military major command to which individual is assigned.

14. DUTY TELEPHONE. Enter individual's duty telephone number.

AUDIOMETRY:

15. REASON FOR CONDUCTING AUDIOGRAM. Enter number in box for reason to complete reference audiogram.
1 - Individual has not yet worked in hazardous noise duty areas and no reference audiogram has been accomplished.
2 - Individual has worked in hazardous noise duty areas but reference audiogram has been lost or was never accomplished.
3 - Individual has worked in hazardous noise duty areas and requires revised reference audiogram following completion of hearing conservation follow-up program.

16. AUDIOMETRIC DATA RE: ANSI S3.6 - 1989. Enter threshold levels determined for this individual at six frequencies in each ear. Results are entered in 5dB increments (e.g., 0, 5, 10, 15, etc). If responses exceed maximum limits of audiometer, enter that limit with plus sign (e.g., 110+).

17. DATE OF AUDIOGRAM. Enter year, month, and day the audiometric test is given. (If January 14, 1999, enter 19990114.)

18. MEETS REFERRAL CRITERIA. Based on the audiometric test results, each DoD component should apply its own criteria.

19. MILITARY TIME OF DAY. Enter four digits for hour of day (24-hour clock) this audiogram is completed (e.g., "0830," "1400," etc.). This field is optional.

20. HOURS SINCE LAST NOISE EXPOSURE. Enter appropriate number of hours prior to this audiogram that individual was last exposed to hazardous noise (e.g., steady noise 85 dBA or greater and/or impulse noise above 140 dBP).

21. EAR, NOSE, AND THROAT PROBLEM AT TIME OF TEST. Enter "1" (NO) if individual has no ear, nose or throat problems at time of test that could be causing a temporary (conductive) hearing loss (e.g., ear canal blocked with ear wax, ear infection, head cold, etc.). Enter "2" (YES) if problem was present and "3" (UNKNOWN) if no way to determine presence of problem.

22. EXAMINER.
a. Name. Enter surname, given name and middle initial of individual operating audiometer.
b. Training Certification Number. Enter audiometric technician training certification number.
c. Service Duty Occupation Code. Enter examiner's service duty occupation code (see Item 10).
d. Office Symbol. Enter complete office symbol where examiner is performing the test.

23. AUDIOMETER.
a. Type. Enter number for type of audiometer used (e.g., "1" for manual type).
b. Model. Enter manufacturer's designation.
c. Manufacturer. Enter name of company that produced audiometer.
d. Serial Number. Enter manufacturer's serial number.
e. Last Electroacoustic Calibration Date. Enter year, month and day (see Item 16) of last electroacoustic determination of this audiometer's performance specifications.

24. PERSONAL HEARING PROTECTION.
a. Type Issued. Enter number for type of hearing protector that the individual was issued (e.g., "2" for triple flange, etc.; if "6 - OTHER," explain in Item 25, "Remarks").
b. Size Earplugs. Enter number for size of earplugs (single or triple flange) used for each ear (e.g., "4" for Large in right ear (R) and "3" for Medium or Regular in left ear (L)).
c. Double Protection Used. Enter "1" in box if earplugs are not routinely worn in combination with noise muffs or a noise-attenuating helmet. Enter "2" if they are routinely worn together.
d. Glasses Worn. Enter "1" in box if eye glasses or goggles are not routinely worn with noise muffs or noise-attenuating helmet.
e. Frequency Glasses Worn. Indicate frequency of use if "2" was entered in Item 24.d. If "1" was entered in 24.d., enter "3" - N/A.

25. REMARKS. Print explanations for any of above items marked "OTHER" and any information considered pertinent. Include the individual's 8-hour TWA noise exposure, when available.

DD FORM 2215 (BACK), JAN 2000

HEARING CONSERVATION DATA

(This form is subject to the Privacy Act of 1974 - use Blanket PAS - DD Form 2005)

1. ZIP CODE/APO/FPO/PAS

2. DOD COMPONENT — A - ARMY; N - NAVY; F - AIR FORCE; M - MARINE CORPS; 1 - OTHER DOD ACTIVITY

3. SERVICE COMPONENT — R - REGULAR; V - RESERVE; G - NATIONAL GUARD; 1 - OTHER

4. SOCIAL SECURITY NUMBER

5. NAME *(Last, First, Middle Initial)*

6. DATE OF BIRTH *(YYYYMMDD)*

7. SEX — M - MALE; F - FEMALE

8. PAY GRADE, UNIFORMED SERVICES

9. PAY GRADE, CIVILIAN

10. SERVICE DUTY OCCUPATION CODE

11. MAILING ADDRESS OF ASSIGNMENT

12. LOCATION - PLACE OF WORK

13. MAJOR COMMAND

14. DUTY TELEPHONE *(Include area code)*

15. AUDIOMETRY — a. PURPOSE: 1 - 90 DAY; 2 - ANNUAL; 3 - TERMINATION; 4 - OTHER

AUDIOMETRIC DATA RE: ANSI S3.6 - 1989	LEFT						RIGHT					
	500	1000	2000	3000	4000	6000	500	1000	2000	3000	4000	6000
b. CURRENT AUDIOGRAM DATE *(YYYYMMDD)*												
c. REFERENCE AUDIOGRAM DATE *(YYYYMMDD)*												
d. SIGNIFICANT THRESHOLD SHIFT (STS) 1 - NO 2 - YES; e. THRESHOLD SHIFT →												

f. REMARKS *(Include exposure data)*

g. TYPE OF PERSONAL HEARING PROTECTION USED — 1 - SINGLE FLANGE (VS1R); 2 - TRIPLE FLANGE; 3 - HAND FORMED EARPLUGS; 4 - EAR CANAL CAPS; 5 - NOISE MUFFS; 6 - OTHER

h. EXAMINER NAME *(Last, First, Middle Initial)* | i. TRAINING CERTIFICATE NO. | j. SERVICE DUTY OCCUPATION CODE | k. OFFICE SYMBOL

l. AUDIOMETER TYPE — 1 - MANUAL; 2 - SELF-RECORDING *(Automatic)*; 3 - MICROPROCESSOR | m. MODEL | n. MANUFACTURER | o. SERIAL NUMBER | p. LAST ELECTROACOUSTIC CALIBRATION DATE *(YYYYMMDD)*

16. FOLLOWUP NO. 1 — a. MINIMUM 14 HOURS NOISE FREE SINCE CURRENT AUDIOGRAM *(See item 15 b.)*

AUDIOMETRIC DATA RE: ANSI S3.6 - 1989	LEFT						RIGHT					
	500	1000	2000	3000	4000	6000	500	1000	2000	3000	4000	6000
b. CURRENT AUDIOGRAM DATE *(YYYYMMDD)*												
c. REFERENCE AUDIOGRAM DATE *(YYYYMMDD)*												
d. SIGNIFICANT THRESHOLD SHIFT (STS) 1 - NO 2 - YES; e. THRESHOLD SHIFT →												

f. EXAMINER NAME *(Last, First, Middle Initial)* | g. TRAINING CERTIFICATE NO. | h. SERVICE DUTY OCCUPATION CODE | i. OFFICE SYMBOL

j. AUDIOMETER TYPE — 1 - MANUAL; 2 - SELF-RECORDING *(Automatic)*; 3 - MICROPROCESSOR | k. MODEL | l. MANUFACTURER | m. SERIAL NUMBER | n. LAST ELECTROACOUSTIC CALIBRATION DATE *(YYYYMMDD)*

17. FOLLOWUP NO. 2 — a. MINIMUM 14 HOURS NOISE FREE SINCE CURRENT AUDIOGRAM *(See item 15 b.)*

AUDIOMETRIC DATA RE: ANSI S3.6 - 1989	LEFT						RIGHT					
	500	1000	2000	3000	4000	6000	500	1000	2000	3000	4000	6000
b. CURRENT AUDIOGRAM DATE *(YYYYMMDD)*												
c. REFERENCE AUDIOGRAM DATE *(YYYYMMDD)*												
d. SIGNIFICANT THRESHOLD SHIFT (STS) 1 - NO 2 - YES; e. THRESHOLD SHIFT →												

f. EXAMINER NAME *(Last, First, Middle Initial)* | g. TRAINING CERTIFICATE NO. | h. SERVICE DUTY OCCUPATION CODE | i. OFFICE SYMBOL

j. AUDIOMETER TYPE — 1 - MANUAL; 2 - SELF-RECORDING *(Automatic)*; 3 - MICROPROCESSOR | k. MODEL | l. MANUFACTURER | m. SERIAL NUMBER | n. LAST ELECTROACOUSTIC CALIBRATION DATE *(YYYYMMDD)*

DD FORM 2216, JAN 2000 PREVIOUS EDITION MAY BE USED.

Reset

INSTRUCTIONS

(Refer to DoD Component Instructions for additional guidance)

PURPOSE: This form is used to record the results of periodic and followup audiometry for individuals routinely exposed to hazardous noise. Before this form is used, a DD Form 2215, "Reference Audiogram," must already be filed in the individual's health record.

1. **ZIP CODE/APO/FPO/PAS.** Enter nine digit ZIP Code/APO/FPO/ PAS of where audiometric test is conducted.

2. **DOD COMPONENT.** Enter letter in box of major organizational subdivision of DoD to which military or civilian individual is assigned. Enter "1" if DoD component is not listed.

3. **SERVICE COMPONENT.** Enter letter in box corresponding to primary subdivision of separate military service in which military is assigned (e.g., Regular (R) - standing military component of armed forces in peace and war; Reserve (V) - component of ready trained personnel for military service when needed, etc.; National Guard (G) - component of National Guard personnel in full-time or part-time status). Enter "1" for all others not listed.

4. **SOCIAL SECURITY NUMBER.** Enter nine digit social security number. If foreign national, enter "FN" in middle two blocks.

5. **NAME.** Enter surname, given name and middle initial of individual being tested.

6. **DATE OF BIRTH.** Enter year, month, day.

7. **SEX.** Enter "M" if male, "F" if female.

8. **PAY GRADE, UNIFORMED SERVICES.** For military personnel only, enter military personnel class and pay level serial number as follows:

O11 - General of the Army/General of the Air Force/Fleet Admiral
O10 - General/Admiral
O09 - Lieutenant General/Vice Admiral
O08 - Major General/Rear Admiral (Upper Half)
O07 - Brigadier General/Rear Admiral (Lower Half)/Commodore
O06 - Colonel (A,F,M)/Captain (N)
O05 - Lieutenant Colonel/Commander
O04 - Major/Lieutenant Commander
O03 - Captain (A,F,M)/Lieutenant (N)
O02 - First Lieutenant/Lieutenant Junior Grade
O01 - Second Lieutenant/Ensign
W05 - Chief Warrant Officer, W-5
W04 - Chief Warrant Officer, W-4
W03 - Chief Warrant Officer, W-3
W02 - Chief Warrant Officer, W-2
W01 - Warrant Officer, W-1
C00 - Cadet/Midshipman
E09 - Sergeant Major/Chief Master Sergeant/Master Chief Petty Officer
E08 - Master Sergeant (A,M)/Senior Chief Petty Officer/Senior Master Sergeant/First Sergeant(A)
E07 - Sergeant First Class/Gunnery Sergeant/Chief Petty Officer/ Master Sergeant (F)/Platoon Sergeant (A)/Specialist-7
E06 - Staff Sergeant/Technical Sergeant/Petty Officer First Class/ Specialist-6
E05 - Sergeant (A,M)/Staff Sergeant/Petty Officer Second Class/ Specialist-5
E04 - Corporal/Sergeant (F)/Petty Officer Third Class/Specialist-4
E03 - Private First Class (A)/Airman First Class/Lance Corporal/Seaman
E02 - Private (PV1)/Airman/Private First Class (M)/Seaman Apprentice
E01 - Private (PV2)/Private (M)/Airman Basic/Seaman Recruit

9. **GRADE, CIVILIAN.** Enter two letters and two numbers of Federal civilian employee rank (e.g., WG05, GS11, etc). Letter entries will be WG, WL, WS, WN, WD or GS. Number entries will be 01 to 18. Enter "1111" if other (e.g., foreign national, contractor, etc.).

10. **SERVICE DUTY OCCUPATION CODE.** Enter code to which military member's duty occupation is assigned (e.g., MOS, SSI, NEC/Rating, NOBC or AFSC in which individual is actually working). Enter number code of civilian job series in which civilian member is actually working (e.g., for a carpenter enter "4607").

11. **MAILING ADDRESS OF ASSIGNMENT.** Enter installation name (and street address for Navy and Marines), unit, office symbol, and ZIP Code/APO/ FPO/PAS of individual's current duty assignment.

12. **LOCATION - PLACE OF WORK.** Enter specific location where individual is routinely exposed to hazardous noise including building number (e.g., Corpus Christi, NAS, Building 1571, Carpenter Shop). For Air Force personnel, enter 12-digit Workplace Identifier Code per AFOSH Std. 161-17.

13. **MAJOR COMMAND.** Enter authorized abbreviation of military major command to which individual is assigned.

14. **DUTY TELEPHONE.** Enter individual's duty telephone number.

15. **AUDIOMETRY.**

a. **Purpose.** Enter number in box for reason to complete audiogram. "1" - First periodic test given 90 days after beginning duties in noise- hazardous area or operation; "2" - Periodic test given at yearly intervals; "3" - Last test given, regardless of noise exposure history, before termination of active duty or employment; "4" - Test at interval for reason not listed above.

b. **Current Audiogram Date.** Enter year, month, day (e.g., if January 31, 2000, enter 20000131) that audiometric test is given and current threshold levels determined for this individual at six frequencies in each ear. Results are entered in 5 dB increments (e.g., 0, 5, 10, 15, etc.). If responses exceed maximum limits of audiometer, enter that limit with plus sign (e.g., 110+).

15.c. **Reference Audiogram Date.** Enter year, month, and day reference test results were obtained. See DD Form 2215, "Reference Audiogram," or other appropriate source. Enter threshold levels in 5 dB increments from reference audiogram.

d. **Significant Threshold Shift (STS).** Enter "1" if no STS is present; enter "2" if STS is present.

STS - NO: See DoD component specific manuals for detailed guidance.

STS-YES. Outlines procedures required when a significant threshold shift present. "Notify Supervisor" - Notify individual's supervisor that significant threshold shift has been found and followup audiogram must be done. "Followup No. 1 After Minimum 14 Hours Noise Free" - Schedule individual for first followup audiogram. They must be instructed to stay in a noise free environment (not to exceed 75 dBA or 120 dBP) for at least 14 hours prior to test. They must be told to avoid environments in which noise levels make it necessary to use raised voice to talk at 1 meter (3 feet) distance. If examinee has obvious ear problem (e.g., earache, draining ear, excessive cerumen buildup), he/she should be examined by physician and followup postponed until after any necessary treatment.

e. **Threshold Shift.** Enter difference between current and most recent reference audiogram for 1000, 2000, 3000 and 4000 Hz. Refer to DoD component manuals for established criteria. Enter "+" to indicate positive shift (poorer hearing) or "-" to indicate negative shift (better hearing) on current audiogram.

f. **Remarks.** Print any information considered pertinent. Include the individual's 8-hour TWA noise exposure, when available.

g. **Type of Personal Hearing Protection Used.** Enter number for type of hearing protection that is routinely used by individual.

h. **Examiner Name.** Enter surname, given name and middle initial of individual operating audiometer.

i. **Training Certificate Number.** Enter audiometric technician training certificate number.

j. **Service Duty Occupation Code.** Enter examiner's service duty occupation code (see Item 10).

k. **Office Symbol.** Enter complete office symbol where examiner is performing the test.

l. **Audiometer Type.** Enter number for type of audiometer used (e.g., "1" for manual type, etc.).

m. **Model.** Enter manufacturer's designation of audiometer.

n. **Manufacturer.** Enter name of company that produced audiometer.

o. **Serial Number.** Enter manufacturer's serial number of audiometer.

p. **Last Electroacoustic Calibration Date.** Enter year, month and day (see Item 15.b.) of last electroacoustic determination of this audiometer's performance specifications.

16. **FOLLOWUP NO. 1.** If significant threshold shift determined on periodic test, record results of first followup audiogram in this section. Mark (X) box to certify "Minimum 14 Hours Noise Free Since Current Audiogram (see Item 15.b.)".

b., c., and e., "Current Audiogram," "Reference Audiogram," and "Threshold Shift" completed in same format as above. Note: Hearing threshold levels entered in 16.c. are the same values as those used in 15.c.

d. "STS - NO" - If no STS noted, enter "1" in box and follow steps in "STS - NO" section.

"STS - YES" - If STS remains following this examination (Followup No. 1), follow service component instructions (e.g., supervisor is notified for the second time, individual is scheduled for Followup No. 2 audiogram, and individual is instructed to stay in a noise free environment (not to exceed 75 dBA or 120 dBP) for a minimum of 14 hours of auditory rest since current audiogram (Item 15.b.)).

e. through m. Enter the required information according to guidelines for entries on periodic audiogram.

17. **FOLLOWUP NO. 2.** If significant threshold shift determined on Followup No. 1, record results of Followup No. 2 in this section. Mark (X) box to certify "Minimum 14 Hours Noise Free Since Current Audiogram (see Item 15.b.)".

b., c., and e., "Current Audiogram," "Reference Audiogram," and "Threshold Shift" completed in same format as above. Note: Hearing threshold levels entered in 17.c. are the same values as those used in 15.c.

d. "STS - NO" - If no STS noted, enter "1" in box and follow steps in "STS - NO" section.

"STS - YES" - If STS remains following this examination (Followup No. 2), enter "2" in box. Refer to DoD component instructions for appropriate patient disposition.

e. through m. Enter the required information according to guidelines for entries on periodic audiogram.

See specific DoD component manuals regarding followup procedures required in addition to those listed above. For example, if the annual test indicates a "negative" threshold shift and is confirmed on the first followup, the reference audiogram may be reestablished at this time without any further followup testing for DA personnel.

DD FORM 2216 (BACK), JAN 2000

BIOLOGICAL AUDIOMETER CALIBRATION CHECK

1. AUDIOMETER

a. MANUFACTURER	b. MODEL	c. SERIAL NUMBER	d. LAST ELECTROACOUSTIC CALIBRATION DATE *(YYYYMMDD)*

2. LISTENER

a. NAME *(Last, First, Middle Initial)*	b. FACILITY	c. LOCATION

3. DATES AND DATA REVIEW				4. HEARING THRESHOLD LEVELS OF TEST FREQUENCIES RE: ANSI S3.6 - 1989											
DATE *(YYYYMMDD)* a.	NAME OF EXAMINER *(Last, First, Middle Initial)* b.	CALIBRATION CHECK c.		LEFT EARPHONE (1)						RIGHT EARPHONE (2)					
		PASS: ± 5dB of Baseline (1)	FAIL: greater than ± 5dB of Baseline (2)	500	1000	2000	3000	4000	6000	500	1000	2000	3000	4000	6000
				a. BASELINE											
				b. PERIODIC BIOLOGICAL CALIBRATION CHECKS											

5. REMARKS

DD FORM 2217, JAN 2000 PREVIOUS EDITION MAY BE USED.

Reset

INSTRUCTIONS

(Refer to DoD Component Instructions for additional guidance.)

PURPOSE: This form is used to record biological/electroacoustic monitor checks of the calibration of one audiometer. Hearing threshold levels of one person tested on this audiometer are recorded as well as notations of any signal distortions and noise transients.

GENERAL: Print all information in ink. Biological audiometer calibration checks will be performed every day the audiometer is used. More frequent intervals (e.g., daily checks) may be required by the DoD component. Start a new form if a different listener is used and/or after the audiometer is re-calibrated.

1. AUDIOMETER.

a. Manufacturer. Enter name of company that produced audiometer.

b. Model. Enter manufacturer's model designation.

c. Serial Number. Enter manufacturer's serial number.

d. Last Electroacoustic Calibration Date. Enter year, month, and day of last electroacoustic determination of this audiometer's performance specifications. If January 31, 2000, enter 2000131.

2. LISTENER.

a. Name. Enter surname, given name and middle initial of individual being tested, i.e., the person listening through earphones of audiometer.

b. Facility. Enter name of installation (e.g., Fort Bliss).

c. Location. Enter state or APO (e.g., TX, etc.).

3. DATES AND DATA REVIEW.

a. Date. Enter year, month, and day (see Item 1.d.) of each biological calibration check.

b. Name of Examiner. Enter surname, given name and middle initial of individual operating audiometer.

3. DATES AND DATA REVIEW *(Continued)*

c. Calibration Check.

(1) Pass: ± 5 dB of Baseline at 500 - 4000 Hertz (Hz) and ± 10 dB at 6000 Hz. Mark (X) this column if periodic biological calibration check is within ± 5 dB of baseline at 500 - 4000 Hz and ± 10 dB at 6000 Hz (e.g., if baseline of 15 dB has been established at 1000 Hz in right ear, any of the following hearing threshold levels obtained on periodic check would require no action: 10, 15, or 20 dB).

(2) Fail: Greater Than ± 5 dB of Baseline at 500 - 4000 Hz and ± 10 dB at 6000 Hz. Mark this column if periodic biological calibration check is greater than ± 5 dB of baseline at 500 - 4000 Hz and ± 10 dB at 6000 Hz (e.g., if baseline of 15 dB has been established at 1000 Hz in right ear, any threshold levels of 5 dB or less or 25 dB or greater would require action). This discrepancy must be accounted for or audiometer should receive an electroacoustic calibration. Refer to DoD component instructions for further guidance.

4. HEARING THRESHOLD LEVELS OF TEST FREQUENCIES.

a. Baseline. After listener has demonstrated test-retest reliability (i.e., if test results of several pre-tests are consistently within ± 5 dB of each other), enter hearing threshold levels of last test results in increments of 5 dB (e.g., 0, 5, 10, 15, etc.).

b. Periodic Biological Calibration Checks. Enter hearing threshold levels in increments of 5 dB. Use a separate line for each calibration check.

5. REMARKS. Enter any comments pertaining to signal distortion or noise transients including date of check. Include additional information on location of the audiometer (e.g., building number and room number) and the type of acoustic test environment (e.g., single-walled, single-person audiometric examination booth, etc.).

DD FORM 2217 (BACK), JAN 2000

Appendix H

Biographical Sketches of Committee Members

Larry E. Humes, Ph.D. (*Chair*), is a professor in the Department of Speech and Hearing Sciences at Indiana University. Past roles have included department chair, director of the Center for Hearing Aid Research and Technical Training, and director of audiology. His research is focused on behavioral issues related to auditory perception, hearing loss, and hearing aids, with a particular interest in age-related hearing loss. He is the principal investigator for a study funded by the National Institute on Aging on Speech-Recognition Difficulties of the Hearing Impaired, and co-investigator for a study funded by the National Institute on Deafness and Other Communication Disorders on the Perception of Speech by Normal and Impaired Listeners. Dr. Humes was a member of the National Research Council Committee on Hearing, Bioacoustics, and Biomechanics (CHABA) (1989–1992) and also served as an advisor to that committee. He is a fellow of the American Speech-Language-Hearing Association and has served as an editor or editorial consultant to publications including *Journal of Speech, Language, and Hearing Research*, *Journal of the Acoustical Society of America*, and *Ear and Hearing*. He holds a Ph.D. in audiology from Northwestern University.

Elliott Berger, M.S., is the senior scientist for auditory research at E•A•R/Aearo Company, where he conducts hearing protector research and development. He has studied hearing protection, hearing conservation, and related topics for more than 25 years. He has authored 14 textbook chapters, written over 60 articles on hearing protection/conservation, and was the principal editor for the fourth and fifth editions of the American Industrial Hygiene Association (AIHA) *Noise Manual*. His principal research has involved evaluation of techniques for measuring and specifying hearing protector attenuation and the limits to their performance caused by the bone-conduction pathways, as well as means for improving the effectiveness of hearing conservation programs. Mr. Berger chairs the American National Standards Institute (ANSI) working group S12/WG11 on hearing protector attenuation and performance, is a fellow of the Acoustical Society of America, past-president of the National Hearing Conservation Association (NHCA), fellow of the AIHA and past-chair of its Noise Committee, a board member of the Council for Accreditation in Occupational Hearing Conservation, and a recipient of the NHCA's Outstanding Hearing Conservationist Award.

Barbara A. Bohne, Ph.D., is professor of otolaryngology (neurobiology) at Washington University School of Medicine. From 1987 though 2003, she was also senior research scientist at the Central Institute for the Deaf in St. Louis. Her research focuses on the anatomy and physiology of the inner ear and the biological mechanisms of noise-related damage to the inner ear. She is the principal investigator for the study Adverse Effects of Noise on Hearing: Basic Mechanisms, which is supported by the National Institute for Occupational Safety and Health. Dr. Bohne is a member of the editorial board for the journal *Hearing Research*. She previously served 8 years on study sections for the National Institutes of Health. She has also served as a member of the National Research Council Committee on Hearing, Bioacoustics, and Biomechanics (CHABA). Dr. Bohne holds a Ph.D. in neurobiology from Washington University.

Anthony Cacace, Ph.D., is associate professor at The Neurosciences Institute of the Department of Neurology at Albany Medical College. From 1993 to 2003, he was an associate professor in the Division of Otolaryngology of the Department of Surgery at Albany Medical College. He also served as director of oto-neurological research, Division of Otolaryngology, Albany Medical College, from 1985 to 2003. Dr. Cacace is a member of the Scientific Advisory Committee of the American Tinnitus Association. He is also an associate editor of the *American Journal of Audiology* and assistant editor for hearing science for the *Journal of the American Academy of Audiology*. He is a fellow of the American Speech-Language-Hearing Association and the American Academy of Audiology. Dr. Cacace has a Ph.D. in communication sciences and disorders/audiology and neuroscience from Syracuse University. His postdoctoral fellowship training in neurophysiology was performed at the Wadsworth Laboratories, New York State Health Department.

Karen J. Cruickshanks, Ph.D., is a professor in the Department of Ophthalmology and Visual Sciences as well as in the Department of Population Health Sciences at the University of Wisconsin School of Medicine. Her research interests are in the epidemiology of age-related sensory disorders, diabetes and its complications, and aging. She has written more than 100 articles on these topics and is the principal investigator for two major studies of the epidemiology of age-related hearing loss. Dr. Cruickshanks serves as frequent advisor or reviewer for the National Institute on Deafness and Other Communication Disorders and as a reviewer for a variety of journals, including the *American Journal of Epidemiology*, the *American Journal of Public Health*, *Ear and Hearing*, and the *New England Journal of Medicine*. She is a member of the Society for Epidemiological Research and the American Epidemiological Society. Dr. Cruickshanks received her Ph.D. in epidemiology from the University of Pittsburgh.

Richard Danielson, Ph.D., is the manager for audiology and hearing conservation at NASA's Johnson Space Center, Houston, Texas. Working for the National Space and Biomedical Research Institute and Baylor College of Medicine, he leads a program aimed at preventing noise-induced hearing loss during space and ground-based missions among astronauts, crew, pilots, and other flight-associated personnel. He is the chair of the Council for Accreditation in Occupational Hearing Conservation and a faculty member of several university graduate programs. Prior to his work at NASA, Dr. Danielson served for 28 years as an audiology officer in the U.S. Army Medical Department, including roles as the director of the Army Audiology and Speech Center at Walter Reed Army Medical Center, director of audiology at Madigan Army Medical Center, and officer-in-charge of an audiology task force in Saudi Arabia. Dr. Danielson holds a Ph.D. in human development and communication sciences from the University of Texas at Dallas.

Judy R. Dubno, Ph.D., is a professor in the Department of Otolaryngology–Head and Neck Surgery of the Medical University of South Carolina (MUSC). She is also a member of the MUSC Center for Advanced Imaging Research. Her research focuses on human auditory system function, with emphasis on the encoding of auditory information in simple sounds and speech and how these abilities change in adverse listening conditions, with age, and with hearing loss. This research is currently funded by the National Institute on Deafness and Other Communication Disorders (NIDCD) of the National Institutes of Health (NIH). Dr. Dubno's other activities currently include serving as a member of the Executive Council of the Acoustical Society of America, the Tinnitus Research Consortium, and the Clinical Research Discussion Group of the NIDCD. She previously served as a member of the National Deafness and Other Communication Disorders Advisory Council of the NIH and the National Research Council Committee on Hearing, Bioacoustics, and Biomechanics (CHABA). She was also president of the Association for Research in Otolaryngology. She is a fellow of the Acoustical Society of America and the American Speech-Language-Hearing Association. She holds a Ph.D. in speech/hearing science from the City University of New York Graduate School and University Center.

George A. Gates, M.D., is the former director of the Virginia Merrill Bloedel Hearing Research Center at the University of Washington, emeritus professor in the Department of Otolaryngology–Head and Neck Surgery in the School of Medicine, and adjunct professor in the Department of Epidemiology in the School of Public Health and Community Medicine. His research includes work on the biology of presbycusis and Ménière's disease, and he has conducted clinical trials in otitis media and Ménière's disease. He is currently the scientific medical director of the Deafness Research Foundation and previously chaired the Noise Subcommittee of the American Academy of Otolaryngology–Head and Neck Surgery and the National Institutes of Health Consensus Conference on Cochlear Implants in 1995. He also served on the Working Group on Speech Understanding and Aging of the National Research Council Committee on Hearing, Bioacoustics, and Biomechanics (CHABA).

Donald Henderson, Ph.D., is a professor in the Department of Communicative Disorders and Sciences at the State University of New York at Buffalo. He also co-founded the Center for Hearing and Deafness of the State University of New York at Buffalo in 1994 and served as its director from 1994 to 2002. He has studied causes of hearing loss over the past 30 years. His current research focus is on the role of free radicals as a factor in hearing loss from noise, drugs, and aging. As a corollary, he is also developing pharmacological techniques for preventing hearing loss by using antioxidants or inhibitors of active hair cell death. He has published more than 150 articles and edited 10 books. Dr. Henderson has a Ph.D. in sensory psychology from the University of Texas.

John H. Mills, Ph.D., is professor in the Department of Otolarynogology–Head and Neck Surgery at the Medical University of South Carolina (MUSC), where he is also director of the Research Division. His primary interests involve age-related changes in hearing in humans, as indicated by psychophysical and electrophysiological methods, and in noise-induced hearing loss. Dr. Mills is the program director and one of the principal investigators for the program in Experimental and Clinical Studies in Presbycusis at MUSC, which is funded by the National Institute on Deafness and Other Communication Disorders. He is a former chair of the National Research Council Committee on Hearing, Bioacoustics, and Biomechanics (CHABA) and also served on its Working Group on Hazardous Exposure to Impulse Noise. He is a fellow of the American Speech-Language-Hearing Association and the Acoustical Society of America and a scientific fellow of the American Academy of Otolaryngology–Head and Neck Surgery. Dr. Mills holds a Ph.D. in audiology from the University of Iowa.

Mark R. Stephenson, Ph.D., is a senior research audiologist in the Division of Applied Research and Technology at the Centers for Disease Control and Prevention's National Institute for Occupational Safety and Health (NIOSH). Before joining NIOSH, he served 20 years as a commissioned officer in the U.S. Air Force (USAF), most recently as associate chief of the Air Force Biomedical Sciences Corps, with worldwide staff management responsibility for all USAF audiology and speech pathology policies and personnel. Dr. Stephenson is a member of the Acoustical Society of America, a fellow of the American Academy of Audiology, a representative to the Council for Accreditation in Occupational Hearing Conservation, and a member and former vice president of the National Hearing Conservation Association. He is an adjunct faculty member at the Ohio State University and at Miami University (Ohio), and he is the recipient of numerous military and civilian awards for his contributions to the fields of audiology and hearing conservation. Dr. Stephenson holds a Ph.D. in hearing science/audiology from the Ohio State University.

Richard Tyler, Ph.D., is a professor in the Departments of Otolaryngology–Head and Neck Surgery and Speech Pathology and Audiology at the University of Iowa. His research focuses on the use of cochlear implants in adults and children, the impact and treatment of tinnitus, and treatments for hyperacusis. Dr. Tyler is an author of more than 100 peer-reviewed journal articles on these topics. He is currently the principal investigator for a randomized trial of counseling and sound therapy treatments, which is funded by the National Institute on Deafness and Other Communication Disorders (NIDCD), and is co-director of the Iowa Cochlear Center Project, which is also funded by NIDCD. Dr. Tyler serves as a member of the Scientific Advisory Committee and is on the Board of Directors of the American Tinnitus Association. He jointly holds a patent for a system and method for reducing tinnitus with electricity. Dr. Tyler previously served as a member of the Working Group on Communication Aids for the Hearing Impaired for the National Research Council Committee on Hearing, Bioacoustics, and Biomechanics (CHABA). He has a Ph.D. in psychoacoustics from the University of Iowa and a clinical degree in audiology from the University of Western Ontario.

Robert B. Wallace, M.D., is professor of epidemiology and internal medicine at the University of Iowa Colleges of Public Health and Medicine. He received the Irene Ensminger Stecher Professorship in April 1999 for cancer- and heart disease-related research. He was formerly head of the Department of Preventive Medicine at the University of Iowa College of Medicine and the director of the University of Iowa Cancer Center. Dr. Wallace's research interests include cancer epidemiology and prevention; the causes and prevention of chronic, disabling diseases among older persons; women's health issues; and risk factors for cardiovascular disease. He is a principal investigator of several large clinical trials. He received his medical degree from Northwestern University School of Medicine. He is an Institute of Medicine member and currently serves as chair of the Institute of Medicine's Board for the Medical Follow-up Agency. He has served on and chaired several other Institute of Medicine studies.